# Learning Materials in Biosciences

Learning Materials in Biosciences textbooks compactly and concisely discuss a specific biological, biomedical, biochemical, bioengineering or cell biologic topic. The textbooks in this series are based on lectures for upper-level undergraduates, master's and graduate students, presented and written by authoritative figures in the field at leading universities around the globe.

The titles are organized to guide the reader to a deeper understanding of the concepts covered.

Each textbook provides readers with fundamental insights into the subject and prepares them to independently pursue further thinking and research on the topic. Colored figures, step-by-step protocols and take-home messages offer an accessible approach to learning and understanding.

In addition to being designed to benefit students, Learning Materials textbooks represent a valuable tool for lecturers and teachers, helping them to prepare their own respective coursework.

Ben Loos • Daniel J. Klionsky

**Editors**

# Autophagy - From Molecular Mechanisms to Flux Control in Health and Disease

 Springer

*Editors*
Ben Loos
Department of Physiological Sciences
Stellenbosch University
Stellenbosch, South Africa

Daniel J. Klionsky
Life Sciences Institute
and Department
of Molecular, Cellular
and Developmental Biology
University of Michigan
Ann Arbor, MI, USA

ISSN 2509-6125     ISSN 2509-6133   (electronic)
Learning Materials in Biosciences
ISBN 978-3-031-88120-6     ISBN 978-3-031-88121-3   (eBook)
https://doi.org/10.1007/978-3-031-88121-3



This Springer imprint is published by the registered company Springer Nature Switzerland AG
The registered company address is: Gewerbestrasse 11, 6330 Cham, Switzerland

If disposing of this product, please recycle the paper.

# Preface

Performing biomedical research in the autophagy field is captivating, with hundreds of laboratories having contributed substantially to fundamental and applied research alike. Over the past 30 years, significant advances in the study of macroautophagy (referred to here as autophagy) and related processes, not only in the life science discipline but also in various biomedical and applied research spheres, have been made. These are increasingly translating into major focus areas of global health concern, including aging, cancer, neurodegeneration, and heart disease. Teaching, and learning autophagy, however, is hard, as it is a complex process. We mean, a truly complex process—approximately 20 proteins are part of the core machinery required to form the autophagosome, and at least that many more act in other aspects of autophagy including cargo recognition. For the student, it must be a challenge to discern which of the many scientific articles are key and should form part of the learning process. Similarly for the teacher, typically a university academic with limited training in teaching, it is not easy to choose suitable material, within a rapidly growing field. Where should one start? What is really important? This was the motivation for this book, to provide a point of departure for the teaching and learning environment, be it the undergraduate or postgraduate student at the university or college, the teaching instructor and university academician, or clinicians and specialists.

What will you learn by reading, studying, and working with this book? By bringing together a large number of experts, clinicians, and scientists, you will have the chance to engage with a framework for teaching and learning autophagy in a rather unique, captivating way, guiding you on a path that begins with molecular mechanisms and ends with flux control in health and disease (you may recognize the title). Each chapter starts with a "what will you learn" introduction, which is followed by a comprehensive section on the material in question (Part 1). Interspersed "take note" sections will assist in emphasizing concepts, while questions (and answers) will put the content into practice. Take note of the combination of schemes and cartoons, often next to real data. This approach is designed to make the topic tangible. In the second part of each chapter (Part 2), 2–3 landmark or cardinal articles are introduced. These are geared to assist the reader in focusing attention. Emphasis is placed on why these particular studies moved the field forward and changed our understanding of autophagy. Which implications do they have, which questions do they raise, and what does really deserve attention? A "guide to answers" and a "take home message" ensure that there are well-defined learning outcomes that can be assessed. Some questions and quizzes provide unique material to engage actively during the teaching or learning experience.

We begin this book uniquely, addressing the students and professors separately, providing techniques for effective learning and interactive teaching of autophagy, followed by an overview of basic vocabulary used to describe the autophagy process. Watch out for the minimum number of terms that are required to participate successfully in an autophagy-focused discussion. This is critical, to assist in a standardized approach for the use of nomenclature, and hence the enabling of subsequent accurate teaching and learning. Strategically, the following chapters introduce the autophagy molecular machinery. Specifically, the reader will first learn how to identify autophagy pathway intermediates in transmission electron microscopy samples, to discern and interpret the morphological features of phagophores, autophagosomes, amphisomes, and autolysosomes. Spend time looking at these micrographs, appreciate the membrane and cargo complexity, and sharpen your skills in recognizing autophagic structures. This chapter will be followed by expanding on microscopy-based techniques, including confocal, fluorescence, super-resolution, and correlative microscopy, to emphasize the dynamic nature of the autophagy pathway, and the key molecular markers that characterize the pathway intermediates. Engage with the functional complexity of the pH environment and its impact on particular fluorochromes. Having laid a foundation for the autophagy pathway and its molecular machinery, the reader will learn about the selectivity of autophagy, how specific cellular components, including damaged organelles or protein aggregates, are targeted and connected to the forming autophagosome and the core autophagosome formation machinery. Do not skip over the comprehensive table summarizing the specific cargo receptors engaging with respective candidate cargo. In the following chapter, the reader will learn about the origins of the autophagosomal membrane, focusing on the primary association with the endoplasmic reticulum but also other membrane compartments. Notice the use of single-molecule imaging to describe and dissect each step of the process in structural and functional detail. In the following chapter, your understanding about autophagy in healthy cells will be put to its first test, where you learn how mutations in core autophagy genes or regulators cause neurodegenerative diseases, and how autophagy-enhancing strategies may act as potential treatments. Take note of the MTOR-dependent and -independent strategies available to control autophagy activity. Maybe at this stage you should consider to page back to the first chapters, that alluded on how to measure autophagy and dissected which cargo receptor would recruit that specific aggregate-prone cargo? Could you rank these autophagy enhancers, selecting a best-suitable drug? Could you quantify the degree of autophagy failure in disease? Could you use CLEM to measure cargo *and* machinery flux?

Knowing that aging is the biggest risk factor for the onset of neurodegeneration, the following chapter will reveal that autophagy dysfunction is one of the central hallmarks of aging. The reader will learn about the mechanisms by which such "disabled autophagy" occurs with aging and the impact of autophagy regulation on lifespan and health span. This will be followed by focusing on autophagy in the regulation of mechanical forces and biochemical responses in renal cells and the importance of autophagy in human skeletal muscle homeostasis and myopathies. Pay attention to the emerging and growing role of histopathology in the assessment of autophagy markers for diagnostics. The next chapter focuses on cardiac myopathies, and the role of autophagy as well as selective mitochondrial degradation through autophagy, i.e., mitophagy, in ischemia-reperfusion injury and cardiovascular disorders. Take note of the early work on the quantification of autophagy activity in the whole heart. What follows is a chapter that focuses on chaperone-mediated autophagy, what it is, how it is regulated, and how it is different from macroautophagy, yet also critical for the cell's protein quality control. Having completed this chapter, your view on the lysosome and its receptors will have changed forever, because you may now be aware of the dynamic crosstalk between protein degradation systems. The last chapter is dedicated to a phenotypic screening routine, which can be employed for the identification of autophagic flux inducers from chemical libraries. Having arrived at the end of the book, the learner will now fully appreciate and accurately contextualize the importance of tools such as fluorescent biosensor cells, utilized with automated liquid handling and robotized bioimaging approaches, to move the autophagy field forward.

It is our hope that this book may become a meaningful resource. May it bring fascination to the learning and teaching environment alike. Happy learning and teaching!

Stellenbosch, South Africa                                                                    Ben Loos
Ann Arbor, MI, USA                                                               Daniel J. Klionsky

# Contents


**How to Learn, and Teach, Autophagy** . . . . . . . . . . . . . . . . . . . . . . . . . . . . . 1
Steven K. Backues, Karen L. Sirum, and Daniel J. Klionsky

**Discerning Autophagy Pathway Intermediates: Transmission
Electron Microscopy Techniques** . . . . . . . . . . . . . . . . . . . . . . . . . . . . . . . . 29
Joanna Biazik, Katri Kallio, Sigurdur Runar Gudmundsson,
and Eeva-Liisa Eskelinen

**Assessment of Autophagy: Correlative and Super-Resolution
Microscopy Techniques** . . . . . . . . . . . . . . . . . . . . . . . . . . . . . . . . . . . . . . 43
Nicola Vahrmeijer, Dumisile Lumkwana, André du Toit, Ben Loos,
and Lize Engelbrecht

**Autophagy Receptors Couple Cargo Destined to Be Degraded
with the Core Autophagy Machinery** . . . . . . . . . . . . . . . . . . . . . . . . . . . . . 71
Hallvard Lauritz Olsvik, Trond Lamark, and Terje Johansen

**The Source of Membrane During Autophagy and the Early
Steps in Autophagosome Formation** . . . . . . . . . . . . . . . . . . . . . . . . . . . . . 101
Nicholas T. Ktistakis

**Autophagy and Neurodegenerative Diseases** . . . . . . . . . . . . . . . . . . . . . . . 127
Antonio Daniel Barbosa, Jennifer E. Palmer, Xinyi Li,
and David C. Rubinsztein

**Autophagy and Intracellular Membrane Dynamics in Aging
and Age-Related Diseases** . . . . . . . . . . . . . . . . . . . . . . . . . . . . . . . . . . . . 171
Satoshi Minami, Shuhei Nakamura, and Tamotsu Yoshimori

**Shear Stress-Dependent Regulation of Autophagy and Metabolism
in Kidney Epithelial Cells** . . . . . . . . . . . . . . . . . . . . . . . . . . . . . . . . . . . . 213
Aurore Claude-Taupin, Nicolas Dupont, and Patrice Codogno

**Autophagy Assessment in Diagnostic Pathology: Focus
on Skeletal Myopathies** . . . . . . . . . . . . . . . . . . . . . . . . . . . . . . . . . . . . . . . . . . 231
Biswarathan Ramani and Marta Margeta

**The Role of Autophagy and Mitophagy in Cardiomyocyte
Ischemic Injury** . . . . . . . . . . . . . . . . . . . . . . . . . . . . . . . . . . . . . . . . . . . . . . . . 257
Roberta A. Gottlieb and Somayeh Pourpirali

**Chaperone-Mediated Autophagy in Health and Disease** . . . . . . . . . . . . . . . . 283
Susmita Kaushik

**A Phenotypic Screening Routine for the Identification
of Autophagic Flux Inducers** . . . . . . . . . . . . . . . . . . . . . . . . . . . . . . . . . . . . . . 311
Marion Leduc, Sabrina Forveille, Allan Sauvat, Giulia Cerrato,
Guido Kroemer, and Oliver Kepp

**Index** . . . . . . . . . . . . . . . . . . . . . . . . . . . . . . . . . . . . . . . . . . . . . . . . . . . . . . . . . 325

# Contributors

**Steven K. Backues** Department of Chemistry, Eastern Michigan University, Ypsilanti, MI, USA

**Antonio Daniel Barbosa** Department of Medical Genetics, Cambridge Institute for Medical Research (CIMR), Cambridge, UK

**Joanna Biazik** Molecular and Integrative Biosciences, University of Helsinki, Helsinki, Finland

Electron Microscopy Unit, Mark Wainwright Analytical Centre, The University of New South Wales, Sydney, NSW, Australia

**Giulia Cerrato** Metabolomics and Cell Biology Platforms, Gustave Roussy Cancer Center, Université Paris Saclay, Villejuif, France

Centre de Recherche des Cordeliers, Equipe labellisée par la Ligue contre le cancer, Université Paris Cité, Sorbonne Université, Inserm U1138, Institut Universitaire de France, Paris, France

**Aurore Claude-Taupin** Université Paris Cité, INSERM UMR-S1151, CNRS UMR-S8253, Institut Necker-Enfants Malades, Paris, France

**Patrice Codogno** Université Paris Cité, INSERM UMR-S1151, CNRS UMR-S8253, Institut Necker-Enfants Malades, Paris, France

**Nicolas Dupont** Université Paris Cité, INSERM UMR-S1151, CNRS UMR-S8253, Institut Necker-Enfants Malades, Paris, France

**André du Toit** Department of Chemistry and Molecular Biology, University of Gothenburg, Gothenburg, Sweden

**Lize Engelbrecht** Central Analytical Facilities, Microscopy Unit, Stellenbosch University, Merriman Avenue, Mike de Vries Building, Stellenbosch, South Africa

**Eeva-Liisa Eskelinen** Institute of Biomedicine, University of Turku, Turku, Finland

**Sabrina Forveille** Metabolomics and Cell Biology Platforms, Gustave Roussy Cancer Center, Université Paris Saclay, Villejuif, France

Centre de Recherche des Cordeliers, Equipe labellisée par la Ligue contre le cancer, Université Paris Cité, Sorbonne Université, Inserm U1138, Institut Universitaire de France, Paris, France

**Roberta A. Gottlieb** Department of Cardiology, Cedars-Sinai Medical Center, Los Angeles, CA, USA

**Sigurdur Runar Gudmundsson** Molecular and Integrative Biosciences, University of Helsinki, Helsinki, Finland

Biomedical Centre, School of Health Sciences, University of Iceland, Iceland

**Terje Johansen** Autophagy Research Group, Department of Medical Biology, UiT The Arctic University of Norway, Tromsø, Norway

**Katri Kallio** Molecular and Integrative Biosciences, University of Helsinki, Helsinki, Finland

FinVector Oy, Kuopio, Finland

**Susmita Kaushik** Department of Developmental and Molecular Biology, Albert Einstein College of Medicine, Bronx, NY, USA

Institute for Aging Research, Albert Einstein College of Medicine, Bronx, NY, USA

**Oliver Kepp** Metabolomics and Cell Biology Platforms, Gustave Roussy Cancer Center, Université Paris Saclay, Villejuif, France

Centre de Recherche des Cordeliers, Equipe labellisée par la Ligue contre le cancer, Université Paris Cité, Sorbonne Université, Inserm U1138, Institut Universitaire de France, Paris, France

**Daniel J. Klionsky** Life Sciences Institute and Department of Molecular, Cellular and Developmental Biology, University of Michigan, Ann Arbor, MI, USA

**Guido Kroemer** Metabolomics and Cell Biology Platforms, Gustave Roussy Cancer Center, Université Paris Saclay, Villejuif, France

Centre de Recherche des Cordeliers, Equipe labellisée par la Ligue contre le cancer, Université Paris Cité, Sorbonne Université, Inserm U1138, Institut Universitaire de France, Paris, France

Institut du Cancer Paris CARPEM, Department of Biology, APHP, Hôpital Européen Georges Pompidou, Paris, France

**Nicholas T. Ktistakis** Signalling Programme, Babraham Institute, Cambridge, UK

**Trond Lamark** Autophagy Research Group, Department of Medical Biology, UiT The Arctic University of Norway, Tromsø, Norway

**Marion Leduc** Metabolomics and Cell Biology Platforms, Gustave Roussy Cancer Center, Université Paris Saclay, Villejuif, France

Centre de Recherche des Cordeliers, Equipe labellisée par la Ligue contre le cancer, Université Paris Cité, Sorbonne Université, Inserm U1138, Institut Universitaire de France, Paris, France

**Xinyi Li** Department of Medical Genetics, Cambridge Institute for Medical Research (CIMR), Cambridge, UK

UK Dementia Research Institute, Cambridge Institute for Medical Research (CIMR), Cambridge, UK

**Ben Loos** Department of Physiological Sciences, Stellenbosch University, Merriman Avenue, Mike de Vries Building, Stellenbosch, South Africa

**Dumisile Lumkwana** Science Technology Platform, Electron Microscopy, Francis Crick Institute, London, UK

**Marta Margeta** School of Medicine, University of California, San Francisco, San Francisco, CA, USA

**Satoshi Minami** Department of Genetics, Graduate School of Medicine, Osaka University, Suita, Osaka, Japan

**Shuhei Nakamura** Department of Biochemistry, Nara Medical University, Kashihara, Nara, Japan

Institute for Advanced Co-Creation Studies, Osaka University, Suita, Osaka, Japan

**Hallvard Lauritz Olsvik** Autophagy Research Group, Department of Medical Biology, UiT The Arctic University of Norway, Tromsø, Norway

**Jennifer E. Palmer** Department of Medical Genetics, Cambridge Institute for Medical Research (CIMR), Cambridge, UK

UK Dementia Research Institute, Cambridge Institute for Medical Research (CIMR), Cambridge, UK

**Somayeh Pourpirali** Department of Biological Science and Technology, Islamic Azad University, Najafabad, Isfahan, Iran

**Biswarathan Ramani** School of Medicine, University of California, San Francisco, San Francisco, CA, USA

**David C. Rubinsztein** Department of Medical Genetics, Cambridge Institute for Medical Research (CIMR), Cambridge, UK

UK Dementia Research Institute, Cambridge Institute for Medical Research (CIMR), Cambridge, UK

**Allan Sauvat** Metabolomics and Cell Biology Platforms, Gustave Roussy Cancer Center, Université Paris Saclay, Villejuif, France

Centre de Recherche des Cordeliers, Equipe labellisée par la Ligue contre le cancer, Université Paris Cité, Sorbonne Université, Inserm U1138, Institut Universitaire de France, Paris, France

**Karen L. Sirum** Department of Biological Sciences, Bowling Green State University, Bowling Green, OH, USA

**Nicola Vahrmeijer** Department of Physiological Sciences, Stellenbosch University, Merriman Avenue, Mike de Vries Building, Stellenbosch, South Africa

**Tamotsu Yoshimori** Health Promotion System Science, Graduate School of Medicine, Osaka University, Suita, Osaka, Japan

Integrated Frontier Research for Medical Science Division, Institute for Open and Transdisciplinary Research Initiatives (OTRI), Osaka University, Suita, Osaka, Japan

# How to Learn, and Teach, Autophagy

Steven K. Backues, Karen L. Sirum, and Daniel J. Klionsky

> **What You Will Learn in This Chapter**
>
> In this chapter we will cover strategies for teaching (what the instructor typically tries to do) and learning (what you must do for yourself). One area of focus is active learning. The concept of constructivism states that you must create the knowledge for yourself in order to learn it. An example would be whether you learn to ride a bike by listening to a lecture or getting on the bike and riding it. You can listen to lectures all day long and you will not really know how to ride a bike. In contrast, a brief introduction followed by practicing riding can get you to an adequate level of competency very quickly. The same is true of essentially any subject, whether it be computer skills, surgery or learning a new language or subject area (such as autophagy). The knowledge cannot simply be transferred from the instructor to the learner. The instructor can guide the learner, and can establish an atmosphere that promotes learning, but only you can do the actual learning. Therefore, this chapter begins with a section of effective strategies that you as a learner can implement yourself. This is followed by a section for instructors that explains how to create a teaching environment that encourages students to actively learn. The final section is an overview of the basic process of autophagy, with a focus on key terminology that will be needed to understand the remaining chapters of this book.

S. K. Backues (✉)
Department of Chemistry, Eastern Michigan University, Ypsilanti, MI, USA
e-mail: sbackues@emich.edu

K. L. Sirum
Department of Biological Sciences, Bowling Green State University, Bowling Green, OH, USA
e-mail: ksirum@bgsu.edu

D. J. Klionsky
Life Sciences Institute and Department of Molecular, Cellular and Developmental Biology, University of Michigan, Ann Arbor, MI, USA
e-mail: klionsky@umich.edu

© The Author(s), under exclusive license to Springer Nature Switzerland AG 2025
B. Loos, D. J. Klionsky (eds.), *Autophagy - From Molecular Mechanisms to Flux Control in Health and Disease*, Learning Materials in Biosciences,
https://doi.org/10.1007/978-3-031-88121-3_1

# 1    Strategies for Learning

## 1.1    Why Bother Reading This Chapter?

If you are holding this book in your hands (or the virtual equivalent), it is probably because a professor told you to buy it for a class you are taking. Good! We think you'll find it a worthwhile investment. But, like anything you buy, it will only pay off if you know how to use it. That's what this chapter is about—how to get the most out of your investment, which is to say, how to use this book correctly so that it can actually help you learn about autophagy. As a bonus, most of the things you'll learn in this chapter apply not only to this book but to any textbook and any class. These strategies will make you a better learner, no matter what class you are taking.

If you are a professor reading this, then this is for you, too! The second part of this chapter ("Strategies for Teaching") has specific tools for the instructor to help you teach your students about autophagy (or any complex subject), but these are predicated on the same principles about how learning works that are presented in the first part. After all, the purpose of teaching is learning. So, we ask you to also read this first section, considering how you as a professor can structure your class to encourage good learning habits in your students, and then the second section will have some additional suggestions for this. But we will start by addressing the students; their learning is ultimately in their hands.

## 1.2    Do I Really Need to Learn How to Learn?

At this point you might be thinking "But I already know how to learn. This sounds like a waste of time." But let us challenge you—do you really? Are you a straight "A" student? If so, you might be right. If you rock every class you take, then you probably already know most of what is in this chapter, though you may still pick up some insights, or learn why what you do is effective. Still, if you get all "A's", you can skip this chapter (though, knowing you, you probably won't). But, if you don't always excel, or you're worried in any way about this class, this chapter might be just what you need.

You might think that after years in the educational system you should know how to learn already—after all, you've been doing it since you were five. Unfortunately, our educational system doesn't always teach students how to learn. Even worse, it sometimes accidentally creates bad habits. Students pick up ineffective strategies, like rereading the chapter over and over, that mostly just waste their time. Or a focus on rote memorization that worked in high school but isn't good enough for a college class—much less an advanced college class on a complex topic like autophagy. Or things like cramming the night before the exam that got you through one class but didn't build a foundation to prepare you for the next. Even though there has been a lot of research that has shown what does and does not work for learning, old habits die hard, and the best information hasn't always made it to the teachers, much less to the students themselves.

The strategies contained in this chapter aren't just someone's idea about how learning might work, or hokey tricks or fads. Instead, these are proven techniques based on educational research, and we've chosen those with the greatest impact, which have been recommended by multiple learning experts. Students who have implemented these strategies have seen their grades improve—often going from failing the first exam to getting an A in the class (McGuire). Professors who have taught their students these strategies have seen the whole class improve, sometimes by a full letter grade [1].

So, are you ready to get started? If so, here's lesson one: you can do it. The ability to learn is not primarily an inborn trait that some people have and others don't. Instead, it's mostly about knowing and applying good learning strategies—something that anyone can do. Also, you don't have to do all of the strategies in this chapter, or do them all right, in order to see the benefit. In fact, trying to do everything here might be a little overwhelming. But if you choose just a few key things that are different from how you currently study and apply those new strategies to this class, you should see a difference in how well you learn and what grade you get.

## 1.3 Strategy 1: Read to Remember

Because this is a textbook, we'll start with reading. Reading is the foundation of our educational system (and civilization itself), and it is the first step in learning because it is usually the way that you encounter new information. (If most of the information in this class is instead delivered in lectures that you watch, that's fine, too—all of the techniques we'll cover also apply to listening to lectures.)

You already know how to read—have known for years. But do you know how to read to remember? Do you know how to read to understand? When you finish a chapter in a textbook, have you learned something that makes sense to you, or is it all just kind of a jumble that you can't remember the gist of the next day? If the latter, then you aren't making very good use of your time. Here's how to do better.

### 1.3.1 Step 1: Previewing

When it comes to remembering what you read, the first key is what you are paying attention to. You won't be surprised to hear that people are better at remembering things that they are paying attention to than things they aren't. For example, in a study where participants read a passage about two countries, they did a better job remembering information about the country they had been told they might visit than the other country described in the same passage. What might be surprising is that participants who had been told to "read for understanding"—in other words, to pay attention to everything—failed to remember much about either country. Even though those "reading for understanding" took their instructions seriously and spent a lot of time reading about both of the countries in the passage, their retention of detail for each country was just as bad as those not paying attention to that specific country, and much worse than those who were paying attention to it

(McCrudden). What this suggests is that you can't just "pay attention to everything", because that's not what *attention* means. You have to go into the reading with some idea of what you want to get out of it, or else you won't get much out of it at all. If you don't give your mind something to focus on, the words will just flow by. With regard to autophagy, consider the difference between these two scenarios: (1) Learn the functions of all of the yeast Atg proteins. (2) Focus on selective autophagy and learn the function of Atg11 and the specific receptors for mitophagy, reticulophagy, pexophagy and the Cvt pathway. There are currently 44 fungal Atg proteins, so the idea of immediately learning the function of each one is not practical. Furthermore, without further guidance, why would you even care to learn this information? In contrast, in the second scenario you decide to focus on five proteins that all function in a related process–cargo recognition and sequestration—a very achievable goal.

But how do you know what you need to pay attention to when you haven't started reading yet? The answer is a technique called *previewing*. It takes only a little bit of time, and has been shown to help you retain the material you read. Here's how it works: before you start reading a passage, spend about 10 minutes previewing it—flipping through, looking at headings, bolded or italicized words, illustrations, things set aside in boxes; anything that can start to give you a feeling for what the reading is about and why it is important [1]. Then, once you have a little bit of sense of what the chapter is about, ask yourself a few *questions* that you want the reading to answer. Following from the example above, you may have noticed in your previewing that selective autophagy was one recurring theme and you decided you needed to know more about this topic.

Coming up with questions that you want to answer by doing the reading is the key trick of previewing, because that is what focuses your attention on what you want to learn. Our mind loves to search for answers to questions, so this is the motivation and focus that it needs (McGuire). The power of asking questions has been experimentally demonstrated, and it works not only for reading but also for listening to lectures. In one study, students in two groups watched a lecture video then took a test on the content covered. One of the groups, however, had been asked some questions regarding the content before they watched the video. That group performed much better on the test afterwards. They did the best on the content that their prequestions had focused on, but even on the content that wasn't the focus they still did better than the students who hadn't gotten any prequestions at all [2].

Maybe you are worried that the questions you come up with won't be the most relevant or important ones. Fortunately, that's not a big concern. You should be able to generate some pretty relevant ones just by glancing through the chapter to see what is emphasized. For example, if you were to preview this chapter, you might notice that this section is called "previewing," and later on, the words "previewing" and "attention" are italicized. So, you might ask yourself, "What is previewing, and what does it have to do with attention?" But even if you don't generate the best possible question, any relevant question is better than no relevant question. If the only question in your mind when you start your reading is "How long is this stupid chapter, and can I get it read before class," you are unlikely to retain much of what you read.

### 1.3.2 Step 2: Active Reading

It's very easy to slip into a passive mode of reading, especially if you are reading something that is dense or complicated. In this mode, the words just slip through your brain without leaving much of an impression—yes, technically you are reading them, but you might as well not be, since you won't remember what you read. Most students know that they need to slow down and pay attention to what they are reading if they want it to do any good. But how do you do that? It's not primarily a matter of willpower, but instead of using the right techniques. Unfortunately, many students use relatively ineffective techniques like underlining, highlighting, or reading the passage over and over again.

What should you do instead? *Paraphrase.* Taking notes on what you read is a key way to cement it into your mind. Simply underlining an important point is not nearly as effective at fixing it in your memory as writing about it, even if it's just a little note jotted in the margins (though for serious reading, it's best to take notes on a separate sheet). Critically, these notes you take should be paraphrases, not a direct copy of what the textbook says. Directly copying something does not involve very much mental processing, and will therefore not help very much with remembering.

How much should you write? This is somewhat an individual preference, but McGuire [1] recommends reading one paragraph at a time and pausing to write a summary sentence after each paragraph. This way of reading is time consuming, without a doubt! But for challenging material it can actually end up saving you time in the long run, because you won't need to spend as much time later on studying. Also, by laying that strong foundation you'll be better prepared for the next level of learning, in the classroom or as you do your homework, so that time won't be wasted either. A minimal approach to writing is to generate an outline as you read the material. Having an outline of a chapter can help you organize your thoughts about the material and can make it easier to recall the information. Again, following from the topic of selective autophagy, an outline might look something like the following:

I. Macroautophagy.
 A. Nonselective.
 B. Selective.
  1. Cytoplasm-to-vacuole targeting.
   (a) Precursor Ape1 propeptide.
   (b) Atg19.
   (c) Atg11.
  2. Mitophagy.
   (a) Atg32.
   (b) Atg11.
  3. Pexophagy.
   (a) Pex3.
   (b) Atg36.
   (c) Atg11.

   4. Reticulophagy.
     (a) Lnp1.
     (b) Atg40.
     (c) Atg11.

This outline can be expanded by including other types of selective autophagy, or by adding in additional components such as kinases.

The second key to active reading, closely related to the first, is this: you need to *think* about what you are reading; not try to memorize it, but think about it critically, and do your best to understand. Understanding something is the best way to remember it; we remember what makes sense to us. Here are some tips for what to think about as you read:

1. Try to *elaborate* and *explain* what you read. When you encounter a new fact, make a point of thinking about why it is important (that's elaborating) and how it connects to other things that you already know (that's explaining) [2]. For example, how does the formation of the phagophore membrane compare with other membrane formation processes that you have learned about? What's the significance of lipids coming via a lipid transfer protein instead of just by membrane fusion? On the topic of selective autophagy, you might wonder how recognition of the cargo is regulated. (If you don't know the answer to any of these questions yet, don't worry—that's in other chapters.)

2. Critically analyze what you read. Does it seem logical? Does it make sense? Don't just trust the authors (even the authors of this book).

3. Pay particular attention to things that seem strange or surprising; these could indicate a misconception you acquired in a previous class, or something you are misunderstanding [3]. Don't let these slide by, but work to make sense of them, as they could be key points for learning. In fact, one of the greatest barriers to learning is incorrect preconceptions that the learner brings with them, and you have to identify these before they can change.

Finally, here are some tips from the authors of the later chapters in this book about what you as a student should particularly be paying attention to:

1. Take note of the cardinal research articles presented in each of the chapters. These will be landmark papers that have shaped what we currently know about autophagy. Not only are they good options for further reading to deepen your knowledge of the topic, but understanding how the original data leads to the development of scientific models is a key part of your development as a scientist.

2. Watch for the comprehensive tables on autophagy induction and control, as that is of central importance to a practical understanding of autophagy.

3. Be aware of the many powerful model systems showcased and how autophagy is the same or different in these organisms.

4. Finally, understand the role of autophagy as basic cellular mechanism versus as a stress response, and appreciate the dynamics of the entire process.

### 1.3.3 Step 3: Understanding Graphs, Figures and Diagrams

Scientific texts are studded with graphs, images and other figures. They come in two major flavors: figures that show data and diagrams that illustrate a concept. One of the biggest mistakes you can make while reading them is to simply skip over the figures and illustrations, assuming that everything you need to know is in the text. This is not the case. Figures are extremely time-consuming to create, so believe me, the authors wouldn't have made them unless they thought that seeing that figure would be critical to your understanding.

Some concepts in science are better summed up in pictures or diagrams than in words. For example, if you need to learn the parts of a biological structure such as a plant or a cell, it would be much easier to do that from a labeled diagram than from a paragraph trying to describe how all of the parts related to each other. Other pictures show more abstract concepts, such as biochemical pathways or the progression of autophagic structures (see Fig. 2 for an example), but again can show this more efficiently in a graphical form than a paragraph.

Diagrams are dense with information, so it takes time to understand them. Pause and study the diagram, asking yourself questions like "what is the main takeaway that this is trying to show me?" and "what do the different colors and symbols mean?" It's important to be aware that the same symbol might mean different things in different figures—for example, an arrow could indicate physical movement, or progression through time, or causality, or that one thing controls another. While ideally the meaning of a figure should be obvious just from looking at it, that is not always the case. However, every figure should have a legend that explains both its purpose and the meaning of the symbols used. In addition, comparing the figure against the text can be a great way to understand both.

Other figures contain data, either quantitative data in tables and graphs, or qualitative data such as pictures. This is because science isn't just a series of logical arguments, but instead data driven, and figures are the way that the data are shown. Although textbooks show much less data than primary literature papers do, they will often include some, because sometimes seeing the data is the only way to fully appreciate a particular fact, and can help it stick in your memory.

Learning how to read graphs is a skill in its own right, but here are a few hints. First, take a look at the axes. What is being measured on each? Do these make sense? Note that sometimes there are a few steps of logic between what is being measured and the conclusions that are drawn—for example, the y-axis might measure fluorescence of a tryptophan residue, but that might actually just be a proxy for a conformational change in a protein that is being monitored. The x axis is often time, but could also be something like position, wavelength, or condition. The rule is that the x axis is the variable that the experimenters themselves are controlling (called the independent variable), such as how long the experiment is run, while the y axis is the outcome that they are measuring (called the dependent variable). However, within this general schema there are an almost infinite number of graphs that can be generated depending on what data are being shown. So, just like with a diagram, take time to make sense of the graph, using the figure legend and the text itself to help you understand it and explaining to yourself what it shows and why that is important.

Often, multiple data sets will be plotted on the same graph. These might be very different data sets, put together to help you compare them (sometimes there are even two different y-axes, one on the left and one on the right, one of each data set). Or, they might be similar data sets but taken under different experimental conditions, to allow you to compare those. Most scientific experiments involve multiple replicates—repeated measurements of the same thing. A graph might show the data for all of the replicates independently, or it might show only summary statistics, such as the mean value plus some error bars to show how much variation there was in it. "N" is used to show how many replicates were included in the analysis (this might be on the graph itself or in the figure legend), and symbols such as asterisks are often used to indicate results that pass a test of statistical significance—again, the details should be in the figure legend. All of these things are worth paying attention to, because they are important for the conclusions of the figure.

The most important question to ask yourself when interpreting a figure is "what are the authors trying to show me with this?" Are they trying to make a point about how one condition is different from another, or how they are the same? Or are they only showing one condition, but want you to see how it varies over time? Sometimes a graph will have more than one main point, in which case you might want to ask yourself which is the largest effect, or seems the most important, and then compare that to what the authors emphasize in the text. If what seems most interesting to you in a figure isn't the main thing that the authors talk about, that doesn't necessarily mean you are wrong—it might mean that you just learned something that they were ignoring. Or it could be that you misunderstood it; either way, it's something that deserves more thought.

As you read the chapters in this book, pay particular attention to the cartoons provided and the data-based figures. This is a unique opportunity to see, side by side, a scheme of the autophagy machinery with real data that supports the conclusions shown in the scheme. Real data, such as an electron micrograph, a super-resolved fluorescence image, a blot or a protein structural motif, can be harder for a novice to interpret than a cartoon, but it is the essence of science, so you should make the most of this structured opportunity to train yourself in data interpretation.

One final word of advice: figures in textbooks are typically generated by illustrators and are designed to look nice. However, it can be helpful if you redraw the figure in a simplified form, emphasizing the points or structures that you consider important. As with writing to help remember the text, redrawing a figure also helps you recall the key points.

### 1.3.4   Step 4: Summarizing What You Read

Once you have completed the reading, take a few minutes to briefly summarize what you've read. Although it may be tempting just to close the book in triumph once you've read the last word, summarizing, like previewing, is a step that doesn't take very much time but can really help you remember what you've read. The goal here isn't to recapitulate the entire chapter, but just to write a *very brief* summary that captures the main points;

this helps you to organize the information in your head and will help you to remember not only those main points but also the details that you wrote down as you were reading [3]. If those notes are like a suitcase of knowledge, the summary is like a handle that will allow you to more easily retrieve the whole batch from your memory.

The summary could be a paragraph or short list of bullet points such as the simple outline mentioned above. As an example, when learning about the (1) morphological stages of macroautophagy (see *Basics of Autophagy–A Common Vocabulary* below) you might have listed, (A) phagophore, (B) autophagosome, (C) autolysosome. That short list of terms can be easy to remember, yet it actually corresponds to a lot of information. After you have completed your reading and finished your outline you can quickly test your recall and comprehension by reproducing the outline and explaining each of the terms either verbally or in writing. If you cannot remember the meaning of a term, you now have a specific question in mind that allows you to immediately go back and read for understanding as described above.

Your summary could also be some sort of *graphical organizer* like a chart or concept map that arranges the information visually. Although making a graphical organizer takes a bit more work than writing a list, multiple studies have shown that this is particularly effective at helping you understand and remember [2]. Some information is better suited to a graphical layout than a written description, and many individuals have good visual memories. Even more importantly, taking information that was presented in written form and transforming it into a graphically organized form requires *processing* that information, and it is well demonstrated that in-depth processing is the key step for moving information into long-term memory [3]. Have you ever tried remembering something, such as a phone number, by simply repeating it over and over again? How well did that work when you tried to remember it a day or week (or even 5 minutes) later? Probably not very well. Instead, a better strategy is to process the number you are trying to memorize, for example by thinking about the relationships between the digits or connecting them to other numbers you already know [4]. The same is true for all information, and the more relevant the processing is to the actual structure of what you are trying to remember, the more effective it will be.

Three common types of graphical organizers are (1) sequence organizers, which show how events are related to each other in time or as steps in a process, (2) hierarchical organizers, which show how concepts are related to each other structurally (e.g., car and bike are both types of vehicles), and (3) matrix organizers, which are particularly good for comparing and contrasting two or more concepts [2]. In some cases, creating a diagram or illustration (a "picture organizer") might be the best way to organize the information—it just depends on what the topic is.

Concept mapping isn't only applicable to summarizing a chapter you've read. You can also use it during the note taking process, or later while studying. It works because it's a good way to process the information, and that's the biggest take-home message of this section: You read to remember by *processing* the information you are reading—elaborating, explaining, and summarizing it, in thoughts, notes, and graphical forms.

## 1.4  Strategy 2: Make Homework Worthwhile

Another resource that you will find in some of the chapters of this book is homework problems. Ah, homework—we can almost see you shudder at the word. The fact is, nobody likes homework—students don't like doing it, and professors don't like assigning it (or grading it). Then why is it that it's still a standard part of most classes? Because it works. You can't expect to get good at something without practice, which is why doing the homework is probably the most important part of preparing for the exam. But, just like every tool, in order to get the most out of it, you have to use it right. Here's how to make your homework work for you, so that you are actually learning from it and not just wasting your time.

### 1.4.1  Step 1: Do the Homework (Even if It's Optional)

It's pretty obvious that you aren't going to get any benefit from the homework if you don't do it. Skipping the homework means that you are missing out on the best opportunity to learn the material for the test, and copying from a friend or finding the answers on the internet is just as useless but more work (not to mention unethical). So, step one is simply this: do the homework.

This is true even if the homework is optional. While most professors recognize the importance of homework, not all choose to assign it. Perhaps it's because they're tired of listening to students complain about homework, or don't want to invest the time that would take them to include it in your grade. Or it could be that it's a strategy for putting your learning in your hands, so that you can decide whether or not to put the effort in. Here's our recommendation: if you want to do well in the class, put the effort in. If you do it right, it will more than pay off.

### 1.4.2  Step 2: Treat It as a Quiz

The first key to getting the most out of the homework is this: treat it as if it were a quiz. What that means is: don't immediately reach for your notes, or the book, or the internet—and importantly, do not reach for the answer (see below)—as soon as you read the question. First, try to answer it without any of those tools. Only if you are unable to answer the question just from memory should you pull out additional resources. If you find that you end up needing those extra resources on most of the questions, that's okay—you shouldn't expect yourself to know everything by heart at this stage in the process (it is, after all, only homework, not the actual test). But it is critical that you try—that before you pull out the mental crutches, you try to answer it just from your mind. This simple strategy is one of the most effective in helping students raise their grades [2]. Also, if you find that you cannot answer a question, you can go back once again and read for understanding. Remember that reading when you have a specific goal in mind, or a question to answer, is a much more effective way to learn the material.

Attempting to answer the question yourself might feel like a waste of time if most of the time you end up needing to go back to the book for help. But it's not—it's actually the most important part of the whole process. It can also be a bit discouraging, so it's tempting just to skip that step, and do the homework with your notes open, glancing back and forth. But please don't—at least if you want the homework to prepare you for the exam.

Why is trying the homework from memory so important? Because the way that you strengthen a memory is by trying to remember. A memory is much like a muscle: in order to keep it strong, you have to use it [4]. Even if you try to remember but fail, that's still helpful—just the act of trying prepares your mind so that when you do look up the information, it will stick.

Trying a problem on your own first isn't only important for simple memorization questions. It's even more important for problem-solving questions where you need to work out an answer. Problem-solving is just like any other skill: the only way to get better at it is to practice. And looking at how someone else solved the problem isn't the same as actually solving it yourself, any more than watching someone else ride a bike is the same as actually getting on and giving it a go [5]. Watching someone else might be a good first step, but most of your time will need to be spent trying it yourself—and the same is true for working problems.

One resource that you don't need to hesitate in using right away for problem solving is a blank piece of paper (or its equivalent). Solving complicated problems requires putting together many pieces of information, but there is only so much we can hold in our working memory at once. Having something to scratch notes on as we think essentially expands our working memory, allowing us to work with more concepts at once [3]. When I was younger, I used to try to do everything in my head, because I thought that writing things down was too much work and that being able to do it all mentally showed I was smart. But as I moved on to more advanced learning in college, I found that a piece of paper was increasingly indispensable, and now I would never try to solve a problem (or hardly even have a conversation) without one in hand.

How long should you try a question on your own before reaching for additional resources? That depends on the type of question. If it's a simple memorization question, such as a definition, you probably don't need to think about it for more than about 30 seconds before accepting that you can't remember and looking up the answer. On the other hand, if it is a problem to figure out, you need to spend some serious time thinking about it, probably at least 5 minutes (you may need to set a timer to force yourself to do that). Then start pulling out other resources, looking for similar examples, etc. But you have to give it a good try yourself first [1].

### 1.4.3   Step 3: Don't Abuse the Answer Key

That brings us to our next point, which is the proper use of an answer key. Problem sets often include an answer key, because that can be a useful tool when used right. But when used wrong, an answer key can be one of the most dangerous weapons for undermining the learning process. And, unfortunately, many students use them in the wrong way.

The right way to use an answer key is to first do all of the homework yourself, and then compare your answers to the key to make sure you have them all right. If yours are wrong, revisit those questions, reworking the problems or reviewing that material to figure out why you are wrong (or why the key is wrong—it happens sometimes). In this way the answer key can serve as a powerful tool to check your understanding and make sure you are on the right track.

The wrong way to use an answer key is to look at the homework problem, decide that it's too hard, and then immediately look up the answer in the key. That's totally useless—it teaches you nothing, just like if you had copied the answer from someone else. An even worse way to use the answer key is to study it, as if memorizing those answers were the key to learning the material. This might be even worse than not doing the homework at all because now you're spending time and energy doing something that isn't going to help. So, please, for your own sake, don't abuse the answer key—don't make your professor regret having given it to you!

### 1.4.4    Step 4: Make Time, Have adequate Space and Find Motivation

Doing the homework right will take a certain amount of time and effort. If you try to get it done in snatched moments in between other things, you won't have the mental resources to do it right (if at all). Instead, you should make a homework plan that features at least three things: A dedicated space, regular times, and some motivation. And, of course, this applies to reading and studying as well!

Having a dedicated place to do your homework, be it a desk in your bedroom or a favorite corner of the library, can go a long way to putting you in the right mindset to get things done. You'll want to remove your favorite distractions (e.g., probably your phone) so that you can focus, and make sure that people leave you alone. Some people need a place that is very quiet (perhaps supplemented with noise-canceling headphones), while others do better with a bit of background noise, as long as it doesn't attract their attention. Whatever it is for you, find that place and make time to go there.

When setting aside dedicated time for homework, it's better to have smaller sessions over multiple days than to try to get it done all at once. Not only will your focus and attention be better if you don't try to take on too much at once, but research has shown that *spaced repetition* leads to better retention than studying straight through. Even taking a four-minute break before reviewing the material a second time leads to better retention, and taking a one-day break is even more effective. Moreover, distributed practice is particularly important for long-term retention—spreading the studying over two days (instead of doing it all in one) leads to modestly better performance on a test one week later, but dramatically better performance when the test comes four weeks later. In other words, spreading your studying out over multiple days instead of doing it all in one cram session should help you do a little bit better on the unit exam, but way better on the cumulative final—and retain it for the next class you have. Distributed practice is key to long-term retention.

Even within a given study session, it's important to take breaks. Shorter bursts of intense work with breaks to relax in between is better for most people than trying to grind on and on. One popular way to organize that is called the "Pomodoro technique" and involves using a timer to create 25 minutes of focused work time, followed by 5 minutes of rewarding yourself with doing whatever you want [6]. You might find that different amounts of time work better for you, or you might prefer to reward yourself with a break after completing a certain amount of work rather than a certain amount of time, as I do. But the principle of interspersing focused work with rewarding breaks is the key.

That brings up our final point, which is the importance of motivation. Ideally, your fundamental motivation is your desire to learn the material. Reminding yourself every so often that the purpose of this effort *is* learning, not just getting through the work, might help you to remember to do it right. It may also be a good idea to build in some other rewards that are a little more immediate and less abstract than learning itself. What sort of reward is totally up to you and what you find rewarding. Just make sure that the magnitude of the reward scales with what you are trying to accomplish, and remember to include both little rewards for frequent effort as well as a bigger one for a job well done in the end.

## 1.5 Strategy 3: Study to Learn

Reading to remember and making homework worthwhile will go a long way to helping you learn the material. But to really master it will take more than that. This shouldn't be any surprise; most students are well aware that in order to do well in a challenging class you can't just do the minimum that is required—you also have to put in time studying for the exams. Unfortunately, many students spend a lot of study time on ineffective study strategies that don't help them learn as much as they could. Let's start by exposing a few of those:

### 1.5.1 Four Study Strategies that Don't Work

Here are four common study strategies that can take up a lot of time but not help very much with learning. Are any of these familiar?

1. *Rereading the chapter (or rewatching the lecture)*—Simply looking at the material again makes it feel familiar, giving a false sense that you know it, but actually does relatively little to help you remember it.
2. *Reviewing the homework*—Looking at the answers to problems you have already solved does almost nothing to help you learn. Reworking the problems from scratch is better, though still not as good as trying new problems you haven't worked on before.
3. *Studying the answer key to the practice test*—This is a terrible waste of a practice test, which can be a very helpful study tool when used right.
4. *Setting aside 10 PM to 4 AM the night before for a cram session.* Not only is a single study session less effective than spaced repetition over multiple days, but rest is also important to learning. You can't do your best on the exam if you are sleep deprived.

If this is how you have been studying, don't feel bad—most students do some or all of these things. In fact, it's good news, because it means you have lots of room to become more efficient. Let's look at some ways to do that.

### 1.5.2 Test Yourself!

One of the major findings of educational research of the past few decades is how powerful testing is at promoting learning. Just the act of taking a test on the material has a major effect on later test performance, often stronger than other study techniques [7]. This is why many professors have increased their use of quizzes. However, it doesn't have to be a formal quiz designed by the professor—you can (and should!) put this principle into practice every time you study. Remember, memory is like a muscle—the more you use it, the stronger it gets. When you reread material you've already learned, your memory isn't having to do any work, so it doesn't get any stronger. But when you take a test or a quiz on the material, now your memory is working as you actively try to remember. For example, one study showed that students who studied the material once then were quizzed on it three times did 50% better on the exam than students who studied the material four times [2]. So test yourself!

### 1.5.3 Flashcards and Free Recall

One classic study technique that relies on this principle is flashcards. When it comes to remembering lots of simple factual information, such as definitions, flashcards are one of the best tools. (The other is *processing*, for example coming up with connections, mnemonics, etc. that help the material make sense. Processing plus recall practice is a winning combination.) Note that it's important to make sure that you use your flashcards in both directions—don't just look at the definition and recall the word, but also look at the word and recall the definition. That way you'll be prepared for whatever comes.

Flashcards are great for memorizing simple material, but what about higher order skills such as problem solving or understanding? Even here the recall principle still works. The best way to get better at problem solving is to practice solving novel problems, which involves recalling the material you've learned and applying it to new situations. The way to understand something is to explain it, repeatedly, from memory—as if answering an essay question on a test or explaining it to a friend or classmate. In fact, *free recall* (that is, just sitting down with a blank piece of paper and seeing how much you can remember about the material) is more effective than cued recall (e.g., filling in a blank). The less prompt you have, the more your memory has to work, and the stronger it gets.

Free recall is perhaps the simplest of all study strategies to implement. All you have to do is try to remember what you have learned, then compare that to your notes to see what you missed. But simple doesn't mean easy; in contrast, free recall is hard work, particularly in the early stages of studying when you haven't mastered things yet. Just remember that, like with any workout, the reason that it is effective is because it is hard—you are making your memory work! Free recall can also be humbling; you'll probably be embarrassed to realize how much you don't remember. But wouldn't you rather be humbled by yourself in your room early in the learning process than on exam day?

### 1.5.4    Study Anytime, Anywhere, Without Missing Anything

Since free recall is so simple, it can be implemented in a number of different ways. When I was a student, I did a lot of studying by simply reviewing the information I needed to learn in my head while I was doing other things like taking a walk or washing dishes. Whatever I couldn't remember I would look up in my notes when I was back at the dorm. Of course this isn't a substitute for dedicated study sessions, but it is a good supplement to it, which can allow you to make the most of your limited time.

A potential pitfall of free recall is that you might miss some things that you forgot that you even needed to remember. To avoid that, here's a strategy for scaffolding your studying of the material vis-a-vis the outline approach mentioned above.

Step 1: Make a list of the major topics that you need to understand. (If your professor provides a study guide, that can be a tool to double check your list to make sure you haven't missed anything.)

Step 2: Commit this list of topics to memory.

Step 3: For each topic, write down what you need to know.

Step 4: To study, first bring the list of topics to mind, and then, for each topic, practice recalling what you need to know about it.

Going back to the specific example above, regarding the morphology of macroautophagy, recalling the first part of the outline would be as simple as writing "1, A, B, C." Then, fill in the minimal words that correspond to this outline, such as "1. Morphology. (A) Phagophore. (B) Autophagosome. (C) Autolysosome." Memorizing that small amount of information primes you to recall the relevant information about the morphological intermediates to answer a question such as, "Describe the key membrane compartments involved in macroautophagy." Without the outline it can be difficult to organize your thoughts to begin to answer the question. With the outline in hand, you have a clear direction to formulate your answer. We recommend practicing these outlines every day, so that you are constantly building the number of outlines you have committed to memory. As you become more comfortable with a particular outline, you may not even need to fill in the words; if you encounter a question about morphology, simply jotting down "1ABC" may be enough to organize your thoughts.

### 1.5.5    Question Everything

Recall means answering questions, and testing yourself means answering questions that you came up with. Deeper, more sophisticated questions will yield better answers that will help you learn the material at a higher level. Your study process should involve asking yourself a lot of questions—not only simple "What is…?" questions, but also more sophisticated questions. Here are some example question stems you might try, as suggested by professor of educational psychology Jeanne Ormrod [3]:

- "Explain why…".
- "How might I use … to …?"
- "What might happen if…?"
- "How are … and … similar? How are they different?"

- "What are some strengths and weaknesses of …?"
- "How might I relate this idea to something I've studied before?"

Another beneficial exercise to help prepare yourself for the exam is to write your own exam questions based on the information in your notes. Try to get inside your professor's head; what sorts of questions might they ask? Then, come back later and answer the questions, treating it like a practice exam. Writing good questions is difficult and trying to come up with your own can help you appreciate the nature of the questions you encounter on an actual exam. For example, you need to balance difficulty with fairness. A good question typically does not ask you to simply "spit back" information you have memorized (unless it is a reading quiz designed to test whether you have done the assigned reading). For example, "Name the morphological intermediates in macroautophagy" is not a very interesting question outside of a reading quiz. In contrast, "Explain the origin of yeast autophagic bodies" requires you to use the morphological information as well as have an understanding of the autophagic process.

Questioning also works well in a study group. Come up with questions for each other, and take turns testing each other. When done among friends, without judgment (it's not about how many you get right now, it's about how much you are learning for the real exam) this can be a very effective study technique. Furthermore, getting back to the point that a major hindrance to learning is the incorrect preconceptions that we bring with us, questioning each other in a group is an effective way to dispel this type of misinformation. For example, providing an incorrect answer based on a preconception, then having it politely refuted is one of the best ways to correct that misinformation in your mind, and even though it may seem a little painful, it is critical to learning.

### 1.5.6  Practice the Exam

Students and professors alike recognize that a practice exam is a powerful exam-prep tool, and thus few students will fail to use it if it is available. However, many students don't know *how* to use the practice exam for maximum effectiveness, and thus get disappointing results.

Fortunately, the right way to use a practice exam is pretty simple: treat it like an exam. Use it late in the studying process, once you've already read for understanding, done the homework, and spent time studying by free recall. Clear some time, put away your notes, and pretend it's the real thing. Do your best, never peeking at the answers, and only afterwards come back to compare your answers to the key to see how well you did. Used in this way, the practice exam is an excellent tool for assessing your current knowledge of the material and finding weak areas that you need to work on more. In addition, the simple act of taking a test will help to build those memories so you'll do better next time.

The wrong way to use a practice exam is to give up too quickly, looking up the answers in the key before you've really tried. Or, worse yet, not even trying to take the exam and only looking at the answer key. Reading the right answers by itself does very little to help you learn. Another wrong way to use the practice exam is to skim through it, telling yourself "Oh yeah, I know how to answer that," but not actually trying it (or checking against the key). Just because a problem looks familiar doesn't mean you actually know how to solve it; you can only learn that by giving it a try. Finally, don't try to use the exam as a study guide. Any professor worth their salt makes the exam different every year, so that you have to learn all of the material, not just the limited number of questions that happen to make it onto that particular exam. So don't make the mistake of assuming that you only need to study the material on the practice exam, or you may have a rude awakening on exam day.

Along these lines, we often employ in-class problem solving as part of our learning strategies for the courses we teach. After allowing groups to work out a problem we ask for a volunteer to provide the group's answer. Whether the answer is right or wrong, we immediately follow up with a question that is at least as important as the original problem, "How did you come up with that?" In other words, what was the reasoning that led to that answer? If the answer was wrong, understanding how it was derived can get to the root of the misunderstanding. Even if the answer was correct, it is not that single answer that matters so much as the method of reasoning that allowed the problem to be solved. Understanding *how* to answer the question, not just the answer to the question, is critical.

## 1.6 Strategy 4: Know What You Know

> For which of the following would you study harder: To make an A on the test, or if you had to teach the material to your classmates?

Dr. Saundra McGuire, then director of the Center for Academic Success at LSU, would ask this question of students, and almost always the answer was "to teach the material." [1]. They were right—experiments have confirmed that most students will study harder when asked to prepare to teach other students than when asked to prepare for a test. More importantly, they also learned more—students asked to prepare to teach did better on the test than students asked to study for the test [4]. In fact, pedagogical research has shown that one of the best ways to learn is for students to teach other students [8].

These results raise an interesting question. Most students care about their grades and want to earn an A on the test. So why would they work harder (and learn better) if instead of studying to get that A, they are studying to teach the material to someone else? Wouldn't you expect the best results when you are focused on doing exactly what you need to do to prepare for the test, instead of some other task? Well, that is true only if you actually know what the best way is to prepare for the test, and what these results show is that many students don't. They know that they want to get an A, but they don't know what they need to do in order to get it. If that is something that you can relate to, then what you need is more *metacognition*.

Metacognition is thinking about your own thinking—being aware of your own learning process so that you can direct it more effectively [1, 3]. It includes:

A. Accurately assessing your current level of skills and knowledge.
B. Understanding what level of skills and knowledge you need to get to in order to achieve your goal.
C. Knowing what strategies to use (and how much time to commit) to get from point A to point B.

Metacognition, just like any other type of skill, is something that can be learned and developed over time. You've probably developed some metacognition just by growing up in the educational system. However, most students have gaps in their metacognitive awareness that hold them back. For example, many students tend to *overestimate* what they already know, while at the same time *underestimating* what they need to know to meet the expectations of the test. Their bubble is burst when they get back their exam grade, but some are resilient in their misconception, just writing it off as a bad day or blaming the exam for being too hard instead of realizing that they don't understand what the professor expects of them.

Another common metacognitive mistake is the fixed mindset, where a student thinks that they are just naturally good (or bad) at a particular topic, and there's not much that they can do to change that. They might think that if they don't understand the material the first time they hear it that it's just beyond them and they aren't going to be able to get it. But in fact, you aren't expected to understand most things right away; challenging material can only be understood through a process of working with it—studying, reviewing, doing homework, and constructing your own understanding. Since we have mentioned "constructing" this is as good a place as any to bring up a brief aside. The term "constructivism" in regard to learning means that you need to create the knowledge yourself in order to truly learn it [9]. This is why watching someone ride a bike is not very effective as a means for you to learn that skill. Similarly, simply reading about autophagy without working through problems that make you use the information you have memorized will not give you true comprehension about the topic.

Finally, some students are well aware of what they know (and don't know) and what they still need to learn, but they just don't know the most effective way to get there. Learning all the material for the exam might seem like an unscalable cliff, but only because they don't have any climbing tools, or don't know about the path that winds up the other side of the mountain. If that's you, then putting into practice some of what you have already read in this chapter should be a great help.

### 1.6.1 Blooms Taxonomy—A Metacognitive Tool

It's likely that you've encountered Bloom's taxonomy before—it's hard to make it through school without acquiring at least a passing recognition of the distinctive pyramid. But do you know what it actually means, or what it is good for? I certainly didn't when I was a student.

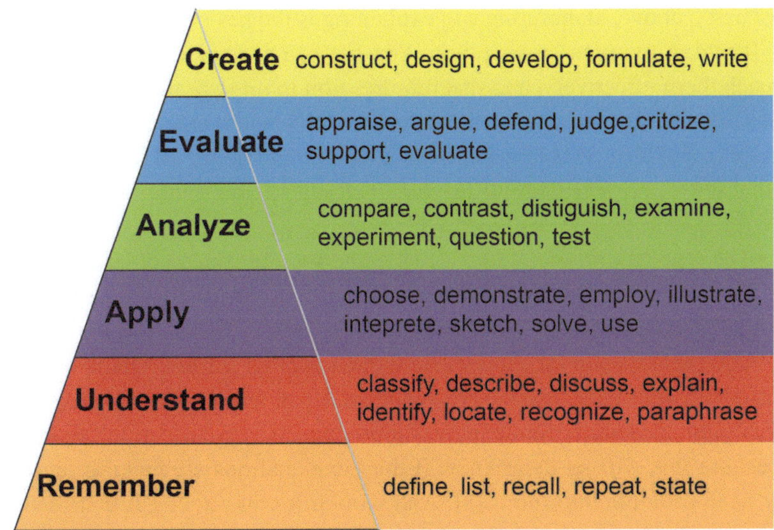


**Fig. 1** Blooms Taxonomy with examples. College level classes expect you to reach into the higher levels of the taxonomy

The basic idea of Bloom's taxonomy is that there are different ways to "know" something—different levels of understanding. While these levels of knowing are not strictly hierarchical, to some extent they are: there's no way to understand something if you don't remember the basic facts about it, for example, and you can't accurately evaluate until you first understand. One of the challenges of Bloom's taxonomy can be figuring out what the different levels actually mean; to help with that, Fig. 1 shows some descriptors for each level of a revised taxonomy developed by Dr. Overbaugh and Dr. Schulz of Old Dominion University [10].

How is Bloom's taxonomy helpful? Because it can help you to compare your current level of knowledge with what is expected for the course. As you move into more advanced courses such as this one, your professors are going to expect you to be operating at higher levels of Bloom's. In lower-level classes or high school, you might have been able to get by just remembering what the teacher said, or maybe with some level of understanding or application. But now you will likely be asked to analyze, evaluate, perhaps even create. If all of your studying has focused on memorizing facts and definitions ("remembering") and then on the exam you are asked to analyze an argument or solve a problem, you probably aren't going to be ready for that. You need to practice what you want to be able to perform [1].

### 1.6.2 Why Teaching Is the Best Way to Learn

How do you get past a "remembering" level of knowledge and begin to operate on higher levels of the taxonomy? We have already gone over a few ways, but there is one strategy that stands out above all of the others: Teach. Teaching something is the best way to learn it, particularly at a deeper level. Teaching is fundamentally a process of explaining things, and, as discussed above, elaborating and explaining is the key strategy that leads to *under-standing*. Teaching requires organization: You have to arrange the ideas in your head in order to teach them effectively, and that requires you to *analyze* them, a higher-level skill on Bloom's. In order to answer questions about what you are teaching, you need to know how to *apply* it, and also be able to *evaluate* what you are being asked. Basically, the type of effort that goes into preparing to teach something is exactly the type of effort that is most effective at helping you really learn [1, 4].

Now maybe you are thinking, "Okay, that's fine for you, since you're a teacher, but I'm not." That doesn't matter. Anyone can learn to teach, and you don't need a formal teaching opportunity to do it. One of the best opportunities to teach to learn is in a study group. You and your study partners can take turns teaching each other different parts of the material. You can also help your peer 'teacher' by being an active learner when you are in that role, asking clarifying questions. Student questions can be very effective feedback for the teacher, because that will let them know if their explanations are making sense and what areas they need to delve into more [4]. And wouldn't you rather get that feedback now, rather than after the exam is handed back?

For that matter, you don't even need a real person to teach in order to get many of the learning benefits from teaching. You can teach your stuffed animals, or line up your rock collection as an audience, or just imagine that you are teaching your family or friends. This isn't quite as good as actually teaching other students because you aren't getting that feedback, but as long as you take the role seriously it still has many of the same benefits. In the section on studying to learn we recommended spending time explaining the material to yourself (from memory). Directing those mental explanations at an imaginary listener can be an even more effective twist on this core learning strategy. And you can do it any-where, anytime that you have a moment free.

### 1.6.3 Using your Time Wisely

Speaking of time: It's no secret that mastering challenging material for a college class takes a lot of time. The techniques in this chapter should save you time, by helping you to use it more effectively, but they still have a significant time requirement. Moreover, they require certain types of time—for example, spaced repetition necessitates that you set aside study time on multiple days, well ahead of the exam. How do you make time for that in your busy life?

The classic strategy for maximizing the use of limited time is to organize it, with a schedule. A key here is that it must be a realistic schedule—your most accurate estimate of how long you need for different tasks. Of course, life is unpredictable, and you shouldn't stress about always keeping your schedule exactly, but if you've only allotted 5 minutes in the morning for shower and breakfast, or 15 minutes for a commute that usually takes at least 20, your schedule will soon fall apart. To make the schedule, think about what you do, and think about your priorities—what do you want to be spending your time doing? Don't forget to make time in your schedule for rest, relaxation, socializing, exercise, self-care and downtime. "All work and no play" isn't sustainable; everyone needs time to rest, and to goof off. These sorts of activities can be used as rewards for working hard and staying focused, and interspersing focused studying with more unstructured time can actually help with problem solving and creativity [6]. But scheduling them helps to make sure that they don't sprawl across your whole day. And remember: Work first, then play.

Another critical thing is to remember that learning does take time, and a class can't be added to a busy schedule without giving up something else. Take stock of your schedule, and think about what adjustments you might need to make. Do you need to give up some other activity, or tell your friends you are busy? Do you need to ask a family member to help with childcare, or cut back your hours at your job? Remember: education is an investment, and it can be an expensive one! So make the time you need to get the most out of it.

### 1.6.4 Start Now!

If you've read this chapter up to this point, you've heard about a lot of strategies that can help you to excel in this class. But none of them will help you at all unless you actually put them into practice. If all you do is *read* this chapter, then you've wasted your time.

You don't need to try everything in this chapter in order for it to make a difference. In fact, it might be more effective to start with a few things that seem most compelling to you. But try something; put it to the test. Not everything works for everyone, but these are the best practices that work for most people, and we are quite confident that at least some of them will make a difference for you.

The best way to ensure that you will actually make a change this semester to improve your learning is to start now. You can practice active reading on the very first chapter (even this one—have you been taking notes?), start your very first homework assignment as soon as you get it and treat it like a test, make a schedule for the semester, and begin thinking like a teacher in order to learn. Don't try it the old way first—you already tried that, in previous classes. Now you have a new set of tools for learning, so give them a spin.

## 2 Strategies for Teaching

While the previous section provides tips for helping students become better learners, in this section we will address how instructors can align their teaching strategies with what is known about how students learn. While much has been written about best practices in teaching, this section will provide some ideas to get you started about how to use this book to promote student engagement.

## 2.1 Interactive Teaching

Effective instructors each may have a different teaching style, or emphasize different concepts and skills, but what they all have in common is that they utilize an interactive engagement approach in the classroom. There is ample evidence that "teaching by telling", i.e.

lecturing the entire class period, is not effective for promoting lasting learning and comprehension [11]. Instructors need to be aware of how students learn (see *Strategies for learning*, above) in order to teach effectively, and they need to teach students how to learn, how to study, and what to do to achieve the learning goals. In order to help students learn, the instructor has to help students think about and apply the material, and this is not accomplished by lecturing, no matter how eloquent and entertaining the speaker [12].

So, what do we do with a class period if we are not lecturing the whole time? Consider how many faculty meetings you have attended where the information presented could have just as readily, and often more efficiently, been conveyed via an email. It is likely that any time that you are talking during class, if it is not answering questions or posing questions for students to answer, it is probably something that students could have attended to on their own time. While the analogy of teacher as coach is often made, think about the following: before practice, football players read and study the manual of plays that they will be working on. The plays are given as new diagrams that might be somewhat confusing, but players come to practice already familiar with the names of the plays and their general idea, even if they do not know exactly what they are supposed to be doing on the field or how to do it. Similarly, students can read the text, and begin to grasp the general ideas, but they may not fully comprehend the material or understand its significance and utility. During practice, the coach runs the team through the plays, correcting players, answering their questions, and giving them feedback on their performance. Similarly, in the classroom, instructors give students practice using the material, explaining key points, and providing students feedback on their understanding. The result is that both the football players and the students know what they need to work on and the drills they must do to attain mastery, after practice and class is over. In other words, they have a scaffold for how to learn.

Structuring a non-lecture based, interactive course in this way requires that the instructor consider the alignment of three key aspects of teaching: (1) learning goals, (2) learning activities, and (3) assessment. Unfortunately, most instructors at the college level are not trained as teachers and do not consider these important points when developing their own courses, instead focusing primarily on factual content.

### 2.1.1 Learning Goals

First the instructor should consider how they want students to be changed by the end of the class period, the end of the unit, or the end of the course. This means not only what the students will know, but what they will be able to DO with that knowledge. This book is filled with amazing information—it has the answers, but what are the questions? Learning goals that go beyond memorization of facts are essential to developing future scientists. For example, as autophagy has gained attention from the general public over the past decade there has been a substantial increase in misinformation available both online and in print. Thus, one goal may be for students to be able to recognize incorrect information that is presented via social media. This goal is slightly different from but complementary to learning the facts about autophagy, and it has a very practical/applied aspect.

## 2.1.2    Learning Activities

With the learning goals in mind, instructors can next consider how to structure class learning activities to give students practice and feedback on their progress towards the goals. These learning activities provide a model for the "workouts" the students will do when they study and continue their learning after class is over (see *Strategies for learning* above), in preparation for a test of their mastery, be it an exam, paper, presentation, or project.

In designing a course, the instructor should plan for what the student does before, during, and after class. The basic structure is to assign students readings to complete before class so they are prepared to utilize the information they read during in-class problem solving activities. Instructors can help students understand what to focus on in their pre-class reading, by providing students with a list of key terms for the student to define, and a set of "reading questions" to answer as they complete the readings. These questions should not be overly complex, but rather, lower-level Bloom's questions (see *Strategies for learning*) that students can answer readily from the assigned readings. In this way the student is prepared to apply the material on the higher Bloom's level questions posed in class, where the instructor is present to guide the students through the complexities.

The class period might start with a short lecture of 7–10 minutes –not more than 20 minutes, given the fact that attention wanes at this time point [8]—followed by a learning activity, and another short lecture. The learning activity is a task that students work on individually, in pairs, in groups, or as a whole class discussion, and often takes the form of questions, asking for ideas or solutions, and importantly, requiring students to justify their answers and provide their reasoning. Students are given the opportunity to demonstrate their conceptual understanding, and to practice the skills defined in the learning goals. These in-class formative assessments address misconceptions and provide real time feedback. Note that "formative" assessments are designed to provide information back to the student, in contrast with "summative" assessments that simply provide a grade. As an example, a learning goal may be that students will have the ability to design controlled experiments. The learning activity to develop this ability would be to have the students design experiments, and get feedback on their design, from each other and from the instructor, perhaps using a checklist of essential design elements. With this feedback, the students can learn how to improve their design.

The in-class learning activities and formative assessments give students feedback on what they know and understand so far, and what they need to work on. Frequent formative assessments are one of the best ways to help students learn [7]. However, students will need to spend additional time on their own after class working to master the material. Some of this additional work will be self-directed, but a set of homework problems is another major tool that you can provide to help students learn. It is important to have a variety of types of questions, in particular including higher level Bloom's questions so that they understand what level they should be engaging with the material. On average, the difficulty of the homework questions should be a little less than the in-class questions were, but a little more than the exam questions will be. In addition, there should also be a range of difficulties, and it is optimal to organize the homework with the easier/lower Bloom's level questions first, to help them build confidence and basic knowledge before moving on to the more in-depth and open-ended questions.

Assigning points for the homework is a significant motivator that indicates to the students how important it is for their learning. The greatest motivation is if the homework is graded; however, simply giving points for completion is sufficient motivation for most students, particularly if they understand that the purpose of the homework is to help them prepare for the exams, and takes less instructor time. In addition to the assigned homework, students should test their own understanding of the material as explained in *Strategies for learning*. As necessary, they should review the pre-class readings and their notes, rework the in-class problems and practice reteaching the material to themselves and others, all in preparation for a culminating summative test of their new understanding.

### 2.1.3 Assessment

While the in-class activities allow the instructor and the students to find out what they understand and where they are in their progression towards the learning goals, at the end of a unit, a summative assessment, such as an exam, measures achievement of the learning goals. In designing exams, instructors need to be sure their tests are in alignment with the learning goals and activities: if you have decided that it's the big concepts that are important, make sure that you are testing the big concepts, not simply recall of details [3]. A test to see if students have learned the elements required to design controlled experiments, would be questions asking them to design experiments, not to simply recall or recognize the items on a checklist of essential design elements.

Teaching by questioning is not a new concept (think "Socrates"), but coming up with good questions is perhaps the most challenging aspect of this approach. Instructors need to minimize excessive jargon and be aware of expert terminology that may initially be confusing to novice learners. Multiple terms that generally mean the same thing, that experts easily recognize, can confuse students, consider for example terms such as transcript versus mRNA, or autophagosome versus autophagic vacuole. Consistent use of the same terms when introducing new concepts, can help to minimize the cognitive load. Similarly, instructors must consider the implicit assumptions in visual representations of cellular constructs and processes. For example, for the novice it is important to be explicit about what arrows or lines and circles represent in diagrams of cellular processes. Does the arrow represent physical movement, perhaps that of a transport vesicle, or is it standing in for a more complicated but otherwise not shown molecular process, such as transcription and translation? Perhaps the arrow is simply pointing to a structure but lacks an explanatory word or phrase. Similarly, lines can cause confusion when diagramming cellular components and attempting to show a structure such as a membrane: we have lipid bilayers, single membranes, and double membranes, so consistency and clarity about what our lines are representing is essential. Does your circle represent a bacterial cell, a mammalian cell, a lysosome, a protein, etc. (Fig. 2)? In addition, even advanced students may need guidance on the meaning of symbols and how to interpret complicated graphs. The

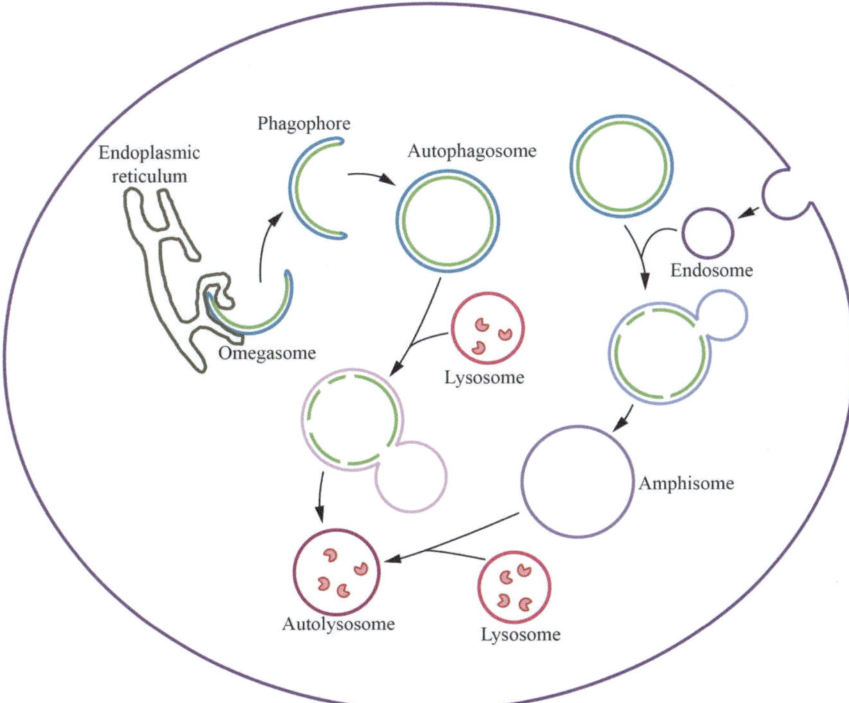


Amphisome: A single-membrane product of an endosome fusing with an autophagosome.
Autolysosome: The degradative organelle resulting from the fusion of an amphisome or autophagosome with a lysosome.
Autophagosome: A double-membrane vesicle that is the completed form of the phagophore.
Lysosome: The degradative organelle (equivalent to the vacuole in yeast).
Omegasome: A portion of the endoplasmic reticulum that nucleates the phagophore.
Phagophore: A transient double-membrane sequestering organelle of macroautophagy.

**Fig. 2** The morphological intermediates of mammalian macroautophagy. Note the presence of many circles that represent different compartments. One key point is that the phagophore and autophagosome have double-membranes, whereas the lysosome, endosome, amphisome and autolysosome all have single membranes. Also note that a typical lysosome is smaller than an autophagosome, in contrast to the situation with a yeast or plant vacuole. An autophagosome may first fuse with an endosome to form an amphisome, which then fuses with a lysosome, or the autophagosome-lysosome fusion may happen directly. A brief definition of each compartment is provided below the schematic

expert through familiarity can be blind to the aspects of an image that make it bewildering to the novice trying to make sense of it for the first time.

In describing biological mechanisms, we often leave students with "black boxes." For example, students can "know" that genotype influences phenotype, that genes determine traits, but they often have a "black box" concerning the mechanism, even though separately they may be able to describe the steps of transcription and translation resulting in a protein [13]. Explaining these black boxes to students, helping them fill in the mechanisms where they are known, and to understand the difference between those "black boxes" that exist because a process is still unknown, will help students develop a deeper conceptual understanding, as well as be better aware of the process of science and the questions remaining to be answered [14].

## 3    Basics of Autophagy–A Common Vocabulary

Although it is often stated that there are more new words in an introductory biology course than in a foreign language course, this is not true in a strict sense [15]. However, "number of new words" is probably too simplistic when it comes to understanding the difficulty of learning a new subject. For example, "escuela" and "Schule" may be new words for "school" to learn in Spanish and German, but they are likely easier for most English-speaking students to learn or comprehend than "phagophore" or "autophagosome". We need to have a common vocabulary in order to communicate; however, it may be best to minimize the number of new terms and focus on the key concepts [16]. With these concerns in mind, we suggest a minimum number of terms that are necessary to carry out a discussion about autophagy with a focus on the mechanism (Fig. 2).

One goal of autophagy is to capture portions of the cytoplasm and deliver them to the lysosome or vacuole. Accordingly, the cell needs to generate a compartment that can carry out the engulfment or sequestration, and that transient structure is called a phagophore. In mammalian cells, the membrane that nucleates the phagophore derives at least in part from a specialized region of the endoplasmic reticulum referred to as the omegasome. The phagophore is a dynamic compartment that expands and eventually seals to form a double-membrane autophagosome. Although most autophagy researchers, especially those new to the field, are familiar with the term "autophagosome," it is important to understand the role of the phagophore. The autophagosome is relatively easy to detect by electron microscopy [17], which is one reason people associate it with autophagy; however, it is essentially a terminal structure and its main purpose is to deliver its contents to the lysosome. In contrast, the phagophore is the compartment that carries out the actual sequestration process. In selective autophagy, receptors attached to specific cargo bind Atg8-family proteins on phagophores. When the sequestration step is completed, the phagophore closes to form the autophagosome. The autophagosome may then fuse with an endosome to form an amphisome, which then fuses with a lysosome, or the autophagosome may fuse directly with a lysosome. The resulting compartment in either case becomes an autolysosome. Autolysosomes are regenerated into lysosomes through a process called autophagic lysosome reformation where membrane tubules bud from the autolysosome and later mature into new lysosomes. To avoid confusion with other processes it is important to differentiate between "phagosome" (the product of phagocytosis) and "autophagosome" (the product of autophagy), as well as between "autolysosome" and "autophagolysosome" (a specialized structure seen only in xenophagy, a specific type of selective autophagy) [18].

As alluded to above, perhaps the greatest confusion with regard to morphology and the molecular events associated with autophagy concerns the differences between the phagophore and autophagosome. Hence, many people refer to the "expansion of" and "sequestration by" the autophagosome or in some cases to recruitment of proteins to this structure. In general, the autophagosome does not expand or sequester cargo as it is already sealed; hence, cargo is subsequently sequestered "within an autophagosome," not "by an autophagosome." This is more than a matter of semantics. For example, the SNARE protein STX17 (syntaxin 17), which is involved in the fusion of autophagosomes with lysosomes, is recruited to autophagosomes [19]. However, recent data [20] indicate an additional role for STX17 in an earlier step of the process. Hence, it is important to distinguish which STX17 population is being discussed, and which autophagic structure is involved. Similarly, Atg8-family proteins are typically recruited to and conjugated at the phagophore, not the autophagosome. Those proteins are present on both sides of the phagophore but are removed from the outer surface of the autophagosome. These two structures can be differentiated in part by the presence of components such as the ATG12–ATG5-ATG16L1 complex, which is located on the phagophore, but not the autophagosome. Correctly interpreting and explaining data requires the use of the correct nomenclature.

**Take-Home Messages**
- There are many evidence-based strategies for helping students learn more efficiently so that they can get the most out of the pre-class readings and the homework, study more effectively, and excel in class. But these strategies only work if used!
- Similarly, instructors can use what is known about how students learn to implement strategies to encourage good learning habits and teach more effectively. The key approach is to teach interactively to facilitate student engagement.
- Finally, a solid grounding in the basic terminology of autophagy presented in this chapter will help to keep the students from getting confused as they delve into the chapters ahead.

*Autophagy is a fascinating topic, and there are many exciting things to be learned from the book. Now you have the learning tools you need to make effective use of the rest of this material—so go and put them into practice!*

## References

1. McGuire SY, McGuire S. Teach students how to learn: strategies you can incorporate into any course to improve student metacognition, study skills, and motivation [Internet]. First. Sterling: Stylus Publishing, LLC; 2015. Available from: https://go.exlibris.link/HWfyqwQ6

2. McCrudden MT, McNamara DS. Cognition in education. New York: Routledge, Taylor & Francis Group; 2018. (Ed psych insights)
3. Ormrod JE. How we think and learn: theoretical perspectives and practical implications. New York: Cambridge University Press; 2017.
4. Schwartz DL, Tsang JM, Blair KP. The ABCs of how we learn: 26 scientifically proven approaches, how they work, and when to use them. 1st ed. New York: W.W. Norton & Company, Inc.; 2016. (Norton books in education)
5. Griswold EN, Klionsky DJ. Not learning how to ride a bike: the lecture approach. Biochem Mol Biol Educ. 2015;43(3):210.
6. Oakley B. A mind for numbers: how to excel at math and science (even if you flunked algebra) [Internet]. New York: Jeremy P. Tarcher; 2014. Available from: https://go.exlibris.link/YBlSXHCD
7. Klionsky DJ. The quiz factor. CBE Life Sci Educ. 2008;7(3):265–6.
8. McKeachie WJ. Teaching tips: strategies, research and theory for college and university teachers. Lexington: Heath; 1986.
9. Gerstman J, Salehi K, Lobo A. Developing a model of student-centred teaching which enhances active engagement. Int J Learn Annu Rev. 2012;18(7):13–30.
10. Overbaugh R, Schultz L. Bloom's Taxonomy Handout [Internet]. Old Dominion University; [cited 2023 Oct 3]. Available from: https://ww1.odu.edu/content/dam/odu/col-dept/teaching-learning/docs/blooms-taxonomy-handout.pdf
11. Freeman S, Eddy SL, McDonough M, Smith MK, Okoroafor N, Jordt H, et al. Active learning increases student performance in science, engineering, and mathematics. Proc Natl Acad Sci. 2014;111(23):8410–5.
12. Deslauriers L, McCarty LS, Miller K, Callaghan K, Kestin G. Measuring actual learning versus feeling of learning in response to being actively engaged in the classroom. Proc Natl Acad Sci. 2019;116(39):19251–7.
13. Newman DL, Coakley A, Link A, Mills K, Wright LK. Punnett squares or protein production? The expert-novice divide for conceptions of genes and gene expression. CBE Life Sci Educ. 2021;20(4):ar53.
14. Haskel-Ittah M. Explanatory black boxes and mechanistic reasoning. J Res Sci Teach. 2023;60(4):915–33.
15. Thonney T. Analyzing the vocabulary demands of introductory college textbooks. Am Biol Teach. 2016;78(5):389–95.
16. McDonnell L, Barker MK, Wieman C. Concepts first, jargon second improves student articulation of understanding. Biochem Mol Biol Educ. 2016;44(1):12–9.
17. Eskelinen EL, Reggiori F, Baba M, Kovács AL, Seglen PO. Seeing is believing: the impact of electron microscopy on autophagy research. Autophagy. 2011 Sep;7(9):935–56.
18. Klionsky DJ, Eskelinen EL, Deretic V. Autophagosomes, phagosomes, autolysosomes, phagolysosomes, autophagolysosomes... wait, I'm confused. Autophagy. 2014;10(4):549–51.
19. Itakura E, Kishi-Itakura C, Mizushima N. The hairpin-type tail-anchored SNARE syntaxin 17 targets to autophagosomes for fusion with endosomes/lysosomes. Cell. 2012;151(6):1256–69.
20. Shen Q, Shi Y, Liu J, Su H, Huang J, Zhang Y, et al. Acetylation of STX17 (syntaxin 17) controls autophagosome maturation. Autophagy. 2021;17(5):1157–69.

# Discerning Autophagy Pathway Intermediates: Transmission Electron Microscopy Techniques


Joanna Biazik, Katri Kallio, Sigurdur Runar Gudmundsson, and Eeva-Liisa Eskelinen



**What Will You Learn in This Chapter**
In this chapter, you will learn guidelines on how to identify macroautophagy/autophagy pathway intermediates in transmission electron microscopy samples prepared using conventional techniques including chemical fixation and plastic embedding. We will describe the typical morphological features of phagophores, autophagosomes, amphisomes and autolysosomes induced by amino-acid starvation in mammalian cells. We will point out what you should keep in mind when interpreting transmission electron microscope images taken of very thin sections cut from much thicker cells and organelles. You will also learn how immuno-electron microscopy of endogenous MAP1 LC3/LC3 can be used to help identification of autophagy pathway intermediates.



J. Biazik
Molecular and Integrative Biosciences, University of Helsinki, Helsinki, Finland

Electron Microscopy Unit, Mark Wainwright Analytical Centre, The University of New South Wales, Sydney, NSW, Australia

K. Kallio
Molecular and Integrative Biosciences, University of Helsinki, Helsinki, Finland

FinVector Oy, Kuopio, Finland

S. R. Gudmundsson
Molecular and Integrative Biosciences, University of Helsinki, Helsinki, Finland

Biomedical Center, School of Health Sciences, University of Iceland, Reykjavik, Iceland

E.-L. Eskelinen (✉)
Institute of Biomedicine, University of Turku, Turku, Finland
e-mail: eeva-liisa.eskelinen@utu.fi

# 1    Introduction

Autophagy is a degradative pathway in which cytoplasmic components, including whole organelles like mitochondria and parts of the endoplasmic reticulum, are first enwrapped into membrane-bound vesicles called autophagosomes, which then deliver their cytoplasmic cargo to lysosomes for degradation and recycling (Fig. 1). Autophagy was originally discovered using electron microscopy [1]. After the autophagy genes and proteins were discovered, it became possible to develop numerous new methods to monitor the activity of the autophagic pathway in cells and tissues [2]. These methods include monitoring the lipidation of the autophagy marker protein LC3 (microtubule-associated protein light chain 3) by immunoblotting [3, 4], as well as quantifying the amount of LC3-positive vesicles in cells by immunofluorescence. LC3 can exist in the cytoplasm as a non-lipidated protein called LC3-I. Once autophagy is induced, LC3 is conjugated to a lipid called phosphatidyl ethanol amine. The lipidated form is called LC3-II, and it associates with the autophagic membranes, including phagophores and the inner and outer limiting membranes of autophagosomes. Thus, LC3 is a marker protein for phagophores and autophagosomes, and to some extent also for amphisomes and even autolysosomes. However, LC3 dissociates from the outer limiting membrane when autophagosomes mature into amphisomes and autolysosomes, and the LC3 that is trapped inside autophagosomes will eventually be degraded together with the cytoplasmic cargo [3, 5].

Tandem fluorescent tagged LC3 can also be used to monitor maturation of autophagosomes to acidic and degradative autolysosomes [6]. This is based on the decrease in pH;

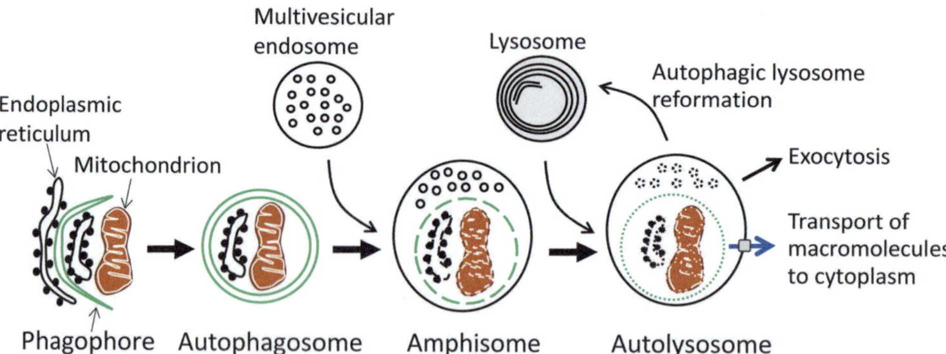


**Fig. 1** Schematic drawing of the macroautophagy pathway when induced by amino acid starvation in a mammalian cell. In many cell types, phagophores form close to rough endoplasmic reticulum and often also segregate pieces of it. Closed autophagosomes can fuse with multivesicular endosomes, forming amphisomes. Fusion of an amphisome or an autophagosome with a lysosome forms an autolysosome. After degradation of the cytoplasmic cargo by lysosomal hydrolases, the degradation products are transported back to cytosol by special pump proteins in the autolysosome limiting membrane. Lysosomes can reform from autolysosomes, and fuse with new amphisomes or autophagosomes. Autolysosomes can also fuse with the plasma membrane and empty their contents to the extracellular space. This is called exocytosis

autophagosomes are neutral while autolysosomes are acidic. The method requires expression of the reporter tandem-tagged LC3 in the cells or tissues. The tandem tag includes both green and red fluorescent protein. In the acidic milieu of autolysosomes, the fluorescence of certain green fluorescent protein (GFP) variants is quenched [7], while the red fluorophores like monomeric red fluorescent protein (mRFP) and mCherry are more acid resistant [6]. Thus, the neutral autophagosomes will show both green and red fluorescence, while the acidic autolysosomes show only red fluorescence.

Despite the numerous new techniques for monitoring autophagy, electron microscopy is still used for this purpose. It is a relatively sensitive technique for detecting the accumulation of autophagy pathway intermediates, especially when combined with careful quantification. The advantage of electron microscopy is that there is no need for the expression of exogenous marker proteins. The disadvantage of electron microscopy is that sample preparation is time consuming compared to many of the newer methods. Further, special equipment and know-how are necessary for sample preparation. In addition, interpretation of the images needs special expertise. This chapter introduces the basic principles that help in identification of autophagic pathway intermediates including phagophores, autophagosomes, amphisomes and autolysosomes in electron microscopy samples prepared using chemical fixation and plastic embedding. The examples we show were prepared using cultured mammalian cells induced for autophagy by 1–2-h serum and amino-acid starvation. Detailed protocols that can be used for sample preparation and quantification of the images have been published elsewhere [8, 9]. The pre-embedding immuno-electron microscopy protocol for LC3 is described in Biazik et al. [10].

## 2 Phagophores Are Curved Membrane Cisterns with a Narrow Intermembrane Space

Phagophores, also called isolation membranes, are the first autophagic structures that can be identified in electron-microscopy samples by morphology. They are rarely seen in thin sections prepared from cells kept under non-starved, basal conditions, but they appear soon when the cells are kept under amino-acid free conditions. Because phagophores are short lived, their number in normal wild-type cells is lower than that of autophagosomes. However, phagophores accumulate in cells if certain autophagy genes have been knocked out or silenced. For instance, cells deficient in the autophagy genes *Atg3*, *Atg5* or *Atg7* accumulate phagophores [11]. Of note, the contrast of the phagophore membranes can be selectively enhanced by using reduced osmium tetroxide during sample preparation. Either potassium ferrocyanide or imidazole can be added as reducing agents during post-fixation [8, 12].

Phagophores are membrane cisterns of varying length and curvature (red arrows in Fig. 2a–e). Their membranes do not have bound ribosomes, which differentiates them from rough endoplasmic reticulum (Fig. 2b–e). The shape of elongated phagophores is typically ellipsoidal or U-shaped (Fig. 2a–e). After closure of the phagophore, the

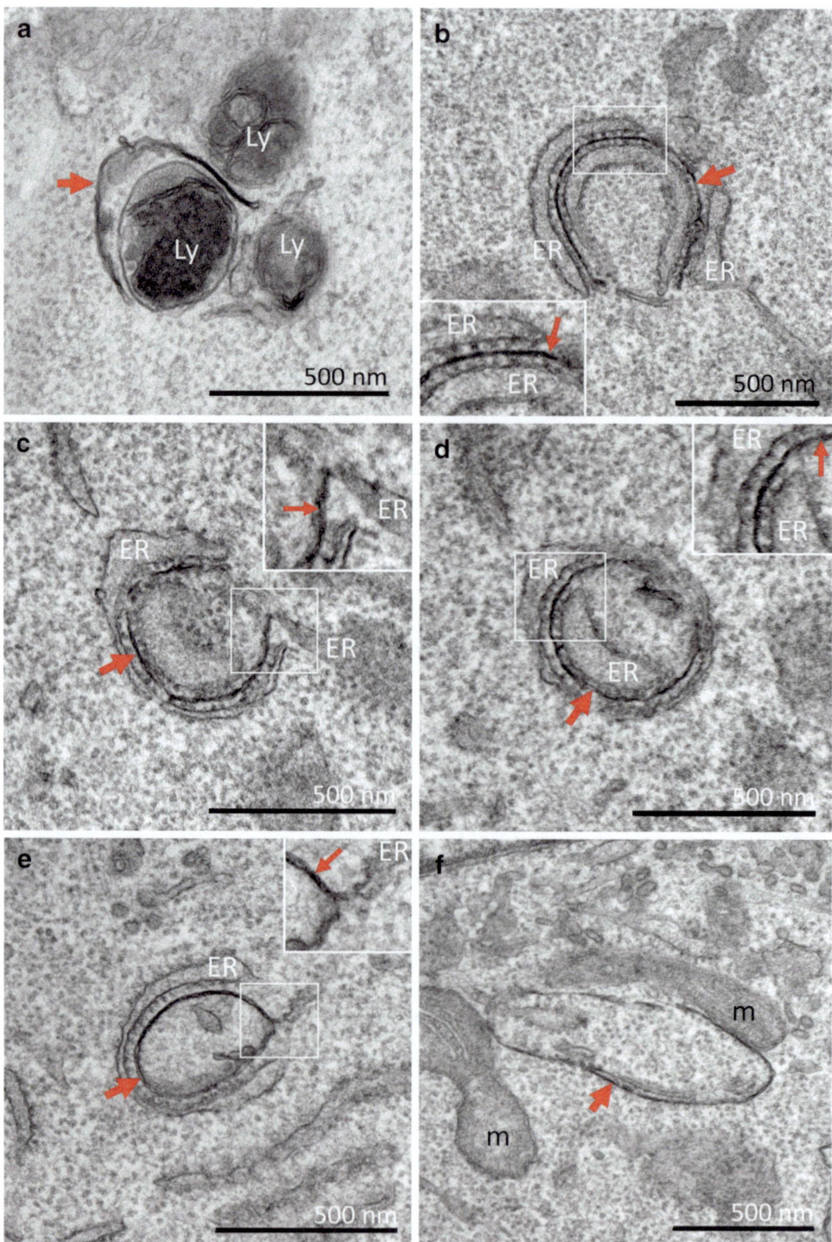

**Fig. 2** Transmission electron microscopy images of phagophores (**a–e**) and a putative nascent autophagosome (**f**) in cultured mammalian cells. Autophagy was induced by amino-acid starvation. The samples were fixed in glutaraldehyde, postfixed in reduced osmium tetroxide, dehydrated in ethanol, and embedded in epoxy resin. The red arrows indicate the phagophore membrane (**a–e**) or the limiting membrane of the putative nascent autophagosomes (**f**). Note the close spatial relationship of the rough endoplasmic reticulum (ER) and the phagophore in panels **b–e**. In panels **c** and **e**, the rim of the phagophore seems to be continuous with ER (inserts). Ly, lysosome; m, mitochondrion

newly-formed autophagosome may have an elongated shape (Fig. 2f). Phagophores are frequently located close to cisterns of endoplasmic reticulum, which in many cell types have ribosomes attached to them, especially when autophagy is induced by amino-acid starvation [13, 14]. Ribosomes are very helpful for identifying cisterns of endoplasmic reticulum (Fig. 2b–e). In many cell types, amino-acid starvation induced phagophores can be lined by cisterns of rough endoplasmic reticulum on both sides [10, 14, 15].

The width of the space between the membranes of phagophores varies depending on the sample preparation protocol, but generally it is significantly narrower than that of cisterns of endoplasmic reticulum (Fig. 2b–e). Careful analysis of samples immobilized using cryofixation, which is able to preserve the native ultrastructure remarkably better than traditional chemical fixation with glutaraldehyde and osmium tetroxide, showed recently that the intermembrane distance of phagophores decreases when the phagophores elongate [16]. However, this phenomenon may not be visible in samples prepared using chemical fixation.

## 3  Autophagosomes Are Spherical, Limited by Two Lipid Bilayers, and Contain Cytoplasmic Cargo

When the phagophore closes, it forms and autophagosome that is limited by two lipid bilayers (Fig. 1). Similar to phagophores, autophagosome membrane do not have bound ribosomes, and their contrast can be enhanced by postfixation in reduced osmium tetroxide [12]. The space between the two limiting membranes may vary depending on sample preparation protocol (Fig. 3a–e). In samples incubated in reduced osmium tetroxide, the two limiting membranes are typically very close to each other, which may look like one thick limiting membrane (Fig. 3d). In samples immobilized by cryofixation, autophagosomes are almost perfectly spherical, and the distance between the inner and outer limiting membranes is very uniform and approximately the same as the thickness of one lipid bilayer [16]. The diameter of autophagosomes varies depending on the cell type. In fibroblasts, the diameter typically varies between 400 nm and 1 μm, while in hepatocytes the size can be considerably larger. In selective autophagy, such as autophagy of intracellular bacteria, the diameter of the autophagosome is determined by the cargo [17]. Similar to phagophores, autophagosomes can localize close to endoplasmic reticulum, which typically has ribosomes attached to it, especially in amino-acid-starved cells (Fig. 3a, b) [10, 14, 15].

In addition to the two limiting membranes, autophagosomes are characterized by the cytoplasmic cargo (Fig. 3). The cytoplasmic cargo is an important feature for identification of autophagosomes in electron microscopy samples. Because autophagosomes have not yet fused with any degradative organelles, the cytoplasmic cargo is still intact. This means that the morphology of the cytoplasmic cargo is similar to that of the cytoplasm that surrounds the autophagosome.

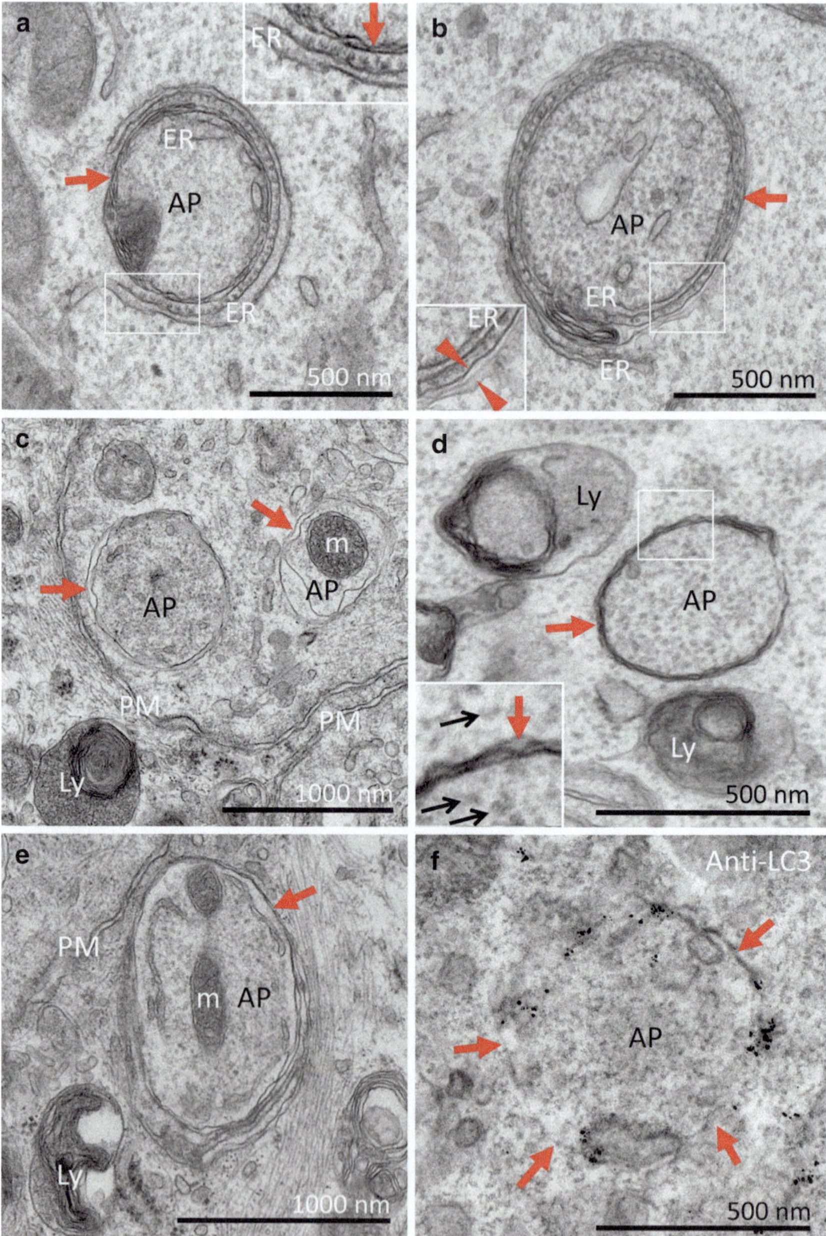





**Fig. 3** Transmission electron microscopy images of autophagosomes (AP). The samples in panels **a–e** were prepared as in Fig. 2. Panel **f** shows pre-embedding immunostaining for endogenous LC3 [10]. The red arrows indicate the limiting membranes of the autophagosomes. Note that the distance between the two limiting membranes varies. The membranes may be tightly attached to each other (panel **d**), or a narrow electron-lucent space may be visible between them (panel **c**). Rough endoplasmic reticulum (ER) surrounds the autophagosome in panel **a**. The red arrowheads in panel

During amino-acid starvation, autophagosomes typically engulf ribosomes and rough endoplasmic reticulum (Fig. 3a–f), and occasionally mitochondria and other organelles (Fig. 3c, e). Especially ribosomes are practically always visible among the cargo in starvation-induced autophagosomes and serve as an important feature for identification (Fig. 3). There are also several types of selective autophagy, in which the phagophores selectively engulf cargo such as mitochondria (mitophagy), endoplasmic reticulum (reticulophagy), peroxisomes (pexophagy), lipid droplets (lipophagy), protein aggregates or condensates (aggrephagy) or intracellular bacteria (xenophagy) [18]. Endogenous LC3 can be localized on autophagosome limiting membranes by immuno-electron microscopy (Fig. 3f).

## 4    Amphisomes Are Born When Autophagosomes Fuse with Multivesicular Endosomes

Autophagosomes often fuse with multivesicular endosomes (also called multivesicular bodies) before fusion with lysosomes (Fig. 1). Multivesicular endosomes are characterized by their intralumenal vesicles, which emerge by invagination and pinching off from the limiting membrane. In electron-microscopy samples, amphisomes are characterized as membrane-bound vesicles that contain small intralumenal vesicles in addition to the cytoplasmic cargo (Fig. 4). The cargo can appear intact, similar to the cytoplasm around the amphisome (Fig. 4a), or partially disintegrated, because amphisomes are acidic (Fig. 4b–e). Ribosomal cargo can still be identified as electron-dense particulate material (yellow asterisks in Fig. 4b–f). Amphisomes typically have one limiting membrane visible in the electron microscopy samples (Fig. 4b–e), but if the fusion with the multivesicular endosome is recent, the two limiting membranes may still be visible (Fig. 4a). Occasionally, several autophagosomes may be seen to have fused with each other and with a multivesicular endosome, forming an amphisomes containing several cytoplasmic 'packages' lined by one lipid bilayer (the inner limiting membrane of the autophagosome) as well as the small internal vesicles delivered by fusion with the multivesicular endosome (Fig. 4a). Endogenous LC3 can still be localized in amphisomes by immuno-electron microscopy (Fig. 4f).

**Fig. 3 (continued)**  **b** insert indicate the inner and outer limiting membranes of the autophagosome. The outer limiting membrane appears abnormally thick likely due to alterations during sample preparation. A cistern of rough endoplasmic reticulum is located inside the autophagosome in panel **b**, very close to the inner limiting membrane. The cytoplasmic cargo includes ribosomes in all panels. The black arrows in the insert in panel **d** indicate some of the ribosomes in cytosol and inside the autophagosome. Autophagosomes in panels **c** and **e** also contain mitochondria (m). Panel **c** shows cytoplasm of three different cells, and panel **e** shows cytoplasm of two different cells; PM indicates the adjacent plasma membranes. Immunostaining for endogenous LC3 in panel **f** shows that the limiting membrane of the autophagosome contains LC3; the black dots are silver-enhanced 1.2-nm gold particles

 

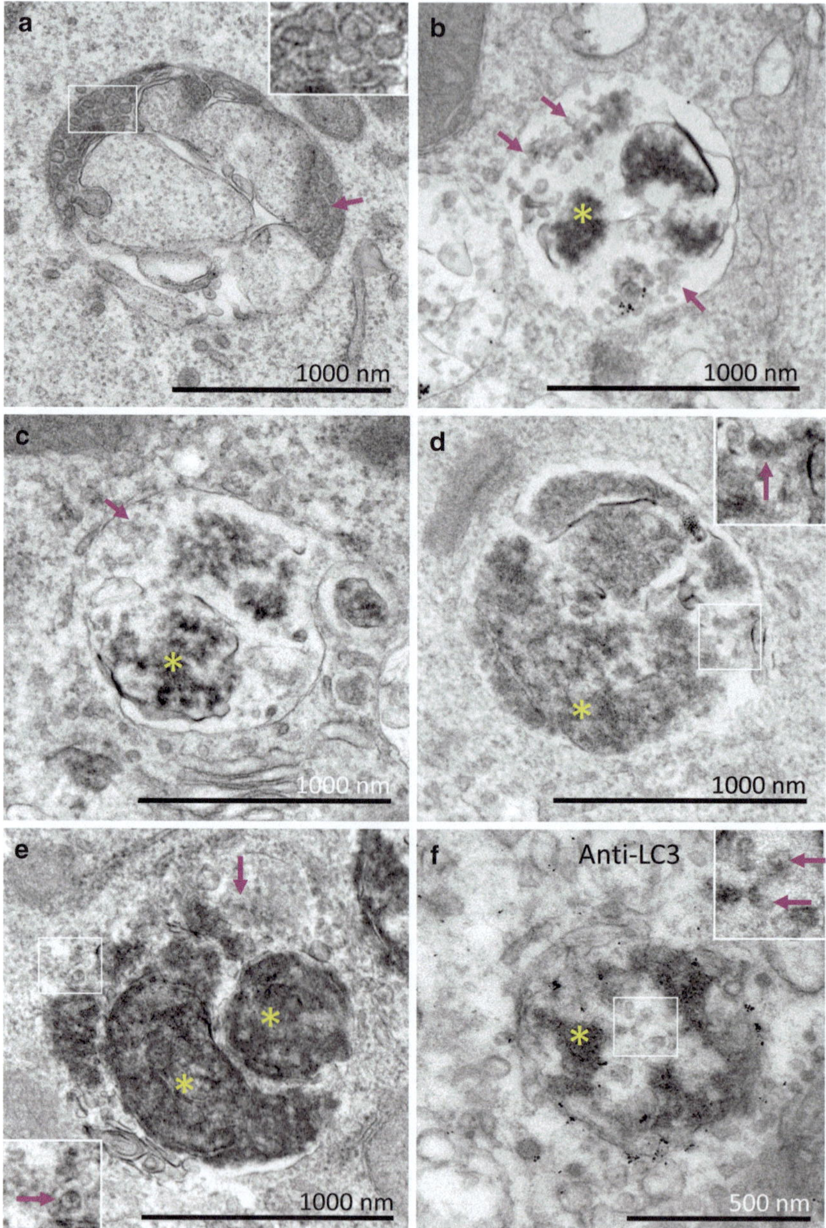

**Fig. 4** Transmission electron microscopy images of amphisomes. The samples in panels **a**–**e** were prepared as in Fig. 2. Panel **f** shows pre-embedding immunostaining for endogenous LC3 [10]. Amphisomes contain both cytoplasmic cargo and small intralumenal vesicles delivered by fusion with multivesicular endosomes. Cytoplasmic cargo is visible as intact ribosomes in panel **a**, and as electron-dense particulate material (partially degraded ribosomes), indicated by yellow asterisks, in panels **b**–**f**. The intralumenal vesicles are highlighted by magenta arrows in all panels, and enlarged in the inserts in panels **a**, **d**–**f**. Immunostaining for endogenous LC3 in panel **f** shows that the amphisome contains LC3; the black dots are silver-enhanced 1.2-nm gold particles

## 5    Autolysosomes Are Born When Autophagosomes or Amphisomes Fuse with Lysosomes

Autophagosomes can fuse directly with lysosomes, or they can undergo sequential fusion with multivesicular endosomes followed by lysosomes. Because lysosomes are full of degradative enzymes, their fusion with autophagosomes and amphisomes initiates the final degradation of the cytoplasmic cargo. In addition to the cytoplasmic cargo, also the inner limiting membrane of the autophagosome is degraded. Compared to the earlier intermediates of the autophagic pathway, autolysosomes are more difficult to identify by morphology in electron microscopy samples, because the characteristic morphological features have been partially degraded (Fig. 5a–f). Autolysosomes are limited by a single (Fig. 5a–d), or sometimes still a double (Fig. 5e, f), limiting membrane. The contents are disintegrated, but for identification of a structure as an autolysosome, the cytoplasmic cargo should still be identifiable. In particular, the ribosomal cargo can still be identified as electron-dense particulate material (yellow asterisks in Fig. 5a–f). In many cases, it is difficult to tell whether a structure is an autolysosome, late endosome or lysosome. Immunoelectron microscopy of a known, degradation-resistant cargo protein such as superoxide dismutase [19] may be used to help identification of the cytoplasmic cargo.

## 6    Autophagic Vacuole or Autophagic Compartment Are General Terms

Sometimes it is not necessary or possible to differentiate all the autophagy pathway intermediates. The terms 'autophagic vacuole' and 'autophagic compartment' can be utilized as general terms to include all intermediates, i.e., phagophores, autophagosomes, amphisomes, and autolysosomes. The term 'initial autophagic vacuole' (initial autophagic compartment) includes phagophores and autophagosomes, while 'degradative autophagic vacuole' (degradative autophagic compartment) includes amphisomes and autolysosomes.

LC3 associates with both phagophore and autophagosome membranes [3, 10, 20]. In autophagosomes, it associates with both of the two limiting membranes. Amphisomes and to some extent also autolysosomes are also LC3-positive, although not as strongly as autophagosomes.

## 7    Take Note of Four Important Points

There are a few issues that you should remember when interpreting images taken using a transmission electron microscope.

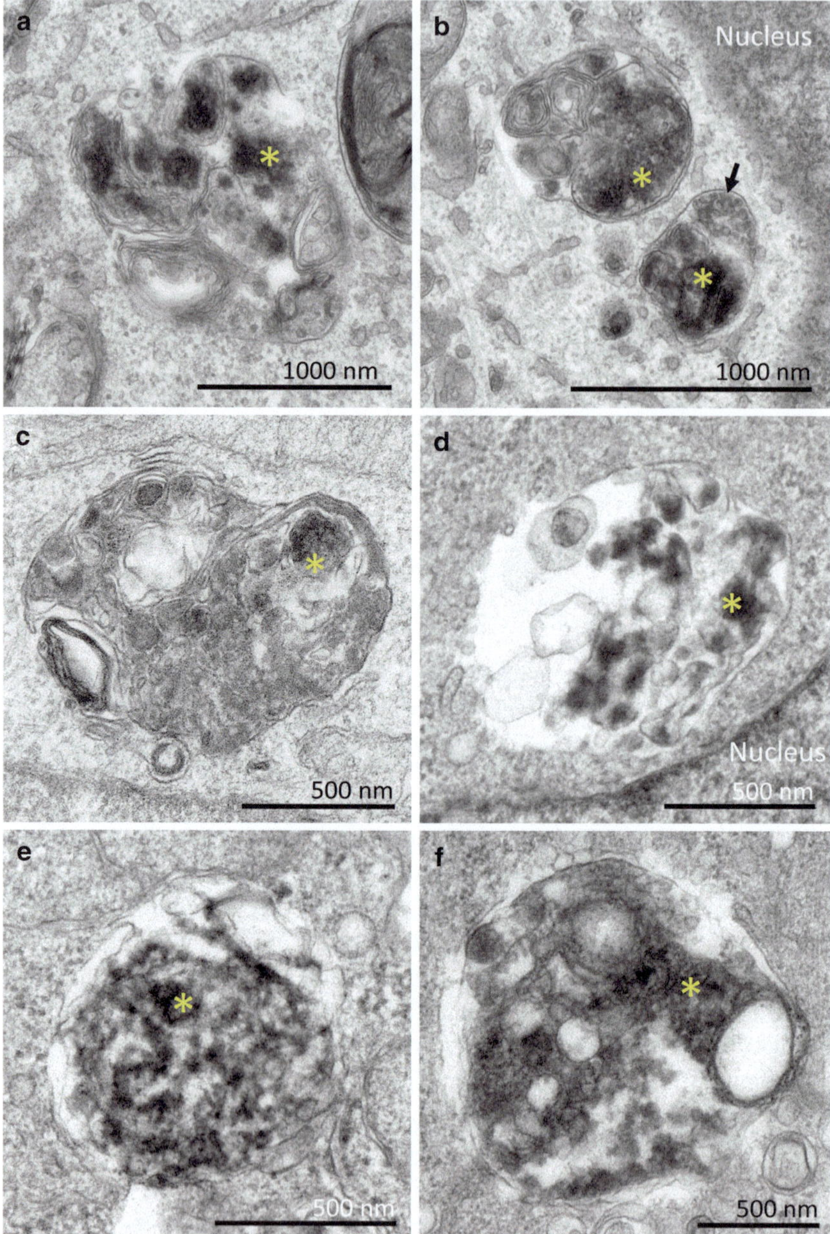

**Fig. 5** Transmission electron microscopy images of autolysosomes. The samples were prepared as in Fig. 2. Autolysosomes contain partially degraded cytoplasmic cargo (panels **a–f**), which in starvation-induced autolysosomes is typically visible as electron-dense particulate material that represents partially degraded ribosomes (yellow asterisks in panels **a–f**). The lower structure in panel **b** also contains small intralumenal vesicles (black arrow), which identify it as an amphisome. The structures in panels **e, f** still have parts of the autophagosome inner limiting membrane visible. It is possible that these structures are amphisomes. However, the thin sections shown in the panels did not hit the small internal vesicles that would clearly identify the structures as amphisomes

First, keep in mind that when looking at thin sections in the transmission electron microscope, you are looking at very thin slices cut from three-dimensional cells and organelles. The typical thin sections are 50–100 nm thick, while a typical mammalian cell has a diameter of tens of micrometers, and the organelles, including autophagy pathway intermediates, have diameters of hundreds of nanometers. This means that in the thin sections, you do not see the whole organelles, only thin slices of them. The appearance of the organelle in the thin section can vary depending on the orientation of the section. For example, a phagophore can look like a closed autophagosome in a thin section, if the section does not hit the open rim. An amphisome can look like an autophagosome or autolysosome, if the section does not hit the small internal vesicles that would identify it as an amphisome (Fig. 5e, f). This may sound confusing. However, in most cases it is acceptable to combine phagophores and autophagosomes in one class and amphisomes and autolysosomes into another class when quantifying the structures. If more specific differentiation between phagophores and autophagosomes or amphisomes and autolysosomes is needed, immunolabelling of phagophore or amphisome markers can be used in either fluorescence microscopy or immuno-electron microscopy.

Second, keep in mind that a double limiting membrane means that autophagosomes are limited by two adjacent lipid bilayers. Occasionally, one lipid bilayer has been mixed with a double membrane in literature [21]. While the double limiting membrane is a good criterion for the identification of autophagosomes, it has some drawbacks. Mitochondria are also limited by two lipid bilayers. However, unlike autophagosomes, the mitochondrial inner limiting membrane is characterized by invaginations called cristae, which are relatively easy to recognize in electron-microscopy images. Depending on the orientation of the thin section, cisterns of endoplasmic reticulum can look like double-membrane limited vesicles containing cytoplasm, which can be mixed with autophagosomes. However, endoplasmic reticulum cisterns are usually studded with ribosomes, which are not present on autophagosome limiting membranes. In addition, the distance between the two lipid bilayers is wider in endoplasmic reticulum than in autophagosomes [16]. Also, the size and shape of autophagosomes can be used as criteria for identification. Autophagosomes are typically spherical, except the situation right after fusion of the phagophore, and in each cell type they have a typical diameter range.

Third, also the appearance of the lipid bilayer can vary in conventional electron microscopy thin sections, depending on the orientation of the section in relation to the bilayer. Occasionally, the lipid bilayer may seem to be abnormally thick, as if it had split into two. An example of this is visible in Fig. 3b insert (lower red arrowhead). The reason for this phenomenon is likely alterations occurring during conventional sample preparation that includes dehydration, plastic embedding and thin sectioning.

Last but not least, pay special attention to the cargo: because autophagosomes form in the cytoplasm, they should always contain cytoplasmic cargo. In some publications, membrane-bound, electron-lucent vacuoles with no visible cargo inside them have been mistakenly called autophagic vacuoles. However, without identifiable cytoplasmic cargo, there is no evidence on the autophagic origin of such empty vacuoles.

**Questions**
1. What are the main morphological criteria of autophagosomes in conventional thin sections?
2. How to differentiate amphisomes and autolysosomes from late endosomes and lysosomes in conventional thin sections?

## 8 Pay Attention to These Articles

Autophagy was originally described using electron microscopy [1], and later on, electron microscopy has been crucial in bringing autophagy field forward. Three-dimensional electron microscopy was used to confirm the close relationship of nascent phagophores with the endoplasmic reticulum [14, 15]. More recently, three-dimensional cryo electron microscopy was used to reveal important details of the morphology of phagophores and autophagosomes [16, 22].

**Take-Home Message**
- When using transmission electron microscopy to identify and quantify autophagy pathway intermediates, pay attention to both the limiting membranes and the cytoplasmic cargo.
- Keep in mind that the electron-microscopy sections are considerably thinner than the diameter of the autophagy pathway intermediates—the sections only show a thin slice that is likely to miss part of the features of the organelles.

**Answers**
1. The double limiting membrane and the morphologically intact cytoplasmic cargo, which frequently includes ribosomes.
2. In many cases, it is difficult to tell whether a structure is an amphisome, autolysosome, late endosome or lysosome. If partially degraded cytoplasmic cargo (such as ribosome clusters) can still be identified among the contents, the structure can be classified as an amphisome or autolysosome. Immuno-electron microscopy of a known, degradation-resistant cargo protein such as superoxide dismutase may be used to help identification of the cytoplasmic cargo.

## References


1. Eskelinen EL, Reggiori F, Baba M, Kovacs AL, Seglen PO. Seeing is believing: the impact of electron microscopy on autophagy research. Autophagy. 2011;7(9):935–56.
2. Klionsky DJ, Abdel-Aziz AK, Abdelfatah S, Abdellatif M, Abdoli A, Abel S, et al. Guidelines for the use and interpretation of assays for monitoring autophagy (4th edition) (1). Autophagy. 2021;17(1):1–382.
3. Kabeya Y, Mizushima N, Ueno T, Yamamoto A, Kirisako T, Noda T, et al. LC3, a mammalian homologue of yeast Apg8p, is localized in autophagosome membranes after processing. EMBO J. 2000;19(21):5720–8.
4. Kabeya Y, Mizushima N, Yamamoto A, Oshitani-Okamoto S, Ohsumi Y, Yoshimori T. LC3, GABARAP and GATE16 localize to autophagosomal membrane depending on form-II formation. J Cell Sci. 2004;117:2805–12.
5. Kirisako T, Ichimura Y, Okada H, Kabeya Y, Mizushima N, Yoshimori T, et al. The reversible modification regulates the membrane-binding state of Apg8/Aut7 essential for autophagy and the cytoplasm to vacuole targeting pathway. J Cell Biol. 2000;151:263–76.
6. Kimura S, Noda T, Yoshimori T. Dissection of the autophagosome maturation process by a novel reporter protein, tandem fluorescent-tagged LC3. Autophagy. 2007;3(5):452–60.
7. Patterson GH, Knobel SM, Sharif WD, Kain SR, Piston DW. Use of the green fluorescent protein and its mutants in quantitative fluorescence microscopy. Biophys J. 1997;73(5):2782–90.
8. Yla-Anttila P, Vihinen H, Jokitalo E, Eskelinen EL. Monitoring autophagy by electron microscopy in Mammalian cells. Methods Enzymol. 2009;452:143–64.
9. Eskelinen EL. Fine structure of the autophagosome. In: Deretic V, editor. Methods in molecular biology. 445: autophagosome and phagosome. Totowa: Humana Press; 2008. p. 11–28.
10. Biazik JM, Yla-Anttila P, Vihinen H, Jokitalo E, Eskelinen EL. Ultrastructural relationship of the phagophore with surrounding organelles. Autophagy. 2015;11(3):439–51.
11. Tsuboyama K, Koyama-Honda I, Sakamaki Y, Koike M, Morishita H, Mizushima N. The ATG conjugation systems are important for degradation of the inner autophagosomal membrane. Science. 2016;354(6315):1036–41.
12. Reunanen H, Punnonen EL, Hirsimaki P. Studies on vinblastine-induced autophagocytosis in mouse liver V. A cytochemical study on the origin of membranes. Histochemistry. 1985;83:513–7.
13. Gudmundsson SR, Kallio KA, Vihinen H, Jokitalo E, Ktistakis N, Eskelinen EL. Morphology of phagophore precursors by correlative light-electron microscopy. Cells. 2022;11(19):3080.
14. Yla-Anttila P, Vihinen H, Jokitalo E, Eskelinen EL. 3D tomography reveals connections between the phagophore and endoplasmic reticulum. Autophagy. 2009;5(8):1180–5.
15. Hayashi-Nishino M, Fujita N, Noda T, Yamaguchi A, Yoshimori T, Yamamoto A. A subdomain of the endoplasmic reticulum forms a cradle for autophagosome formation. Nat Cell Biol. 2009;11(12):1433–7.
16. Bieber A, Capitanio C, Erdmann PS, Fiedler F, Beck F, Lee CW, et al. In situ structural analysis reveals membrane shape transitions during autophagosome formation. Proc Natl Acad Sci USA. 2022;119(39):e2209823119.
17. Nakagawa I, Amano A, Mizushima N, Yamamoto A, Yamaguchi H, Kamimoto T, et al. Autophagy defends cells against invading group A Streptococcus. Science. 2004;306:1037–40.
18. Vargas JNS, Hamasaki M, Kawabata T, Youle RJ, Yoshimori T. The mechanisms and roles of selective autophagy in mammals. Nat Rev Mol Cell Biol. 2023;24(3):167–85.
19. Rabouille C, Strous GJ, Crapo JD, Geuze HJ, Slot JW. The differential degradation of two cytosolic proteins as a tool to monitor autophagy in hepatocytes by immunocytochemistry. J Cell Biol. 1993;120:897–908.

20. Mizushima N, Yamamoto A, Hatano M, Kobayashi Y, Kabeya Y, Suzuki K, et al. Dissection of autophagosome formation using Apg5-deficient mouse embryonic stem cells. J Cell Biol. 2001;152:657–67.
21. Eskelinen EL, Kovacs AL. Double membranes vs. lipid bilayers, and their significance for correct identification of macroautophagic structures. Autophagy. 2011;7(9):931–2.
22. Li M, Tripathi-Giesgen I, Schulman BA, Baumeister W, Wilfling F. In situ snapshots along a mammalian selective autophagy pathway. Proc Natl Acad Sci USA. 2023;120(12):e2221712120.

# Assessment of Autophagy: Correlative and Super-Resolution Microscopy Techniques


Nicola Vahrmeijer, Dumisile Lumkwana, André du Toit, Ben Loos, and Lize Engelbrecht



**What Will You Learn in This Chapter?**

Macroautophagy/autophagy is a critical protein degradation pathway that preserves cellular and metabolic homeostasis [1]. It is therefore no surprise that its dysfunction is associated with many human pathologies, impacting directly cellular fate. Although much is known about the molecular machinery that governs the autophagy pathway, how to best monitor autophagy and how to precisely quantify its activity has remained more challenging. This is, at least in part, due to the highly dynamic rearrangement of membranes involved in the autophagy process and the degradation of proteinaceous cargo [2]. In this chapter, we highlight the various techniques currently employed to assess autophagy, focusing primarily on correlative light and electron microscopy and super-resolution microscopy. Context will be provided with regards to complimentary approaches. We emphasize the dynamic nature of the autophagy system, which marker proteins to utilize [3–5], and may be most important, how to discern between steady state protein abundance levels, the major autophagy pathway intermediates, their pool sizes and the autophagy activity, or flux. The techniques highlighted range from biochemical approaches such as western blotting to microscopy techniques, including,


(continued)


N. Vahrmeijer · B. Loos (✉)
Department of Physiological Sciences, Stellenbosch University, Stellenbosch, South Africa
e-mail: bloos@sun.ac.za

D. Lumkwana
Science Technology Platform, Electron Microscopy, Francis Crick Institute, London, UK

A. du Toit
Department of Chemistry and Molecular Biology, University of Gothenburg, Gothenburg, Sweden

L. Engelbrecht
Central Analytical Facilities, Microscopy Unit, Stellenbosch University, Stellenbosch, South Africa

various fluorescence-based microscopy approaches, super-resolution structured illumination and correlative light and electron microscopy. Section 2 of the chapter will place emphasis on the importance and outcome measures of each technique, whether it reveals the dynamic nature of autophagy or whether it can integrate machinery and cargo measures. By providing examples where autophagy is functional and dysfunctional, the reader will be able to grasp how to best assess the autophagy pathway at basal levels and in diseased states and how to approach scenarios where autophagy activity requires to be offset. Furthermore, you will learn how to use these techniques so that they are well aligned with the given research question, since some techniques are highly labor intensive, and some performed best at single cell level. Finally, this chapter highlights how to discern not only between autophagy activity, but also how to integrate the respective cargo turnover, using a model system of Alzheimer disease.

*Take note*: Applying the most suitable technique or method that best answers your research question is critical as it may dictate the metric and therefore the type of answer you may get. Hence, choosing the technique carefully requires understanding of what it really measures.

## 1 Why Focussing on Autophagy Monitoring?

Before we begin, it is important to highlight why it is fundamental to monitor autophagy accurately, and why it could be fatal, in terms of derived interpretations, if not careful. Firstly, it is important to be reminded that the autophagy pathway is anchored in an energetic feedback loop that integrates substrate metabolite sensing, i.e. monitoring intracellular amino acid availability and ATP consumption, i.e. monitoring the energetic charge of the cell [6] (Fig. 1a). Both these characteristics are core to governing cell viability and cell function, and are hence relying on the basal autophagy activity, the turnover of primarily long-lived proteins at steady state. Given cell-type dependent metabolic demands, it is no surprise that different cell and tissue types are characterized by an inherent basal autophagy activity [9]. Secondly, it is equally important to note that changes in autophagy activity, typically an upregulation, is the cell's first stress response to metabolic perturbations [10], and is intricately linked to the induction and execution of cell death subroutines, such as apoptosis or necrosis (Fig. 1b). Therefore, autophagy activity, when modulated, can impact the positioning of the cell's 'point of no return', i.e. the position that discerns cell death induction from execution, and hence a 'still living' and an 'already dead' cell [7]. Increased autophagy typically delays this point, thereby changing the cell's susceptibility to die. Therefore, measuring autophagy activity accurately has implications in revealing, governing and controlling cellular fate, with obvious translational value. However, it is

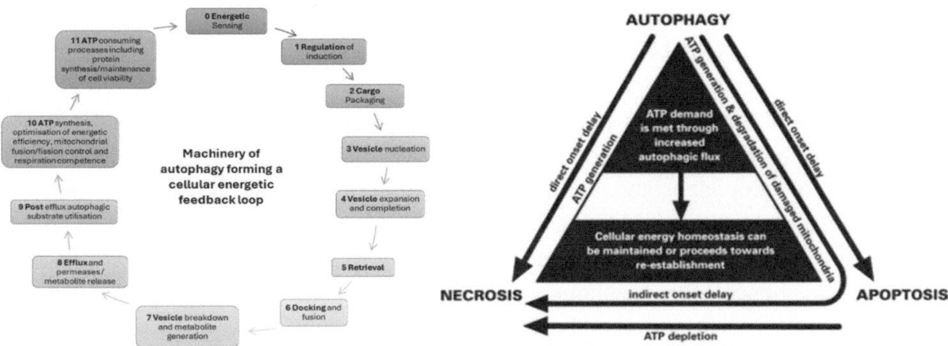



**Fig. 1** Autophagy, metabolic feedback and energy homeostasis. Autophagy, metabolic feedback and energy homeostasis. (**a**) The molecular machinery of autophagy regulation can be depicted as integral part of a metabolic feedback loop. The sequence of events begins with its role in energetic sensing, which regulates the induction, whereas post efflux and permease release the sequence extends to metabolite substrate utilization, ATP synthesis and energetic efficiency as well as protein synthesis/ATP consumption. In this way, the metabolic feedback loop is fully complete and indicates the role of autophagy in tuning the degradation of proteins with the cell's metabolic and energetic demands. (**b**) The metabolic demands of a cell can be well understood when assessing ATP-dependent processes. A functional relationship exists between these very processes that consume ATP, and autophagy. When ATP demands are not met, or autophagy capacity is not sufficient, the cell may be sensitized to apoptosis or necrosis induction. This depends on the magnitude of ATP loss and the severity of the metabolic injury. Autophagy, when enhanced, induces a direct and indirect onset delay of apoptosis and necrosis through ATP generation as well as the removal of damaged mitochondria and apoptotic stimuli. Energy homeostasis is hence more likely to be maintained. Autophagy is hence engaged in a complex metabolic feedback loop that is critical for cellular energy homeostasis [7, 8]

now very well known that the abundance of autophagosomes is by no means an indication of autophagy activity [11–14]. In fact, an increased abundance of autophagosomes may be a result of either enhanced autophagosome synthesis, typically associated with improved cellular function, or due to decreased autophagosome clearance, often associated with lysosomal dysfunction and overall autophagy impairment. Both are, in terms of metabolic contribution for the cell, at the opposite of the spectrum of cell survival—enhanced autophagy function, or cell death—autophagy dysfunction. Therefore, it is critical to be able to discern between autophagy pathway intermediate abundance, and autophagy activity. As we move through this chapter it may hence be valuable to ask, whether the given technique is able to quantify the abundance of specific autophagy proteins or autophagy pathway intermediates, or whether it quantifies the autophagy activity or flux, i.e. the rate of protein degradation through the entire autophagy pathway? If not, it may be of 'fatal' consequences.

## 1.1    Electron Microscopy: Identifying the Unique Ultrastructural Morphology

A first glimpse of the complex morphology of the autophagy pathway intermediates was achieved by using electron microscopy (EM) [15–19]. Indeed, this method led to the coining of the term autophagosome, and the identification of the morphological hallmarks of this organelle, such as the double membrane structure with cytosolic material [20, 21]. In contrast, the autolysosome would contain partially degraded cytoplasmic and organelle material and is hence often less homogeneous in nature. The ultrastructural detail that is enabled through EM allows researchers to differentiate between autophagy pathway intermediates, i.e. autophagosomes, autolysosomes and lysosomes in their complexity and different sizes, with great precision (Fig. 2). It also allows us to observe the nature and type of cargo being degraded. However, since no direct molecular identity is revealed, classification of the pathway intermediates requires major experience and technical expertise to accurately define the features and correctly identify each intermediate. On many occasions was autophagy incorrectly identified [22, 23]. Despite this shortcoming, electron microscopy remains one of the most accurate methods currently used in the autophagy research field.

*Take note*: Electron microscopy allows for the visualization of ultrastructural detail of pathway intermediates at a very high resolution. However, poor sample preparation may result in artifacts and can lead to loss of sample integrity. Moreover, organelles cannot always be identified with absolute certainty, since no molecular identity is revealed or depicted. This leaves an element of subjectivity when analyzing electron micrographs.

> *Question 1: What advantage does electron microscopy offer in terms of observing autophagosome morphology?*
>
> *Question 2: Given is a series of micrographs captured at different stages of the autophagy pathway, arrange them in chronological order. What are the key morphological features that aid in distinguishing between these stages?*

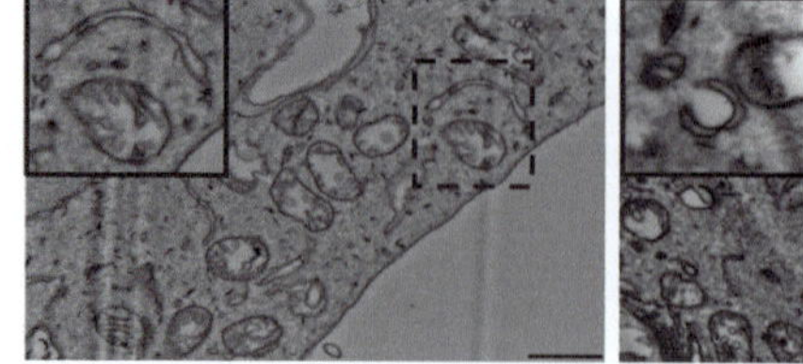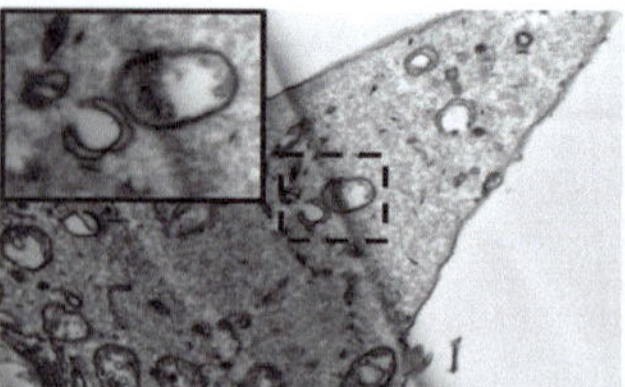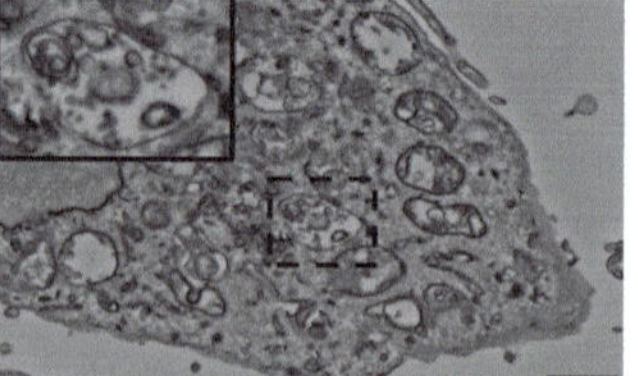

**Fig. 2** The autophagosome and its cargo. Electron microscopy enables the visualization of intracellular organelles with high resolution, enabling the identification of delicate organelle and membrane structures. Here the autophagosome synthesis with the phagophore membrane (**a**), a maturing autophagosome (**b**) and a heterogeneous cargo-containing autolysosome (**c**) can be identified. Scale bar: 1 μm

## 1.2    Western Blotting: Protein Abundance and Molecular Footprint

Imaging data is very often contextualized with western blotting, to provide additional context. Here we briefly summarize key features and pitfalls of this technique. Western blotting is a semi-quantitative tool often used to assess changes in protein expression or abundance [24]. Various ATG (autophagy related) proteins play an invaluable role in regulating the molecular machinery of the autophagy pathway. One of the most widely assessed proteins of interest that are blotted for, is MAP1LC3/LC3, a member of the Atg8 family, that is recruited in its lipidated form (LC3-II) to the phagophore membrane. LC3-II remains associated with the completed autophagosome. Therefore, the abundance of LC3-II correlates with the abundance of autophagosomes, and can hence, provides an indication of the magnitude of the protein present (Fig. 3a). A second protein, SQSTM1/ p62 (sequestosome 1) is an autophagy receptor, and is often assessed, since it is degraded with the proteinaceous cargo it had recognized and sequestered [25]. In addition, it is necessary to follow a second marker because—as mentioned above with regard to autophagosomes—an increase in LC3-II can reflect either an increase in autophagy or a block in a late stage of the process.

Indeed, LC3 is the protein that 'leads a golden thread' throughout multiple techniques employed to assess the autophagy pathway. When assessing LC3 levels using western blotting, it is important to note that LC3 is expressed as different isoforms, mainly LC3A, LC3B and LC3C. Each of these isoforms correlates with an increase in the number of autophagy-related structures found within distinctly different regions of the cell. LC3A

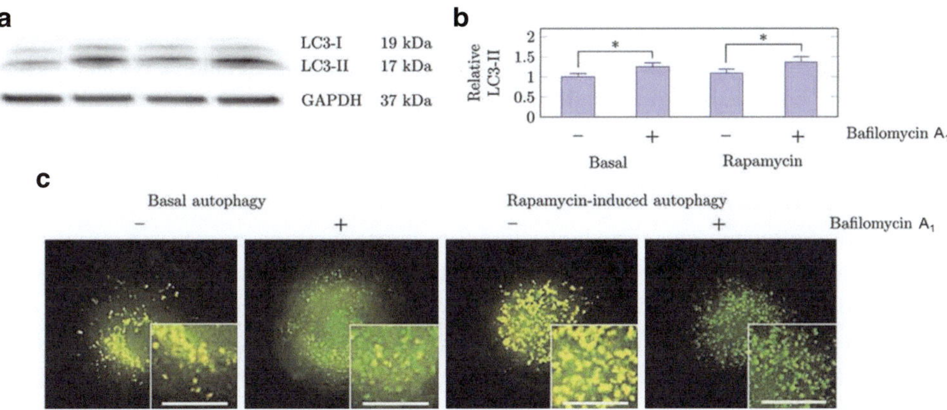


**Fig. 3** Methods used to assess autophagy activity. Western blot analysis (**a**) enables the quantification of LC3-II abundance. When performed in the absence and presence of an autophagosome/lysosome inhibitor, the degree of LC3-II accumulation is indicative of the autophagy flux. Here, basal autophagy activity (left panel) and induced autophagy flux (right panel) through the treatment with 25 nM rapamycin is shown, and relative LC3-II abundance depicted in a bar graph (**b**). Vesicle-based analysis of fluorescence micrographs (**c**) makes it possible to visualize the abundance change of autophagosomes (green), autolysosomes (yellow) and lysosomes (red). Scale bar: 20 μm [13]

and LC3C are primarily expressed in the nucleus of the cell while LC3B is strongly expressed in the cytoplasm [26]. Furthermore, LC3 is expressed as either LC3-I, which is cytosolic, or LC3-II, its lipidated form, which is recruited to the phagophore itself, decorating both sides of the membrane. The conversion of LC3-I to LC3-II therefore serves as an indirect indication of the abundance of autophagy-related structures present within a protein lysate [26–28].

The abundance level of the protein BECN1 often reflects the rate at which autophagy is initiated [29], while elevated SQSTM1 levels signify either an increase in cargo to be degraded or a disruption in the autophagy pathway resulting in impaired autophagosome turnover [30]. In addition to its value in analyzing changes in protein expression, western blotting can be used to indirectly assess autophagy structure formation by assessing the conversion of cytosolic LC3-I to membrane bound LC3-II, which can easily be observed because LC3-II migrates faster when proteins are separated on an SDS-PAGE compared to LC3-I [31]. The steady state formation and degradation results in a constant turnover of LC3-II protein. As a result, the ratio between LC3-II and LC3-I serves as a representative snapshot of autophagy-related structures within a cell at any point in time. To gain a better understanding of autophagy activity under various conditions, autophagy inhibitors and autophagy inducers need to be used to accurately measure autophagy activity [14].

### 1.2.1 Advantages

Through western blotting researchers can detect changes in the level of protein abundance under various conditions. The detection of these changes highlights why this method remains crucial when performing autophagy research, as even minor changes in protein expression could have significant implications further downstream on the initiation and execution of this pathway. Furthermore, western blots allow researchers to assess a wide range of sample types such as cultured cells and tissues. A major advantage is that multiple proteins can be assessed in a single blot as a result of the difference in the molecular weight of proteins of interest.

### 1.2.2 Limitations

Western blots often present with a high variability, hence subtle changes in sub-populations of cells can be masked. Antibody cross-reactivity can arise especially in cases where various isoforms exist. Careful thought should go into the experimental design and extra care should be taken to ensure appropriate controls and inclusion of standards, the use of high-quality antibodies, that have been verified in knockout cells. Variability can also arise due to changes in the way that buffers are made up and stored, as well as sample preparation as it has been shown that certain buffers result in the loss of protein, leading to inaccurate results [10]. Last, while western blotting provides valuable insights into the protein levels at play, it fails to capture the dynamic nature of autophagy in real-time. Per se, western blotting does not measure a rate. Hence, when measuring autophagy activity, cell lysates must be prepared derived from cells or tissues that were exposed to saturating concentrations of an autophagosome/lysosome fusion inhibitor. In this way, deductions can be made with regards to the basal or changed autophagy activity. Insights into single-cell level autophagy behavior is hence not possible using western blot analysis.

### 1.2.3   Conclusion

Western blotting serves as an important tool for assessing autophagy due to its ability to quantify the relative abundance of key ATG proteins. Its variability presents limitations to sensitively describe minor changes in protein abundance. When combined with other techniques, western blotting strongly contributes to the understanding of autophagy and its molecular machinery.

*Take note*: Care should be taken to choose an appropriate sample buffer for your specific sample and question, as certain buffers may result in incomplete protein extraction, post-translational modification and even protein degradation.

*Question 3*: *Which proteins are mostly used to assess autophagy through means of western blotting and what is the significance of these proteins?*

## 1.3   Fluorescence Microscopy: Shedding Light Onto a Dynamic Process

From the previous sections, it becomes clear that current methods used to detect and analyze autophagy are static in nature and are representative of a given time point of this highly dynamic process. This brings to light the need for a technique which allows for accurate and precise analysis of cellular and molecular events in real-time under live conditions. Fluorescence microscopy has emerged as an invaluable and transformative tool in autophagy research due to its ability to visualize the entire autophagy processes with high spatial and temporal resolution with a high degree of specificity [32, 33]. Steady state autophagy flux and changed, i.e. enhanced or decreased autophagy activity can be assessed through means of tagging subcellular targets with fluorochromes or fluorescent proteins, enabling to discern autophagosomes, autolysosomes and lysosomes, as well as specific cargo or receptor proteins and organelles [34]. A molecular identity is being provided, with a high degree of precision.

Multiple fluorescently tagged autophagy-related proteins have been developed to allow researchers to track autophagic structures in live cells, providing insights into autophagy initiation, maturation and progression and fusion (Fig. 4). Fluorescently labelled LC3 tagged with, for example, green fluorescent protein (GFP) allows for the numerical quantification of LC3-positive structures in a given cell over time, capturing the temporal dimensions of autophagy, revealing its highly dynamic and responsive nature. The three fluorescent probes that are most often used in autophagy research are: GFP-LC3 [35], mRFP-GFP-LC3 [36] and GFP-LC3-RFP-LC3ΔG [37]. Note that it is important to denote these chimeras as, for example, GFP-LC3 as opposed to LC3-GFP. The C terminus of LC3 is proteolytically processed by ATG4, such that LC3-GFP would quickly be converted to

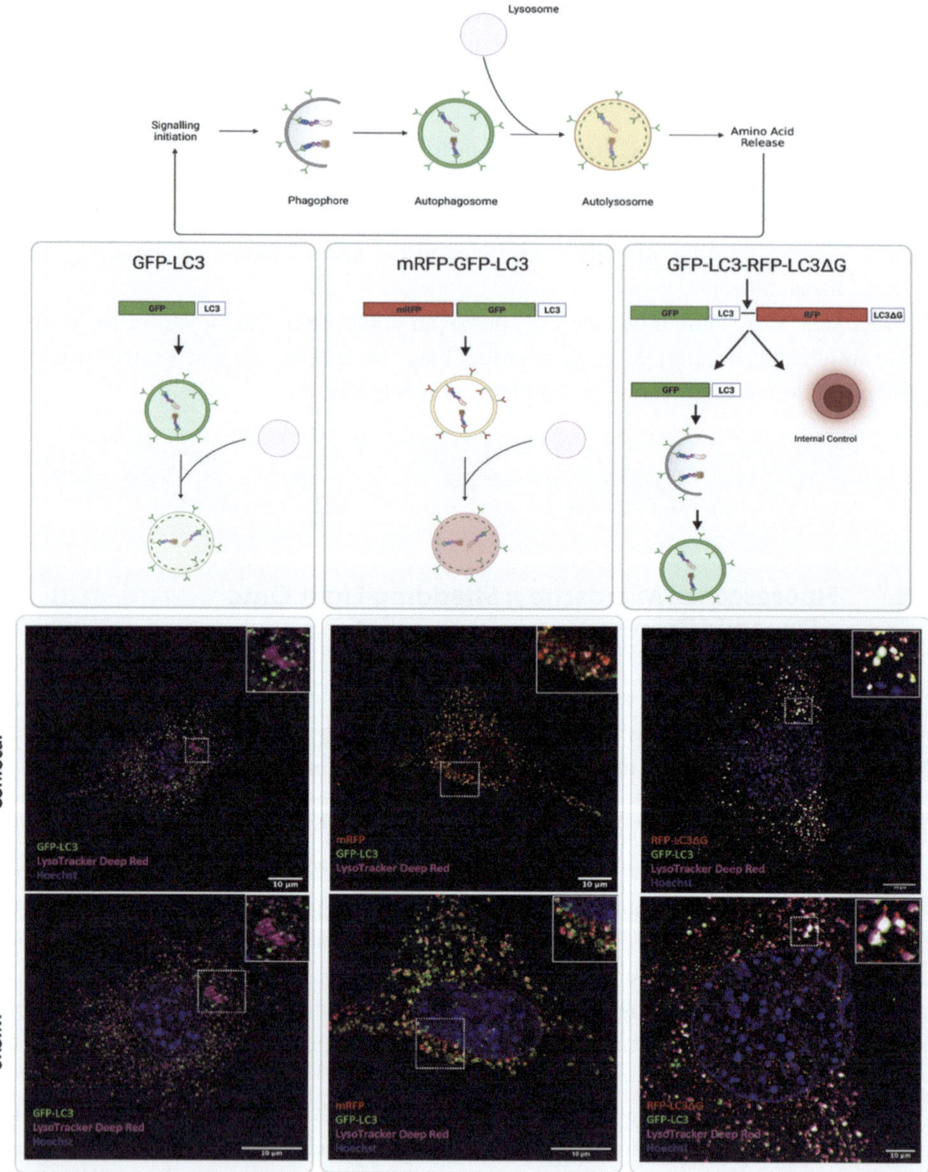


**Fig. 4** Three generation autophagy probes. The autophagy process (A) can be monitored by using fluorescence proteins co-expressed with LC3, in the presence of an inherent binding probe for labelling lysosomes. Here, GFP-LC3 together with a LysoTracker (left panel), a mRFP-GFP-LC3 probe in the presence of a LysoTracker (middle panel) and a RFP-LC3ΔG-GFP-LC3 probe in the presence of a LysoTracker (right panel) have been acquired in confocal (C) and SR-SIM (D) mode. Not only does this approach allows the identification of all three pathway intermediates, i.e. autophagosomes, autolysosomes and lysosomes, but also enables to observe the fusion zone between autophagosomes and lysosomes with greater precision and hence quantitative power. With enhanced resolution, signal overlap becomes equally better resolved and hence more precise

LC3 and free GFP in the cytosol; this can be the basis for assaying ATG4 activity, but obviously cannot be used to monitor LC3 location by microscopy.

While investigating the fate of LC3 during autophagy and its localization to the autophagosome membrane, LC3 fluorescently tagged with GFP is still popularly used in the field of autophagy research today. GFP-LC3 specifically labels phagophores and autophagosomes, with a loss of GFP signal upon fusion with the lysosomal compartment due to its acidic and degradative nature. The pH-sensitive nature of GFP with subsequent emission signal quenching makes it possible to exploit the functional properties of the autophagy pathway intermediates [38]. To fully elucidate all three intermediates of the autophagy pathway, i.e. autophagosomes, autolysosomes and lysosomes, other fluorescent probes such as LysoTrackers have to be used in combination with the LC3 fluorescent probe (Fig. 4, left panel). Due to the loss in fluorescence of GFP upon lysosomal fusion, Kimura et al. (2007) developed a tandem mRFP-GFP-LC3 fluorescent construct which makes it possible to discern between autophagy pathway intermediates and to distinguish between autophagosomes before and after fusion with lysosomes [11]. The autophagosome membrane is positively labelled by both fluorescent proteins, but upon fusion, the GFP portion of the construct loses its fluorescence while the RFP part of the chimera continues to emit fluorescence in the red spectrum. Therefore, an autophagosome is represented by the colocalization of the GFP and RFP signal which appears yellow while an autolysosome will only fluoresce red (Fig. 4, middle panel). This approach allows for pool size count, i.e. the quantification of the entire number or pool of autophagosomes, autolysosomes and lysosomes, to be performed with high precision. However, a lysosomal inhibitor is still required if autophagy activity is to be assessed, since otherwise only steady state values are extracted. Lysosomal inhibitors such as bafilomycin $A_1$ interfere with the lysosomal function and prevent the degradation of autophagosomes [39].

Due to the complexities faced by researchers when measuring autophagy activity, Kaizuka et al. [37] developed a fluorescent construct, GFP-LC3-RFP-LC3$\Delta$G, that reports on autophagy activity based on a ratio-metric signal, making it possible to recognize regions with different fluxes with ease [40]. The plasmid is designed in a manner that during the process of autophagosome formation, ATG4 cleaves the construct into equal amounts of GFP-LC3 and RFP-LC3$\Delta$G. The GFP-LC3 chimera undergoes lipidation and is recruited to the phagophore membrane, and subsequently degraded by the acidic compartment of the lysosomes. The RFP-LC3$\Delta$G chimera is unable to undergo lipidation, and remains stably in the cytoplasm of the cell (Fig. 4, right panel). Therefore, autophagy flux can be assessed by measuring the signal ratio between GFP and RFP. Ratio-metric imaging thus makes it possible to rapidly discern changes in autophagy flux. This fluorescent construct hence allows researchers to rapidly assess autophagy activity and to screen large numbers of autophagy enhancing or inhibiting drugs for their properties [37]. However, performing pool size analysis and pathway intermediate counts remains a challenge and care should be taken since flux is not associated with pool size, hence a ratio may be similar even though pool sizes could differ substantially. Therefore, when designing an experiment and considering the type of probe to use for the analysis of autophagy, it is crucial to weigh up the advantages and disadvantages of each and to consider how each probe will best answer the research question at hand.

*Take note*: Live cell cultures should be maintained as close to the optimal physiological conditions as possible through maintaining the temperature, humidity and gas at optimal levels at all times and ensuring the culture media is properly buffered. An on-stage incubator coupled to the fluorescence microscope is required in order to maintain the culture throughout the live cell imaging experiment [41]. This is of importance, since the autophagy system is the first line of defense and will respond rapidly to metabolic perturbations or changes in temperature [42].

### 1.3.1   Advantages

Fluorescence microscopy offers real-time visualization of the autophagy process of live cells in three dimensions, allowing researchers to capture the dynamic nature of the process from its initiation to autophagosome fusion with lysosomes and subsequent degradation. Due to the nature of the probes used to assess autophagy and the possibility to label multiple cellular structures simultaneously, different proteins and organelles can be localized at the same time allowing for a more comprehensive analysis and understanding of the process on a sub-cellular level. Current developments in image analysis software, the progression of machine learning integration and incorporation of artificial intelligence in image processing allows for the quantification of many parameters that contribute to the autophagy flux.

### 1.3.2   Limitations

Prolonged exposure of live cell probes to light can lead to toxicity and photobleaching, affecting the autophagy response, cell viability and fluorescence intensity. Whereas fluorescence microscopy provides excellent spatial resolution, especially when using a confocal laser scanning microscope, it does not necessarily resolve structures at the nanoscale. This is due to the physical resolution limit of light, a limitation which is overcome through super-resolution techniques which we discuss next in this chapter [43]. Non-specific binding of antibodies or fluorescent probes and spillover of the fluorescence spectrum when using multiple labels can lead to false-positive results, highlighting the importance of experimental setup and controls for validation. Endogenous cellular fluorescence, overexpression artifacts and auto-fluorescence might interfere with the signal generation, demanding background subtraction and post-image processing.

### 1.3.3   Conclusion

Fluorescence microscopy allows for the visualization and localization of multiple proteins and organelles of interest on a subcellular level, in real-time and in three dimensions, allowing researchers to follow the entire process of autophagy. Protein abundance levels can be monitored by measuring the mean fluorescence intensity in a specific cell or region. Fluorescence microscopy and image post-processing further allows for the counting of autophagosomes, autolysosomes and lysosomes, which in turn reveals important aspects of the process. The likelihood of interaction between various proteins and structures of interest, such as recruitment, receptor engagement and degradation can be assessed and quantified using colocalization analysis.

*Take note*: When using fluorescence microscopy as a tool to analyze the autophagy pathway, careful consideration should go into the selection of fluorescent probes. Additional precautions should be taken during the experimental setup and execution, to guarantee minimal contribution of autofluorescence at the desired wavelengths and to prevent any bleed-through between different emission channels.

> **Question 4: Analyze the provided cartoon of a cell with autophagy activity.*
> (a) *Identify which of the micrographs are representative of cells that have been treated with a lysosomal inhibitor such as bafilomycin A₁.*
> (b) *Quantify the pool size of each pathway intermediate, i.e., how many autophagosomes, autolysosomes and lysosomes can be identified?*

## 1.4    Super-Resolution Microscopy and the Autophagosome: Pushing Optical Boundaries

As the demand for higher resolution intensified, super-resolution microscopy techniques were developed, for the first time breaking the limit of resolving power, typically governed by the wavelength of excitation light used, and the numerical aperture (NA) of the objective employed [44]. Super-resolution techniques, including structured illumination microscopy (SIM) and direct stochastic optical reconstruction microscopy (dSTORM) enabled researchers to visualize autophagic structures with unprecedented detail, at sub-diffraction limit, with the added benefit of labelling specific cellular structures or proteins [40, 45–47]. SIM requires superimposing of a grid pattern of illumination onto the sample, which leads to the generation of Moiré fringes. These are in turn represented as a Fourier transform frequency plot. When acquiring a number of such patterns in multiple rotations while light shifts in phase, the Fourier transform is reconstructed, generating a higher frequency feature containing image that is super-resolved. The resulting micrograph is characterized by a resolving power of 80–100 nm, hence double of what a confocal system would typically generate (Fig. 4, lower panels). This becomes important not only for applications where organelles or membrane structures need to be acquired at higher resolution, but also when overlays with electron micrographs are required [48]. In dSTORM, an entirely different approach is being followed. Here, only a small fraction of the fluorophores present in the sample emit fluorescence at a time, enabling the precise localization of the molecule in the sample, using the center point of the point spread function. This effect of fluorophores undergoing switching between fluorescent 'on' and 'off' states, manifests in a molecular blinking frequency. Blinking events are acquired using a time-lapse acquisition protocol, drift corrected and processed, achieving a resolving power of 20–30 nm [49]. These methods allowed researchers to push the boundaries associated with optical resolution and unraveled the finer intricacies at play within autophagy related structures. Due to the increased resolution under which structures of interest can be viewed, the accuracy of measurements is improved substantially, and the interaction between different role players can be brought to light. STORM furthermore allows for the measurement of molecular density distribution in a specific location, a measurement which is not achievable by either confocal microscopy or electron microscopy [50, 51]. Moreover, clusters of proteins can be acquired and quantified in this manner, with very high localization precision. When assessing protein degradation processes, this approach can be highly insightful.

To achieve super-resolution, the acquisition process is lengthy and limits the application of live cell imaging. For some of the super-resolution techniques extensive optimization is required, as only very specific buffer conditions allow for such 'on' or 'off' state of molecules to be enabled. Finally, extensive post-acquisition processing, including drift correction, is required to produce the final super-resolved image. It is also important to note that super-resolution microscopes are far less accessible for many and although recent developments in super-resolution microscopy improved the temporal resolution together with improvement of spatial resolution, such microscope systems are scarce.

*Take note*: The use of SR-SIM allows researchers to visualize cellular structures with a high-resolving power, however with an increase in resolving power comes a decrease in imaging speed. The contrast between signal and background is also of utmost importance, so it is important to select very bright fluorescent labels and eliminate as much background as possible. Moreover, higher resolution may impact the way how overlapping signal is being visualized. Signal that may have occupied the same pixel at a resolving power of, e.g., 200 nm, may, when generated using SR-SIM, occupy less space when a resolution of 80 nm is achieved, and hence, colocalization may now be less or even lost. Hence, overall colocalization analysis will differ, but become more precise, when using super-resolution techniques.

## 1.5  Correlative Light and Electron Microscopy (CLEM): Bridging the Gap Between Molecular/Organelle Identity and Ultrastructural Backdrop

For an improved understanding of the molecular processes underpinning the autophagy pathway, a multimodal approach is desirable where both the molecular identity of structures as well as their fine, intricate ultrastructural and morphological detail can be revealed. Correlative light and electron microscopy (CLEM) emerged as a transformative technique, revolutionizing our understanding of autophagy by overcoming the shortcomings of fluorescence and electron microscopy as independent modalities. Through their integration, CLEM allows researchers to correlate the molecular identity of structures observed when using the fluorescence microscope with the ultrastructural detail derived through electron microscopy [50, 52, 53].

## 1.6    Autophagy with Multidimensional Detail

The requirement for fixation and the typically restricted field of view in EM limits the information of the biology gained, specifically in terms of kinetics in the spatiotemporal context. Fluorescence microscopy, on the other hand, can provide very precise information associated with specific proteins, such as LC3, albeit molecular context and resolving power is limited. CLEM seeks to provide an unparalleled view of cellular processes by combining the power of light microscopy's molecular specificity and large field of view with superior spatial resolution against the backdrop of the complete sample and its ultra-structural information offered by electron microscopy [54]. This synergy allows researchers to correlate and position the molecular identity of specific structures (Fig. 5c), such as autophagosome and autolysosome, as revealed by fluorescent markers (Fig. 5b), with the ultrastructural characteristics of the specific sample, revealed only by electron microscopy

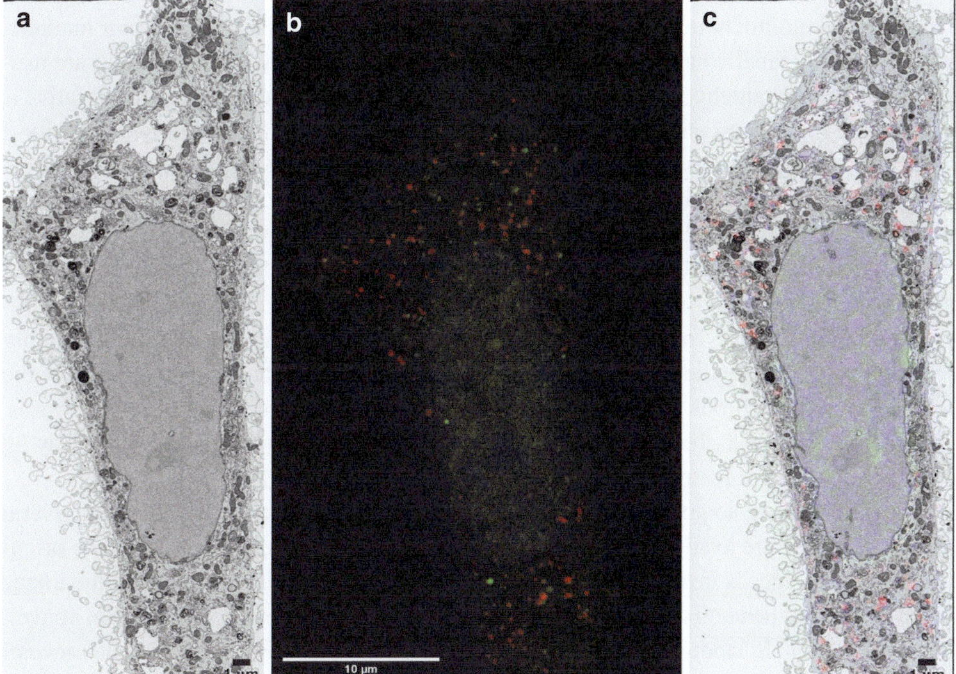

**Fig. 5** CLEM with SR-SIM reveals autophagy compartment with high resolution in the fluorescence mode, projected to a highly resolved ultrastructural cellular context. (**a**) SEM micrograph, (**b**) SR-SIM micrograph and (**c**) Overlayed image of the images seen in **a** and **b**. GT17 neuronal cell treated with 10 μM Spermidine for 8 h induces phagophore recruitment of GFP-LC3, supporting autophagy induction. SR-SIM confirmed that GFP-LC3 binds to the inner and outer membrane of the autophagosome and therefore the portion on the inner membrane is subsequently degraded, while RFP-LC3ΔG remains in the cytosol, serving as an internal control. The EM makes it possible to appreciate the nature of the cargo in terms of electron density and complexity [55]

(Fig. 5a). In doing so, CLEM introduces a multidimensional perspective that breaks traditional boundaries, enabling precise confirmation of structures and their identity and complexity from 2 dimensional workflows to three-dimensional volume CLEM.

## 1.7　The CLEM Workflow: An Integrated Expedition

The CLEM workflow relies on the meticulous combination of light and electron microscopy preparation, acquisition, processing, and analysis, through strategies to locate the same structure or cell in both modalities. Sample preparation for fluorescence microscopy and electron microscopy differs significantly, so it is vital to optimize the workflow. Often, cells are cultured in culture dishes that contain a coordinate grid system and optional fiducial markers, which need to be retained after fluorescence microscopy acquisition has been performed, during sample preparation and resin embedding, so that the same cell or structure can be identified when transferring the sample to an electron microscope. Some structures, like mitochondria can be particularly well utilized as natural fiducial markers, if appropriately labelled and retained for fluorescence microscopy. While samples are usually fixed in glutaraldehyde for subsequent electron microscopy acquisition, this fixative is detrimental for fluorescence properties, but the risk to compromise membrane integrity if glutaraldehyde fixation is not used should be minimized. Hence, samples are often fixed with EM-grade formaldehyde prior to fluorescence microscopy acquisition, followed by fixation in glutaraldehyde, before dehydration and resin embedding takes place, in preparation for electron microscopy. Another major challenge for CLEM, especially when using monolayer cell cultures, is ultramicrotome sectioning. While the conventional ultramicrotome protocol is rather 'forgiving' if a number of sections are removed before good quality sections are obtained, the single layer of cells which is present at the cutting edge of the resin block requires good quality sections to be picked up from the first sections onwards. If this is not achieved, the cell of interest for CLEM could easily be cut away halfway by the time sections are ready for collection.

Following image acquisition on both the fluorescence and electron microscope, several software options are available for correlation and for producing overlays. Often the orientation of the images is inverted (confocal microscopes are typically inverted and the image is acquired from beneath the sample, while in an SEM, the image is acquired from above). Several identifiable landmark spots need to be assigned on each image, sometimes several hundreds, and with the help of the software the two images will be rotated and fitted. This is a tedious task, but well-worth the effort since the overlayed micrograph carries substantial information (Fig. 5c).

With the development of volume electron microscopy, the challenge of ultramicrotome sectioning can be overcome, but a protocol that uses buffers with a higher presence of heavy metals is required, to produce adequate contrast during electron microscopy acquisition. During this process, the fluorescence microscopy workflow will include z-stacking, i.e. acquiring multiple images of a cell or region of interest, moving through focal planes in the image stack. If this step is performed using structured illumination microscopy, the

CLEM data set will be of high quality. Next, following resin embedding, a small resin block can be placed into the electron microscope adapted for 3D imaging. If the microscope houses an in-chamber ultramicrotome, an image at the face of the resin block is acquired, a section is generated, and a subsequent image is acquired. This technique is tedious and requires long hours of microscope acquisition time, often several days, but the result is usually a high-resolution construct of the sample or organelle of interest, such as a GFP-LC3-positive structure, and reveals not only molecular identity but also highest ultrastructural detail in 3D in the same image data set. With appropriate calibration, the volume of specific structures, e.g., autophagosomes, can now be determined.

*Take note*: While CLEM is an invaluable technique that offers insights into both molecular identity and ultrastructural detail, its execution necessitates specific equipment, time, expertise and experience. The meticulous alignment and identification of regions of interest are imperative to effectively link observations made between the distinct modalities.

**Question 5: How does correlative light and electron microscopy (CLEM) address the limitations brought about by each of the two microscopy techniques when used in isolation?*
**Question 6: Compare the techniques fluorescence microscopy, SR-SIM and CLEM with one another.*
  *(a) What are the advantages of using CLEM over other techniques in studying cargo clearance?*

## 1.8    Conclusion: A Journey of Discovery

The evolution of analytical methods in autophagy research has been an exciting journey, with microscopy techniques at the center stage, each technique providing a unique and distinctive characteristic. The combination of multimodal, multiscale imaging approaches, underpinned with deep learning-based image analysis pipelines will with no doubt contribute to better document not only autophagy activity and pathway intermediates, but also to further dissect the role of the molecular machinery, its kinetics and the respective cargo recruitment and clearance properties in health and disease. This will assist in creating increasingly translational value to monitor autophagy across scale and time.

*Take note*: Embrace the multidimensional nature of autophagy and its assessment. No single technique provides a comprehensive picture of the autophagy process. An integrative, multi-modal, multi-scale and multi-dimensional approach generates picture with well-defined and measurable quantitative output that captures the properties of the autophagy pathway.

**Question: Why is it important to adopt a multi-faceted approach when assessing autophagy?*

## 2 Measuring Autophagosome Flux and Its Application in Disease: Introduction to the 2 Articles

### 2.1 Measuring Autophagosome Flux [13]

#### 2.1.1 Introduction

It becomes clear that the assessment of autophagy is complex and that an accurate quantification of autophagy flux or activity requires a careful experimental design with application of suitable techniques. Although it had been shown that autophagy flux assessment requires the use of an autophagosome/lysosome inhibitor, such as bafilomycin $A_1$, to assess the accumulation of LC3-II using western blotting [56], a quantitative metric for autophagy activity remained unclear. Based on the theoretical framework of metabolic control analysis, where the pathway (autophagy pathway) along which there is flow of material (proteinaceous cargo that had been sequestered by autophagosomes) is clearly distinguished from the quantitative measure of this flow (autophagy flux), a new method was proposed [11] to define and measure autophagy activity. Built on this concept, an accurate assessment of single cells with their complete pool size of autophagosomes, autolysosomes and lysosomes was implemented, to reveal the autophagosome flux, according to a rate, with the appropriate unit autophagosomes/h/cell, which can be abbreviated as A/h/cell [11, 13] (Fig. 6).

#### 2.1.2 How Did They Move the Field Forward?

Based on the conceptual framework for defining autophagosome flux, emphasis was placed on the dynamic equilibrium that exists between the formation and degradation of autophagosomes. This clarified several aspects that were, at the time, confusing to many researchers in the field. (1) When both the autophagosome synthesis rate and autophagosome degradation rate are equal, the autophagosome pool size remains unchanged. The autophagy pathway would then be in a so-called steady state. (2) The steady state can however change, for example as the cell is enhancing autophagy, the cell's autophagy activity transits to a new steady state. We now know that the feedback integration of metabolites and the autophagy initiation process can be rapid and is highly dynamic [57]. The cell is hence constantly tuning its autophagy activity to its metabolic demands. (3). The abundance of autophagosomes in a functional cellular system does by no means imply autophagosome flux. A low autophagosome abundance and pool size (or in a western blot a faint LC3-II band) can go hand in hand with a low flux or high flux, and likewise, a large autophagosome pool size (or strong LC3-II band) could be associated with a low flux or high flux [11, 13]. Hence, abundance of LC3-II does not infer the autophagosome flux. Instead, the autophagy activity is inherent to the cell type and the response of the cell in its metabolic context. This newly described definition laid down essential foundational groundwork for measuring basal autophagy activity and induced autophagy activity in cells and tissues of clinical relevance.

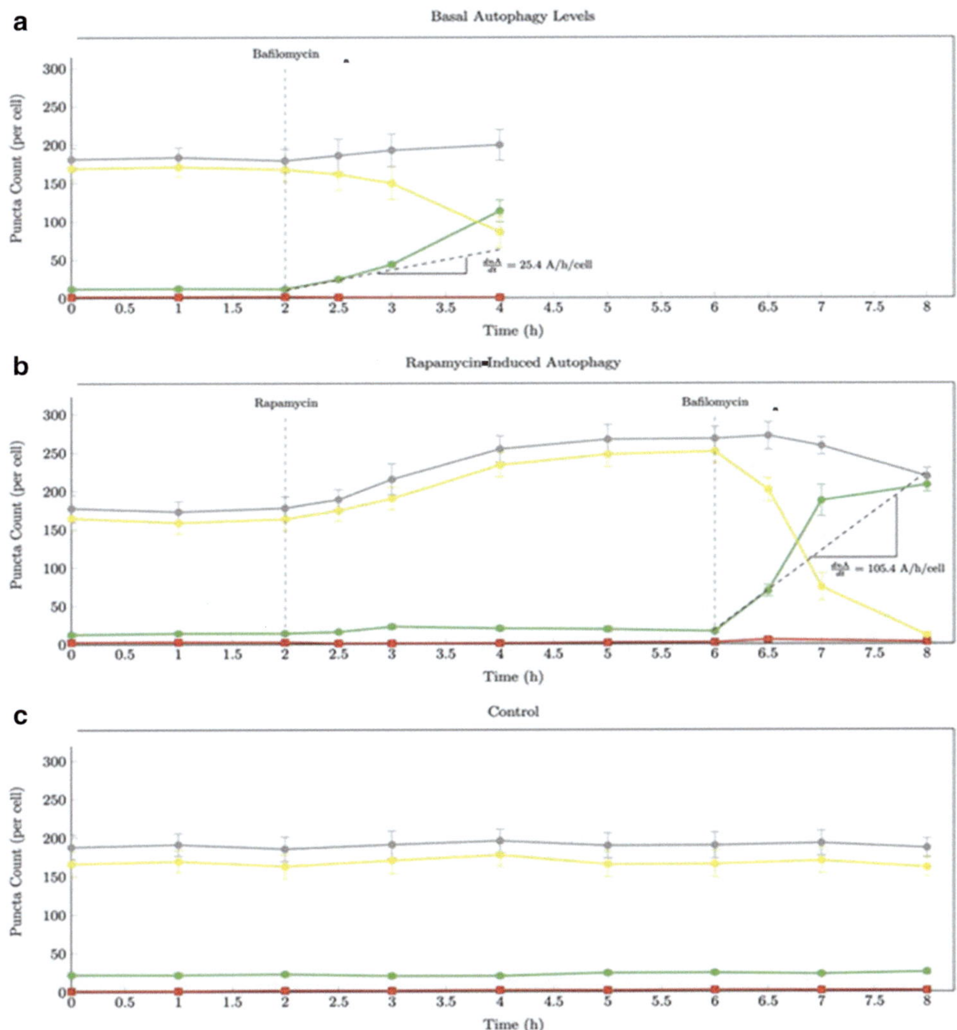


**Fig. 6** The autophagosome (green), autolysosome (yellow) and lysosome (red) pool size plotted over time. Pool sizes under basal conditions (**a**). The slope of the progress curve of autophagosomes upon bafilomycin $A_1$ exposure indicates the basal autophagosome flux (25.4 autophagosomes per hour per cell (A/h/cell)). Pool sizes under induced conditions (**b**). The slope of the progress of autophagosomes following rapamycin treatment, followed by bafilomycin $A_1$ exposure indicates the induces autophagosome flux (105.4 A/h/cell). When the autophagosome synthesis rate equals the autophagosome degradation rate, the total number of autophagosomes per cell remains the same (**c**). The cellular system is in a steady state flux. It is important to note that the autophagosome pool size does not infer the autophagy activity or flux [13]

### 2.1.3 Emerging Questions and Implications

The methodological insights derived from this work immediately prompts questions about altered autophagosome flux in pathology and autophagy dysfunction. Degenerating cells typically decline in autophagy activity, while cancer cells are often characterized by heightened autophagy flux (in this specific context also termed 'autophagy addiction'). How does dysregulated autophagosome flux contribute to disease pathogenesis? Can modulation of flux be exploited as a novel means of therapeutic intervention? How can this methodology be adapted for relevance in clinical studies? Can it be transferred to an in vivo-based approach [6]. Can the measurement of autophagosome flux aid in disease diagnosis, prognosis or therapeutic strategies? Theoretically and practically, all cells and tissues can now be catalogued according to their basal autophagy activity, and their magnitude of response in flux when exposed to autophagy enhancing or inhibiting drugs.

### 2.1.4 What Deserves Attention?

It becomes clear that fluorescence-based microscopy techniques, coupled to 3D live cell imaging, quantitative image analysis pipelines and high throughput, will continue to play a central role to advance the field of autophagy and our abilities to monitor this process accurately. It becomes also evident, that an increasing need exists for a multi-dimensional approach to autophagy analysis, for example SR-SIM or CLEM, in order to provide a comprehensive understanding of the autophagy process that can extend to proteinaceous cargo turnover and autophagosome size, in a precise and quantitative manner.

## 2.2 Investigating the Role of Spermidine in a Model System of Alzheimer Disease Using Correlative Microscopy and Super-Resolution Techniques [55]

### 2.2.1 Introduction

Autophagy enhancing drugs, such as spermidine, have received increasingly attention due to their potential therapeutic benefits in autophagy flux control. However, whether the concentration of a given drug impacts the magnitude of autophagy flux change, had thus far been unclear. By using a model system of Alzheimer disease, where APP (amyloid beta precursor protein) is being overexpressed and overburdens the autophagy machinery with proteinaceous cargo, this article demonstrates the value of combining cutting-edge microscopy techniques, such as SR-SIM and CLEM together with an innovative novel method in which autophagosomes were 3D rendered (Fig. 7). This approach enabled precision quantification of the autophagy system and marks a step forward in our understanding of autophagy, Alzheimer disease and the potential role of modulating autophagy using spermidine.

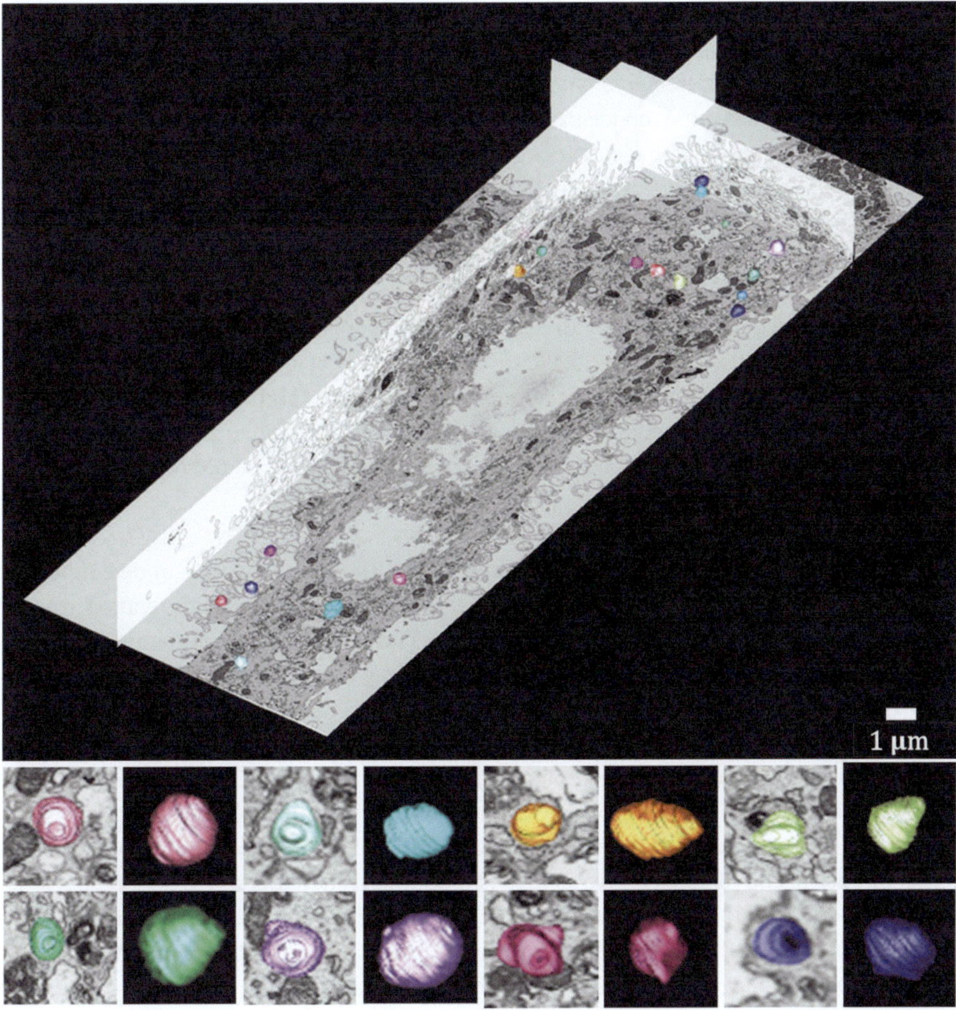

**Fig. 7** Rendering single autophagosomes. When correlative light and electron microscopy (CLEM) is performed, the fluorescence modality indicates the specific molecular identity of the structure, such as a GFP-LC3-positive, but LysoTracker-negative, structure, i.e., an autopghagosome. Volume electron microscopy, here performed through focussed ion beam scanning electron microscopy (FIB-SEM), allows then the exact segmentation of autophagosomes, verified (i.e., ground truth) through the CLEM approach. 3D segmentation enables a highly accurate volumetric assessment of the autophagosomes, while enabling a further characterisation of the cargo, in terms of electron density and complexity. Further fluorescence microscopy enables the dissection of specific cargo contribution. It is important to note that both, autophagy activity and autophagosome size contribute to a given cargo turnover [55]

### 2.2.2 How Did They Move the Field Forward?

This study revealed that indeed autophagy flux is induced in a concentration-dependent manner, which is only detectable when using sensitive metrics, beyond western blotting. It was uncovered that a low concentration of spermidine induces autophagosome formation capable of large volume clearance. Through employing CLEM, the study could link changes in autophagosome formation, surface area, volume and cargo in the presence of spermidine. This approach shed light on the critical aspect that autophagy flux metrics are one side of the 'functional autophagy' equation, cargo clearance is the other. Although the importance of both autophagosome size and flux as critical mechanistic factors had previously been stressed, [58] this study dissected these factors in a mammalian model system in a proteotoxic context.

The innovative use of cutting-edge microscopy techniques highlights the value of combining different imaging modalities, such as SR-SIM with volume EM, to gain a better understanding of the cellular process and its modulation for therapeutic benefit.

### 2.2.3 Emerging Questions and Implications

This study demonstrated that not all techniques are sufficiently sensitive in detecting differences in autophagy activity. It confirmed the distinct and context-dependent role of autophagy modulating drugs *in vitro* and *in vivo*. The results show that the cell type, the nature of the injury, the region of the injury and the drug concentration utilized, need to be equally considered when performing drug screening. Which drug concentration can most favorably offset the degree of autophagy dysfunction? How can autophagy modulation be best aligned with aggregate-prone protein cargo levels? Here, dSTORM was employed to quantify the clusters of APP aggregation, but how could this be achieved in the complex context of brain-specific spatiotemporal autophagy? How could this be characterized in terms of pharmacodynamics and kinetics? Is blood-derived autophagy assessment a suitable window for systemic autophagy function assessment? This study's findings provide evidence for the translational potential of spermidine as a therapeutic intervention for the treatment of neurodegenerative diseases since its molecular target was clearly assessed. Could similar approaches pave the way towards novel therapy?

### 2.2.4 What Deserves Attention?

The integration of microscopy techniques employed, from life-cell-based autophagy flux quantification, to dSTORM, EM, volume EM and CLEM, in context of western blot-analysis data deserves attention. Moreover, the value of morphometrics and quantitative image analysis in whole cells and whole tissues becomes clear. The effect of autophagy enhancement through spermidine on the autophagy pathway, its molecular machinery and cargo highlights the importance of a multi-modal and multi-scale acquisition and analysis approach. Applications with this approach, especially when linked to behavioral studies assessing cognitive function have the potential to not only reshape our understanding of autophagy modulation, but also assists in defining the arsenal of most suitable cutting-edge techniques to monitor autophagy.

**Take Home Message**

- Microscopy approaches provide fundamental insights when studying the autophagy machinery and form an integral part of autophagy research.
- Electron microscopy provides highly resolved ultrastructural detail of organelles, such as the double-membraned autophagosome or the electron-dense lysosome.
- Fluorescence microscopy enables dynamic image acquisition in the living cell and makes it possible to detect a very specific biological event. Fluorescent probes reveal the molecular identity and therefore allow confirmation of structures, such as discerning between an autophagosome or autolysosome.
- Super-resolution microscopy makes it possible to acquire a fluorescence signal below the typical resolving limit of 200 nm, enabling precision localization of single molecules.
- Correlative light and electron microscopy brings together both the ultrastructural detail and backdrop as well as the molecular identity and its localization, based on the specificity of the fluorescent probe.
- Three-dimensional acquisition approaches provide image data in the z dimension, enabling precise subsequent analysis, such as volumetric measurements, in the three-dimensional space.

**Answer Guide to Questions**

*Question 1: What advantage does electron microscopy offer in terms of observing autophagosome morphology?*

1. Electron microscopy offers an advantage in terms of ultrastructural morphological detail due to its high resolving power. This technique allows researchers to visualize ultrastructural detail of autophagosomes, including their size, shape, and interactions with cellular components and cargo targeted for degradation. Electron microscopy provides researchers with the ability to capture and visualize subtle structural alterations, such as the presence of a double-membraned vesicle, characteristic of the autophagosome, electron density and complexity of sequestered cargo. This enables confirmation of the presence of autophagosomes or autolysosomes, and hence provides a deeper understanding of the autophagy process at the subcellular level.

*Question 2: Given is a series of micrographs captured at different stages of the autophagy pathway, arrange them in chronological order. What are the key morphological features that aid in distinguishing between these stages?*

2. The presence of a phagophore membrane in early stages and fully formed double-membraned structures makes it possible to distinguish between phagophores and autophagosomes. Fusion between a lysosome and an autophagosome to form an autolysosome results in the breakdown of the double

(continued)

membrane as well as an increase in the electron density and heterogeneity of cargo and can thus be distinguished by these features.

*Question 3: What are the main proteins of interest when assessing autophagy through means of western blotting and why are these proteins of significance?*

3. The main proteins of interest when assessing autophagy through western blotting include LC3 and SQSTM1. LC3 exists in two forms: LC3-I which is the cytosolic form and LC3-II which is lipidated and recruited to the phagophore membrane. LC3-II is specifically associated with transient phagophores and autophagosomes, hence its abundance in a protein lysate is suggestive of the abundance of autophagosomes. However, one must be careful because the increase in LC3-II levels may indicate either an enhanced autophagosome synthesis rate or an impaired autophagosome degradation. The exposure to bafilomycin $A_1$ will make it possible to discern the autophagy activity, and hence the function of the autophagy system. In contrast, SQSTM1 serves as a receptor protein recruiting proteinaceous cargo/substrates to be selectively degraded. Elevated SQSTM1 levels often imply a disruption in autophagy degradation. Therefore, monitoring changes in LC3 and SQSTM1 levels through western blotting offers valuable insights into autophagy induction, progression and degradation, as well as possible dysregulation, enabling researchers to decipher the autophagic function of a given cellular system under varying conditions.

*Question 4: Analyze the provided cartoon micrographs (Fig. 8).*

(a) *Identify which cell in the given micrograph had been treated with a lysosomal inhibitor, such as bafilomycin $A_1$.*

(b) *Quantify the pool size of autophagosomes and autolysosomes.*

(c) *Quantify the basal and the induced/enhanced autophagosome flux.*

(d) *What is the correct unit for autophagosome flux?*

4. (a) *A and D*

(b)     *Autophagosome pool size:*

   *A: 6 autophagosomes/cell B: 4 autophagosomes/cell C: 3 autophagosomes/cell D: 10 autophagosomes/cell*

   *Autolysosome pool size:*

   *A: 3 autolysosomes/cell B: 9 autolysosomes/cell C: 5 autolysosomes/cell D: 3 autolysosomes/cell*

(c)     *Basal flux: 3 A/h/cell*

*Induced flux: 6 A/h/cell*

(d)     *Autophagosomes/hour/cell; A/h/cell*

*Question 5: How does correlative light and electron microscopy (CLEM) address the limitations of each of the single microscopy techniques involved?*

(continued)

(a) *Compare and contrast the techniques employed for fluorescence microscopy, i.e. confocal microscopy and SR-SIM versus CLEM for studying autophagy.*

(b) *What are the advantages of using CLEM over other techniques in studying cargo clearance*

5.  (a)  Confocal microscopy offers improved resolution than traditional wide-field microscopy due to the emitted light collected only in the focal plane. Since light emitted above and below the focal plane does not contribute to the image, it is suitable for imaging thicker specimens. SR-SIM provides super-resolution capabilities, not only in x/y but also in the z dimension, enabling the visualization of finer structures with enhanced resolving power and detail, but it requires specialized equipment and data processing. Confocal microscopy uses a pinhole to eliminate out-of-focus light, which results in improved optical sectioning, while SR-SIM achieves super-resolution by spatially modulating the excitation pattern and reconstructing a higher-resolution image from the acquired data.

(b)	CLEM combines the strengths of both light microscopy and electron microscopy, providing a comprehensive image data set in terms of function/biology as well as resolution and ultrastructural context. Fluorescence microscopy allows for the precise identification of specific proteins and organelles since it employs labelling strategies for a specific molecular identity. For example, GFP-LC3 or LAMP1-YFP can be easily visualized. Electron microscopy provides highly resolved ultrastructural detail of all structures present, without any specific molecular identify, and hence reveals the overall ultrastructural backdrop of the cell or sample. Therefore, the ability provided by CLEM to register the molecular identity with the ultrastructural detail positions it as a powerful technique for studying the autophagy compartment, the autophagy pathway intermediates, and the respective cargo involved.

*Question 6: Why is it important to use several techniques and methods to assess autophagy?*

6.  A multi-faceted approach is favored to comprehensively monitor autophagy, since different techniques reveal distinct aspects of the complex cellular process. Protein abundance derived in whole lysate using western blotting is a powerful tool, especially if combined with visualization approaches. Quantitative, fluorescence-based image analysis, preferably using a 3D approach to optically section through the sample with great precision, can allow a more precise, and more accurate approach to quantify parameters associated with the autophagy molecular machinery. Autophagy pathway intermediates can be counted, and autophagy activity quantified. Electron microscopy approaches can confirm the nature of size, shape and make up of autophagosomes, autolysosomes and lysosomes, as well as the nature complexity of cargo. No single method in isolation can provide a comprehensive picture of the autophagy pathway.

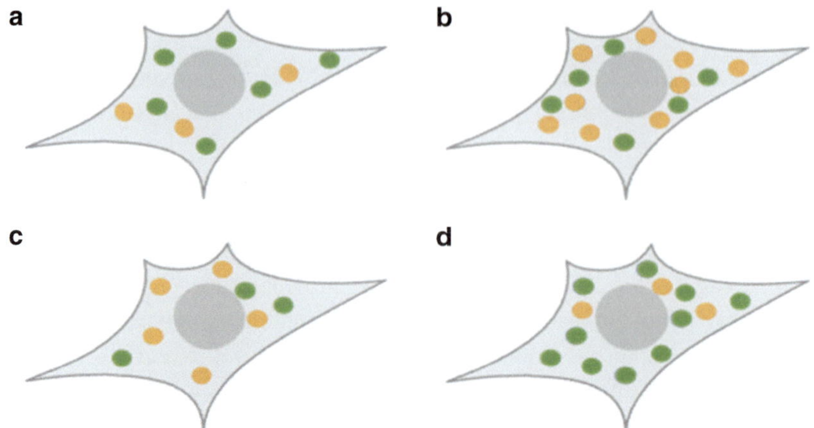

**Fig. 8** Cells depicting autophagosomes (green) and autolysosomes (yellow), at basal and induced autophagy activity, in the absence and presence of Bafilomycin A1. a) Control cells, b) bafilomycin A1 treated cells, c) rapamycin treated cells, d) cells treated with rapamycin, followed by bafilomycin A1 exposure

## References


1. Levine B, Klionsky DJ. Development by self-digestion. Dev Cell. 2004;6:463–77. https://doi.org/10.1016/S1534-5807(04)00099-1.
2. Klionsky DJ, Emr SD. Autophagy as a regulated pathway of cellular degradation. Science (1979). 2000;290:1717–21. https://doi.org/10.1126/science.290.5497.1717.
3. Klionsky DJ. The molecular machinery of autophagy: unanswered questions. J Cell Sci. 2005;118:7–18. https://doi.org/10.1242/jcs.01620.
4. Mizushima N. Methods for monitoring autophagy. Int J Biochem Cell Biol. 2004;36:2491–502. https://doi.org/10.1016/j.biocel.2004.02.005.
5. Yoshii SR, Mizushima N. Monitoring and measuring autophagy. Int J Mol Sci. 2017;18:1865. https://doi.org/10.3390/ijms18091865.
6. Loos B, Klionsky DJ, Du Toit A, Hofmeyr J-HS. On the relevance of precision autophagy flux control *in vivo* – points of departure for clinical translation. Autophagy. 2020;16:750–62. https://doi.org/10.1080/15548627.2019.1687211.
7. Loos B, Engelbrecht A-M, Lockshin RA, Klionsky DJ, Zakeri Z. The variability of autophagy and cell death susceptibility. Autophagy. 2013;9:1270–85. https://doi.org/10.4161/auto.25560.
8. Loos B, Engelbrecht AM. Cell death: a dynamic response concept. Autophagy. 2009;5(5):590–603.
9. Mitra S, Tsvetkov AS, Finkbeiner S. Protein turnover and inclusion body formation. Autophagy. 2009;5:1037–8. https://doi.org/10.4161/auto.5.7.9291.
10. He C, Klionsky DJ. Regulation mechanisms and signaling pathways of autophagy. Annu Rev Genet. 2009;43:67–93. https://doi.org/10.1146/annurev-genet-102808-114910.
11. Loos B, du Toit A, Hofmeyr J-HS. Defining and measuring autophagosome flux—concept and reality. Autophagy. 2014;10:2087–96. https://doi.org/10.4161/15548627.2014.973338.
12. Lumkwana D, du Toit A, Kinnear C, Loos B. Autophagic flux control in neurodegeneration: progress and precision targeting—where do we stand? Prog Neurobiol. 2017;153:64–85. https://doi.org/10.1016/j.pneurobio.2017.03.006.

13. du Toit A, Hofmeyr J-HS, Gniadek TJ, Loos B. Measuring autophagosome flux. Autophagy. 2018;14:1–12. https://doi.org/10.1080/15548627.2018.1469590.

14. du Toit A, De Wet S, Hofmeyr J-H, Müller-Nedebock K, Loos B. The precision control of autophagic flux and vesicle dynamics—a micropattern approach. Cells. 2018;7:94. https://doi.org/10.3390/cells7080094.

15. Eskelinen E-L. Maturation of autophagic vacuoles in mammalian cells. Autophagy. 2005;1:1–10. https://doi.org/10.4161/auto.1.1.1270.

16. Fengsrud M, Roos N, Berg T, Liou W, Slot JW, Seglen PO. Ultrastructural and immunocytochemical characterization of autophagic vacuoles in isolated hepatocytes: effects of vinblastine and asparagine on vacuole distributions. Exp Cell Res. 1995;221:504–19. https://doi.org/10.1006/excr.1995.1402.

17. Lawrence BP, Brown WJ. Autophagic vacuoles rapidly fuse with pre-existing lysosomes in cultured hepatocytes. J Cell Sci. 1992;102:515–26. https://doi.org/10.1242/jcs.102.3.515.

18. Punnonen EL, Autio S, Kaija H, Reunanen H. Autophagic vacuoles fuse with the prelysosomal compartment in cultured rat fibroblasts. Eur J Cell Biol. 1993;61:54–66.

19. Punnonen E-L, Pihakaski K, Mattila K, Lounatmaa K, Hirsimki P. Intramembrane particles and filipin labelling on the membranes of autophagic vacuoles and lysosomes in mouse liver. Cell Tissue Res. 1989;258:269–76. https://doi.org/10.1007/BF00239447.

20. Deter RL, Baudhuin P, de Duve C. Participation of lysosomes in cellular autophagy induced in rat liver by glucagon. J Cell Biol. 1967;35:C11–6. https://doi.org/10.1083/jcb.35.2.C11.

21. de Duve C, Wattiaux R. Functions of lysosomes. Annu Rev Physiol. 1966;28:435–92. https://doi.org/10.1146/annurev.ph.28.030166.002251.

22. Barth S, Glick D, Macleod KF. Autophagy: assays and artifacts. J Pathol. 2010;221:117–24. https://doi.org/10.1002/path.2694.

23. Eskelinen E-L. To be or not to be? Examples of incorrect identification of autophagic compartments in conventional transmission electron microscopy of mammalian cells. Autophagy. 2008;4:257–60. https://doi.org/10.4161/auto.5179.

24. Klionsky DJ. Autophagy revisited: a conversation with Christian de Duve. Autophagy. 2008;4:740–3. https://doi.org/10.4161/auto.6398.

25. Kang R, Zeh HJ, Lotze MT, Tang D. The Beclin 1 network regulates autophagy and apoptosis. Cell Death Differ. 2011;18:571–80. https://doi.org/10.1038/cdd.2010.191.

26. Koukourakis MI, Kalamida D, Giatromanolaki A, Zois CE, Sivridis E, Pouliliou S, Mitrakas A, Gatter KC, Harris AL. Autophagosome proteins LC3A, LC3B and LC3C have distinct subcellular distribution kinetics and expression in cancer cell lines. PLoS One. 2015;10:e0137675. https://doi.org/10.1371/journal.pone.0137675.

27. Chittaranjan S, Bortnik S, Gorski SM. Monitoring autophagic flux by using lysosomal inhibitors and western blotting of endogenous MAP1LC3B. Cold Spring Harb Protoc. 2015;2015:pdb.prot086256. https://doi.org/10.1101/pdb.prot086256.

28. Gómez-Sánchez R, Pizarro-Estrella E, Yakhine-Diop SMS, Rodríguez-Arribas M, Bravo-San Pedro JM, Fuentes JM, González-Polo RA. Routine Western blot to check autophagic flux: cautions and recommendations. Anal Biochem. 2015;477:13–20. https://doi.org/10.1016/j.ab.2015.02.020.

29. Liang XH, Jackson S, Seaman M, Brown K, Kempkes B, Hibshoosh H, Levine B. Induction of autophagy and inhibition of tumorigenesis by beclin 1. Nature. 1999;402:672–6. https://doi.org/10.1038/45257.

30. Bjørkøy G, Lamark T, Brech A, Outzen H, Perander M, Øvervatn A, Stenmark H, Johansen T. p62/SQSTM1 forms protein aggregates degraded by autophagy and has a protective effect on huntingtin-induced cell death. J Cell Biol. 2005;171:603–14. https://doi.org/10.1083/jcb.200507002.

31. Mizushima N, Yoshimori T. How to interpret LC3 immunoblotting. Autophagy. 2007;3:542–5. https://doi.org/10.4161/auto.4600.
32. Geng J, Baba M, Nair U, Klionsky DJ. Quantitative analysis of autophagy-related protein stoichiometry by fluorescence microscopy. J Cell Biol. 2008;182:129–40. https://doi.org/10.1083/jcb.200711112.
33. Kim S, Choi S, Kang D. Quantitative and qualitative analysis of autophagy flux using imaging. BMB Rep. 2020;53:241–7. https://doi.org/10.5483/BMBRep.2020.53.5.046.
34. Rai S, Manjithaya R. Fluorescence microscopy: a tool to study autophagy. AIP Adv. 2015;5:084804. https://doi.org/10.1063/1.4928185.
35. Kabeya Y. LC3, a mammalian homologue of yeast Apg8p, is localized in autophagosome membranes after processing. EMBO J. 2000;19:5720–8. https://doi.org/10.1093/emboj/19.21.5720.
36. Kimura S, Noda T, Yoshimori T. Dissection of the autophagosome maturation process by a novel reporter protein, tandem fluorescent-tagged LC3. Autophagy. 2007;3:452–60. https://doi.org/10.4161/auto.4451.
37. Kaizuka T, Morishita H, Hama Y, Tsukamoto S, Matsui T, Toyota Y, Kodama A, Ishihara T, Mizushima T, Mizushima N. An autophagic flux probe that releases an internal control. Mol Cell. 2016;64:835–49. https://doi.org/10.1016/j.molcel.2016.09.037.
38. Morishita H, Kaizuka T, Hama Y, Mizushima N. A new probe to measure autophagic flux in vitro and in vivo. Autophagy. 2017;13:757–8. https://doi.org/10.1080/15548627.2016.1278094.
39. Mauvezin C, Neufeld TP. Bafilomycin A1 disrupts autophagic flux by inhibiting both V-ATPase-dependent acidification and Ca-P60A/SERCA-dependent autophagosome-lysosome fusion. Autophagy. 2015;11:1437–8. https://doi.org/10.1080/15548627.2015.1066957.
40. Lumkwana D, Engelbrecht L, Loos B. Monitoring autophagy using super-resolution structured illumination and direct stochastic optical reconstruction microscopy. Methods Cell Biol. 2021;165:139–52.
41. Cole R. Live-cell imaging. Cell Adhes Migr. 2014;8:452–9. https://doi.org/10.4161/cam.28348.
42. Dokladny K, Myers OB, Moseley PL. Heat shock response and autophagy—cooperation and control. Autophagy. 2015;11:200–13. https://doi.org/10.1080/15548627.2015.1009776.
43. Heintzmann R, Ficz G. Breaking the resolution limit in light microscopy. Brief Funct Genomic Proteomic. 2006;5:289–301. https://doi.org/10.1093/bfgp/ell036.
44. Hell SW, Sahl SJ, Bates M, Zhuang X, Heintzmann R, Booth MJ, Bewersdorf J, Shtengel G, Hess H, Tinnefeld P, Honigmann A, Jakobs S, Testa I, Cognet L, Lounis B, Ewers H, Davis SJ, Eggeling C, Klenerman D, Willig KI, Vicidomini G, Castello M, Diaspro A, Cordes T. The 2015 super-resolution microscopy roadmap. J Phys D Appl Phys. 2015;48:443001. https://doi.org/10.1088/0022-3727/48/44/443001.
45. Abdollahzadeh I, Hendriks J, Sanwald JL, Simons IM, Hoffmann S, Weiergräber OH, Willbold D, Gensch T. Autophagy-related proteins GABARAP and LC3B label structures of similar size but different shape in super-resolution imaging. Molecules. 2019;24:1833. https://doi.org/10.3390/molecules24091833.
46. Man H, Zhou L, Zhu G, Zheng Y, Ye Z, Huang Z, Teng X, Ai C, Ge G, Xiao Y. Super-resolution imaging of autophagy by a preferred pair of self-labeling protein tags and fluorescent ligands. Anal Chem. 2022;94:15057–66. https://doi.org/10.1021/acs.analchem.2c03125.
47. Wang H, Fang G, Chen H, Hu M, Cui Y, Wang B, Su Y, Liu Y, Dong B, Shao X. Lysosome-targeted biosensor for the super-resolution imaging of lysosome–mitochondrion interaction. Front Pharmacol. 2022;13:865173. https://doi.org/10.3389/fphar.2022.865173.
48. Heintzmann R, Huser T. Super-resolution structured illumination microscopy. Chem Rev. 2017;117:13890–908. https://doi.org/10.1021/acs.chemrev.7b00218.
49. Henriques R, Griffiths C, Hesper Rego E, Mhlanga MM. PALM and STORM: unlocking live-cell super-resolution. Biopolymers. 2011;95:322–31. https://doi.org/10.1002/bip.21586.

50. Karanasios E. Correlative live-cell imaging and super-resolution microscopy of autophagy. Methods Mol Biol. 2019;1880:231–42.
51. dos Santos Á, Rollins DE, Hari-Gupta Y, McArthur H, Du M, Ru SYZ, Pidlisna K, Stranger A, Lorgat F, Lambert D, Brown I, Howland K, Aaron J, Wang L, Ellis PJI, Chew T-L, Martin-Fernandez M, Pyne ALB, Toseland CP. Autophagy receptor NDP52 alters DNA conformation to modulate RNA polymerase II transcription. Nat Commun. 2023;14:2855. https://doi.org/10.1038/s41467-023-38572-9.
52. Ligeon L-A, Barois N, Werkmeister E, Bongiovanni A, Lafont F. Structured illumination microscopy and correlative microscopy to study autophagy. Methods. 2015;75:61–8. https://doi.org/10.1016/j.ymeth.2015.01.017.
53. Takahashi S, Saito C, Koyama-Honda I, Mizushima N. Quantitative 3D correlative light and electron microscopy of organelle association during autophagy. Cell Struct Funct. 2022;47:22071. https://doi.org/10.1247/csf.22071.
54. Pope I, Tanner H, Masia F, Payne L, Arkill KP, Mantell J, Langbein W, Borri P, Verkade P. Correlative light-electron microscopy using small gold nanoparticles as single probes. Light Sci Appl. 2023;12:80. https://doi.org/10.1038/s41377-023-01115-4.
55. Lumkwana D, Peddie C, Kriel J, Michie LL, Heathcote N, Collinson L, Kinnear C, Loos B. Investigating the role of spermidine in a model system of Alzheimer's disease using correlative microscopy and super-resolution techniques. Front Cell Dev Biol. 2022;10:819571. https://doi.org/10.3389/fcell.2022.819571.
56. Rubinsztein DC, Cuervo AM, Ravikumar B, Sarkar S, Korolchuk VI, Kaushik S, Klionsky DJ. In search of an "autophagomometer". Autophagy. 2009;5:585–9. https://doi.org/10.4161/auto.5.5.8823.
57. Sahani MH, Itakura E, Mizushima N. Expression of the autophagy substrate SQSTM1/p62 is restored during prolonged starvation depending on transcriptional upregulation and autophagy-derived amino acids. Autophagy. 2014;10:431–41. https://doi.org/10.4161/auto.27344.
58. Jin M, Klionsky DJ. Regulation of autophagy: modulation of the size and number of autophagosomes. FEBS Lett. 2014;588:2457–63. https://doi.org/10.1016/j.febslet.2014.06.015.

# Autophagy Receptors Couple Cargo Destined to Be Degraded with the Core Autophagy Machinery

Hallvard Lauritz Olsvik, Trond Lamark, and Terje Johansen

**What Will You Learn in This Chapter**

In this chapter you will learn about selective macroautophagy/autophagy in mammalian cells and how specific cellular components, such as damaged or surplus organelles or protein aggregates are selectively degraded by autophagy. You will learn about selective autophagy receptors; how they select cargo to be degraded, how they connect to the forming autophagosome via the MAP1LC3/GABARAP proteins and how they also connect to the core autophagosome formation machinery. All this enables the generation of an autophagosome directly around the selected cargo. The autophagosome fuses with lysosomes forming autolysosomes where the content is degraded by lysosomal hydrolases which are active in the low pH environment. The degradation products are transported out of the autolysosome and used as energy sources and building blocks for synthesis of new macromolecules like proteins, nucleic acids, sugars and lipids. Dysregulation of selective autophagy has been implicated in the pathogenesis of several major diseases, including neurodegenerative diseases, cancer, and metabolic disorders.

## 1 Selective Autophagy

Macroautophagy (autophagy hereafter) is an evolutionarily conserved process for lysosomal degradation of intracellular components involving the sequestration of the cargo to be degraded in double-membrane vesicles 0.5–1 µm in diameter called autophagosomes [94]. The forming autophagosomes appear as crescent-like membrane structures, called phagophores. These expand and close upon themselves to form autophagosomes. Autophagy serves two main purposes: removal of toxic and surplus components, and to secure a supply of energy and building blocks upon nutrient deprivation. Autophagy was

H. L. Olsvik · T. Lamark · T. Johansen (✉)
Autophagy Research Group, Department of Medical Biology, UiT The Arctic University of Norway, Tromsø, Norway
e-mail: hallvard.olsvik@uit.no; trond.lamark@uit.no; terje.johansen@uit.no

for a long time considered a non-selective bulk degradation process of cytoplasmic material [93]. However, even before Christian de Duve had coined the term autophagy [20], Ashford and Porter had observed mitochondria selectively encapsulated within a double-membrane structure by electron microscopy [5]. In hindsight, this was clearly an example of selective autophagy targeting mitochondria [52], now referred to as mitophagy. The discovery of the selective autophagy receptors (SARs) provided the mechanistic underpinning for the formation of autophagosomes around the targeted cargo, often labelled for destruction by ubiquitin chains [45, 46]. The SARs carry out a basal, constitutive selective autophagy serving a quality control purpose which is increased upon stress such as nutrient starvation, oxidative stress and infections [62]. This quality control process is especially important for long-lived cells, like neurons.

In the baker's yeast, *Saccharomyces cerevisiae*, phagophores form at one unique site called the phagophore assembly site (PAS). The PAS is attached to the big, sole lysosome in yeast called the vacuole. In mammalian cells, there are instead multiple PAS and many small lysosomes [84, 146]. In mammalian cells the autophagosomes for the most part grow out of the endoplasmic reticulum (ER) in structures called omegasomes because they resemble the Greek letter omega [6]. When discussing selective autophagy this difference is relevant since either what is to be degraded is moved to the PAS or the PAS is formed on the sequestered cargo. As we go along it will become clear that in both yeast and man there is evidence that SARs can bind to proteins in the core machinery, and these are located at the PAS. What is the core machinery governing formation of autophagosomes composed of?

## 2    Autophagy Core Machinery

Before we delve into the mechanisms of selective autophagy, we need to introduce the components of the autophagy core machinery required for formation of autophagosomes. Of the more than 40 *ATG* genes and proteins about 18 make up the evolutionarily conserved core autophagy machinery that help nucleate, initiate and expand the phagophore and form the autophagosome [80].

The mammalian cell core autophagy machinery is organized into six functional units with (1) the ULK1/ULK2 protein kinase complex, (2) the class III phosphatidylinositol 3-kinase complex I (PtdIns3K-C1), (3) ATG9-containing vesicles, (4) the ATG2A/ATG2B lipid transfer complex including WIPI (WD repeat domain, phosphoinositide interacting) proteins, (5) the ATG12–ATG5-ATG16L1 complex, and (6) the Atg8-family proteins lipid conjugation system including the ATG4A to ATG4D proteases [84, 88] (Fig. 1).

The ULK1/ULK2 complex consists of the protein kinases ULK1 or ULK2 and the scaffold proteins RB1CC1/FIP200, ATG13 and ATG101, and its activation is required during nucleation and initiation of the phagophore. Studies in yeast have revealed that the single PAS is formed by liquid-liquid phase separation (LLPS) to form a fluid like condensate (droplet) due to multivalent interactions between Atg1 (ULK homolog), Atg13 and

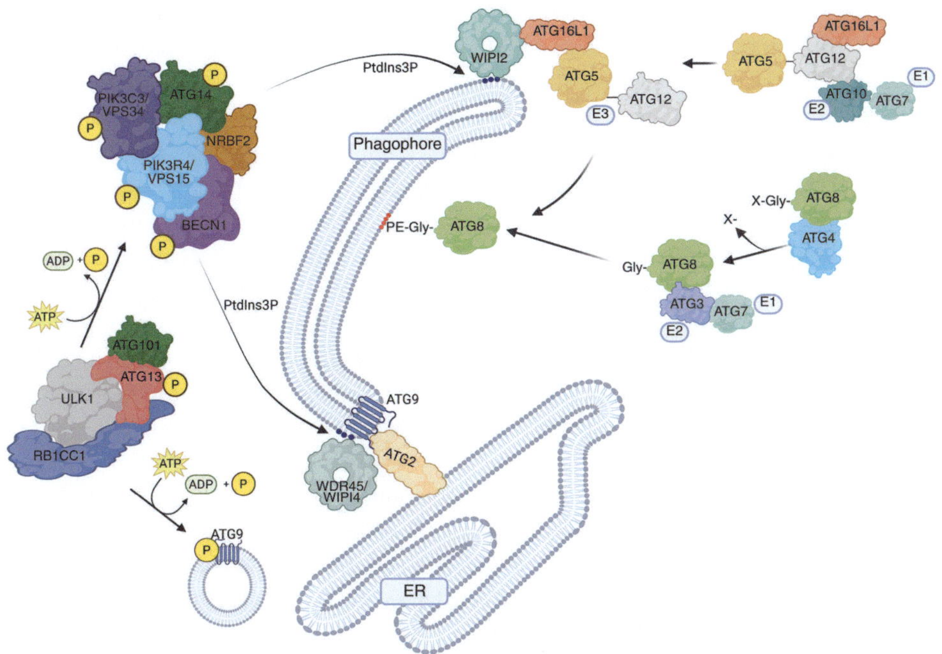


**Fig. 1** Six functional units of the core autophagy machinery. (1) The ULK1 protein kinase complex comprised of ULK1, RB1CC1, ATG13 and ATG101 recruits and activates by phosphorylation the (2) class III phosphatidylinositol 3-kinase complex 1 (PtdIns3K-C1) comprised of PIK3C3/VPS34, PIK3R4/VPS15, BECN1, ATG14 and NRBF2. This complex generates PtdIns3P on the phagophore membrane (indicated by blue dots). (3) Vesicles containing the transmembrane protein ATG9 are recruited to the phagophore where ATG9 scrambles lipids across the two membrane leaflets. (4) The ATG2A/ATG2B lipid transferase complex channels lipids synthesized in the endoplasmic reticulum to the phagophore via ATG9 and binds to both ATG9 and to WDR45/WIPI4 to transfer lipids to the growing tip of the phagophore. (5) The E3-like ATG12–ATG5-ATG16L1 complex, including the E1-like enzyme ATG7 and the E2-like enzyme ATG10 is part of, (6) the Atg8-family protein lipid conjugation system, which includes the protease ATG4, the E1-like enzyme ATG7 and the E2-like enzyme ATG3. (Illustration created with BioRender.com). ATG8: mammalian Atg8-family protein

Atg17 (homologous to part of RB1CC1) [27, 138]. In mammalian cells multiple PAS form from the ER upon amino acid starvation. Calcium transients on the ER membrane induce multivalent interactions between ATG13 and RB1CC1, and this leads to LLPS and assembly of the activated ULK complex [148]. When the ULK complex is assembled, the HORMA domains of the ATG13-ATG101 dimer bind to ATG9A transmembrane proteins [87, 104]. ATG9 is recruited on small vesicles emanating from the Golgi apparatus and helps define the sites (i.e., the PAS) where autophagosomes are formed [38, 76]. The ULK complex subsequently recruits the class III phosphatidylinositol 3-kinase complex I (PtdIns3K–C1) consisting of scaffold proteins ATG14, BECN1, PIK3R4/VPS15, NRBF2 and the lipid kinase PIK3C3/VPS34 [7, 39]. This complex labels the phagophore membranes with phosphatidylinositol-3-phosphate which then acts as a handle for binding of the WIPI proteins to the growing membranes. There are four WIPI proteins (WIPI1, WIPI2, WDR45B/WIPI3 and WDR45/WIPI4) [113]. WDR45/WIPI4 helps recruit the large, rod-shaped lipid-channeling proteins ATG2A and ATG2B to the PAS. ATG2A/ATG2B forms an inter-membrane lipid transfer complex where the C-terminal end of ATG2A/ATG2B is bound to WDR45/WIPI4 and ATG9A at the phagophore membrane and the N-terminal end is associated with VMP1 and TMEM41B in the ER [29, 127]. ATG9A is a lipid scramblase which helps establishing equilibration of lipids across the membrane bilayer of the phagophore membrane [70, 74]. VMP1 and TEMEM41B are lipid scramblases that re-equilibrate lipids between the ER membrane leaflets [29]. The complex of VMP1, TMEM41B (lipid scramblases), ATG2A/ATG2B (lipid channel) and ATG9A (lipid scramblase) transports phospholipids synthesized in the ER to the phagophore membrane to allow its growth and expansion to form the autophagosome [88] (Fig. 1).

An intriguing finding upon elucidating the core autophagy components and their roles in autophagosome formation was the discovery of two ubiquitin-like conjugation systems involved in autophagosome formation. A covalent coupling of ATG5 to ATG12 gives the ATG12–ATG5 conjugate that binds to ATG16L1 to form the ATG12–ATG5-ATG16L1 complex [78, 79]. This complex is recruited to the PAS by the binding of ATG16L1 to WIPI2 [23]. The other conjugation system attaches the phospholipid phosphatidylethanolamine (PE) to the C-terminal glycine of Atg8-family proteins [47, 55]. ATG7 serves as the E1 enzyme for both ATG12 and Atg8-family proteins. ATG10 acts as the E2 enzyme for ATG12 and can directly recognize ATG5 and conjugate it to ATG12. ATG3 acts as the E2 enzyme for Atg8-family proteins. As we will see below, the ATG12–ATG5-ATG16L1 complex acts as an E3 ligase to target the conjugation of Atg8-family proteins to PE on the phagophore membrane [84, 88] (Fig. 1).

## 3    Atg8-Family Proteins and LIR or AIM Motifs

In our initial studies of SQSTM1/p62 as an autophagy receptor we could show that SQSTM1 binds to Atg8-family proteins via a short sequence motif we called the MAP1LC3/LC3-interacting region (LIR) motif [98]. To understand how autophagy receptors work in selective autophagy we need to look into the Atg8-family of proteins and the LIR motif which is often called the Atg8-family interacting motif (AIM) in yeast and plants. Atg8-family proteins are small ubiquitin-like proteins with two extra N-terminal α-helices hovering above the ubiquitin fold [46]. The number of Atg8-family protein members varies from one in yeast, two in the nematode model organism *Caenorhabditis elegans*, two in the fruit fly *Drosophila*, seven in humans, and nine or more in plants. In humans seven genes encode the Atg8-family proteins: *LC3A*, *LC3B*, *LC3B2*, *LC3C*, *GABARAP*, *GABARAPL1* and *GABARAPL2* [105]. Interestingly, rodents lack LC3C [43]. LC3A comes with two different N-termini due to alternative splicing.

Atg8-family proteins are synthesized as precursors that need to be cleaved by ATG4 proteases to expose a C-terminal glycine that is conjugated to PE with the ATG12–ATG5-ATG16L1 complex working as an E3 ligase [34]. Atg8-family proteins are conjugated to both membrane surfaces of the phagophore membrane, but those that are anchored in the outer membrane surface are mostly deconjugated by ATG4 proteases during closure or maturation of autophagosomes. Atg8-family proteins are sometimes conjugated to other membranes and can also be conjugated to phosphatidylserine [22, 26]. The function of the Atg8-family proteins anchored in the phagophore membrane is to act as adaptors that recruit other proteins. Virtually all proteins that are recruited to Atg8-family proteins on the phagophore membrane use a LIR/AIM motif to interact. Different sets of proteins are recruited to the inner and outer membrane surfaces. Proteins involved in autophagosome formation or in the transport or fusion of autophagosomes are recruited to the outer membrane surface. In mammals this includes several of the core autophagy complexes such as the ULK1/ULK2 [2] and PtdIns3K-C1 [10] complexes. Mammalian core autophagy proteins have an interesting binding preference for the GABARAP subfamily [2, 10]. This possibly reflects that the GABARAP subfamily is the evolutionarily oldest subfamily and therefore used for autophagosome formation. Membrane formation occurs at the rim of the phagophore membrane. Core autophagy proteins are thought to be recruited to this site via LIR interactions to facilitate the growth of the phagophore membrane. Autophagy receptors that perform selective autophagy are bound to Atg8-family proteins anchored in the inner membrane surface. An efficient docking of the autophagy receptor and associated cargo onto the membrane requires multiple LIR/AIM-Atg8-family protein interactions. A tight sealing is essential because it allows the phagophore to grow on a cargo coated with autophagy receptors, and this ensures that the cargo is efficiently engulfed by the forming autophagosome. This type of mechanism also explains why the autophagy receptor is always degraded along with the cargo it has selected for degradation. The Atg8-family protein binding preference varies between different mammalian autophagy receptors involved in selective autophagy, and proteins such as SQSTM1 and NBR1 have a broad binding preference indicating that all mammalian Atg8-family proteins may participate in docking of cargos in selective autophagy. However, engulfment of larger cargos like bacteria or mitochondria require large amounts of the Atg8-family proteins that are engaged in cargo docking and the Atg8-family proteins may therefore be a limiting factor in selective autophagy.

The LIR/AIM motif is an evolutionarily conserved motif of around 10 residues and is used in all types of selective autophagy [105] (Fig. 2). In the canonical structure, the LIR/AIM motif folds as a β-strand and becomes an extension of β-sheet 2 in the Atg8-family proteins. Early studies identified a W-x-x-L core LIR motif [40, 89, 98]. This motif was later modified into a more general $\theta_0$-$X_1$-$X_2$-$\tau_3$ motif where $\theta$ is an aromatic amino acid residue (W/F/Y; Trp/Phe/Tyr) and $\tau$ a hydrophobic amino acid (L/I/V; Leu/Ile/Val). Another important feature of LIR/AIM motifs is the presence of a variable number of negatively charged residues (Glu/Asp) or phosphorylatable residues (Ser/Thr). Negatively

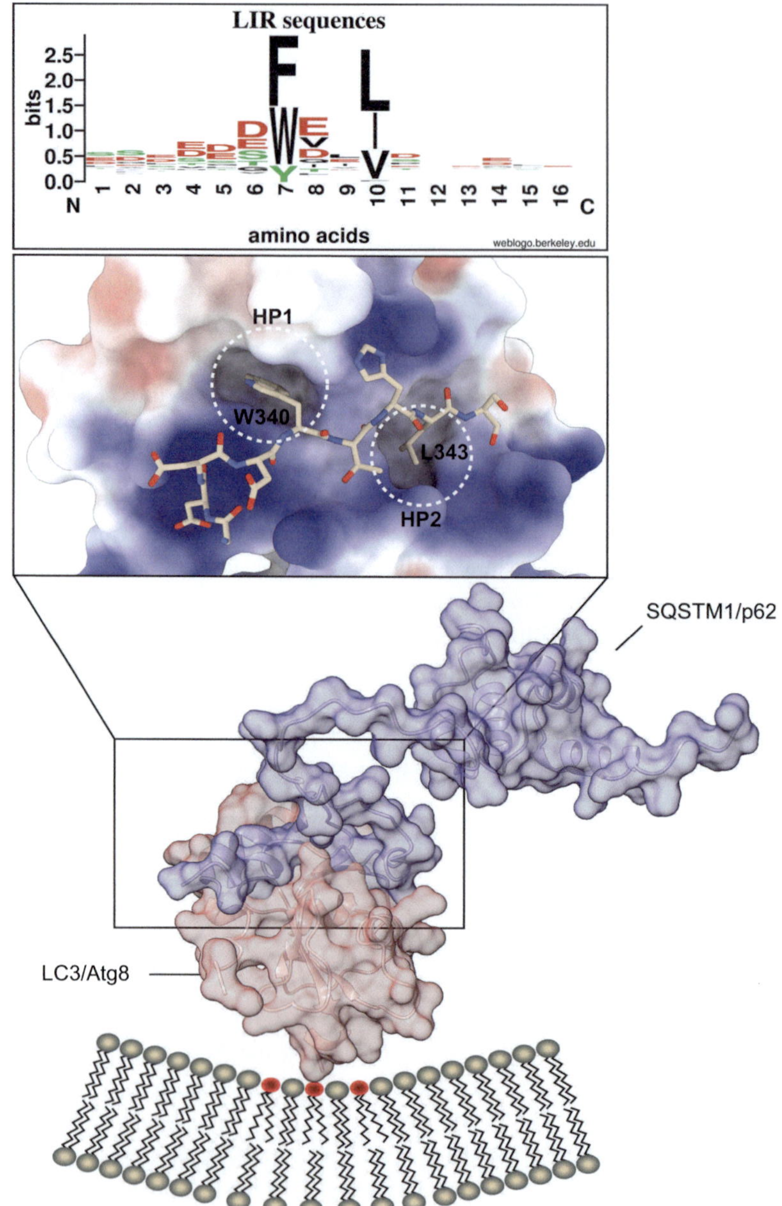



**Fig. 2** The SAR SQSTM1 binds to LC3/Atg8 on the phagophore via the LIR motif. (Top) A sequence logo made using WebLogo [19], shows the representation of amino acid residues at 16 consecutive positions from 117 LIR motifs with the core LIR motif starting at residue 7 with the conserved aromatic residues W, F or Y. Acidic residues (red letters) are preferred in the three positions preceding the conserved aromatic residue. Phosphorylatable residues, Ser (S), Thr (T) and Tyr (Y), are shown in green letters. (Bottom) A 3D structure of LC3B (pink) and SQSTM1 (purple) predicted using AlphaFold2 [77]. The part of SQSTM1 shown encompasses amino acids 329–440, which includes the LIR and UBA domains. (Middle) Inset shows a close-up image of the structure of the SQSTM1 LIR (amino acids 337–344), with W340 docking into the first hydrophobic pocket (HP1), and L343 docking into the second hydrophobic pocket (HP2) in LC3B. LC3B is conjugated to PE (red dots) in the phagophore membrane on the opposite side of where it binds to the SAR SQSTM1

charged residues are often seen at positions −1, −2 or −3 relative to the core motif, but they can also be found in the core motif at positions +1 or +2 or C-terminal to the core motif. The LIR motif interacts with a so-called LIR docking site (LDS) in the Atg8-family proteins that consists of two hydrophobic pockets; HP1 and HP2. HP1 is occupied by the side chain of the aromatic residue (W, F or Y), while HP2 is occupied by the side chain of the hydrophobic residue (L, I or V). This is shown in Fig. 2 for SQSTM1 LIR docking to the LDS in LC3B. In addition, the affinity of the LIR-LDS interaction is strongly increased by electrostatic interactions formed between negatively charged residues in the LIR motif and positively charged residues in the LDS of the Atg8-family protein [46]. These interactions are more variable and contribute to the specificity of LIR interactions. A few identified LIR/AIM motifs are non-canonical in the sense that they are helical or dock into only one of the two hydrophobic pockets in the LDS. One example is the LIR in the autophagy receptor CALCOCO2/NDP52. CALCOCO2 has an atypical core motif (LVV) that docks into HP2 only, and it binds specifically to LC3C [130].

**Question**
1. What are the two main purposes served by autophagy?
2. What is the main function of the Atg8-family proteins bound to the phagophore membrane?
3. How do the Atg8-family proteins recruit selective autophagy receptors (SARs) and other interacting proteins?

## 4    Selective Autophagy Receptors (SARs)

The use of selective autophagy receptors (SARs), which can be soluble or membrane bound, distinguishes selective autophagy from the nonselective, bulk autophagy [45]. The first mammalian SAR discovered, SQSTM1, was defined as such because it could recognize ubiquitinated protein aggregates (aka cargo) via its C-terminal ubiquitin-associated (UBA) domain and connect this cargo to the inner membrane surface of the phagophore membrane by interacting with LC3 via its LIR motif [11]. Selective autophagy is usually classified according to the cargo or substrate that is targeted for degradation (see Table 1). The SARs can be grouped into two groups: soluble SARs that in many cases recognize ubiquitin-labeled cargo, and a large group of membrane-bound SARs [62].

The yeast biosynthetic cytoplasm-to-vacuole targeting (Cvt) pathway was discovered by Dan Klionsky's group before the discovery of the first mammalian autophagy receptors. The Cvt pathway has continued to act as a very important model for the field of selective autophagy [68]. The Cvt pathway is used to transport the zymogens of vacuolar aminopeptidase 1 (prApe1), the vacuolar aspartyl aminopeptidase Ape4 and the vacuolar α-mannosidase Ams1 to the yeast lysosome called the vacuole [140]. Atg19 acts as a

**Table 1** Selective autophagy in mammals classified according to the cargo or substrate

| Pathway | Cargo or Substrate | Autophagy receptors | References |
|---|---|---|---|
| Aggrephagy | Protein aggregate | SQSTM1/p62, NBR1, OPTN, TAX1BP1, CCT2 | [11, 56, 58, 69, 98, 107] |
| Clockophagy | Circadian clock protein BMAL1/ARNTL | SQSTM1 | [141] |
| Cvt pathway (yeast) | Ams1, prApe1 | Yeast Atg19, Atg34 | [68, 85] |
| Reticulophagy | ER | RETREG1, SEC62, RTN3, CCPG1, ATL3, TEX264, CALCOCO1 | [4, 14, 15, 28, 33, 50, 91, 115] |
| Ferritinophagy | Ferritin | NCO4A | [25, 71] |
| Fluidophagy | Condensates (SQSTM1/p62 bodies) | SQSTM1 | [1, 11] |
| Golgiphagy | Golgi apparatus | CALCOCO1, YIFP3 and YIFP4 | [37, 92] |
| Glycophagy | Glycogen | STBD1 | [44] |
| Lipophagy | Lipid droplets | SQSTM1 | [131] |
| Lysophagy | Lysosome | SQSTM1, TAX1BP1, TRIM16 | [13, 57] |
| Midbody autophagy | Midbody rings | SQSTM1, NBR1, TRIM17 | [41, 73, 100] |
| Mitophagy Ub-dependent | Mitochondria | CALCOCO2, OPTN, SQSTM1, TAX1BP1, AMBRA1 | [36, 64, 116, 126, 135] |
| Mitophagy Ub-independent | Mitochondria | BNIP3L, BNIP3, FUNDC1, BCL2L13, FKBP8, PHB2, NLRX1, AMBRA1, cardiolipin, ceramide, NIPSNAP1/NIPSNAP2 | [9, 16, 35, 66, 90, 101, 112, 116, 132, 145] |
| Nuclear lamina autophagy | Nuclear lamina | LMNB1 | [24] |
| Pexophagy Ub-dependent | Peroxisome | NBR1, SQSTM1 | [21] |
| Pexophagy Ub-independent | Peroxisome | BNIP3L | [134] |
| Virophagy (viral xenophagy) | Viral components | TRIM5, SQSTM1, OPTN | [3, 72, 97] |
| Xenophagy | Bacteria | CALCOCO2, SQSTM1, OPTN, TAX1BP1 | [119, 122, 133, 147] |
| Zymophagy | Secretory granule | SQSTM1 | [32] |
| Ribophagy | Ribosomes | NUFIP1, SQSTM1 | [67, 137] |

receptor in this transport pathway and was as such the first SAR discovered [65, 111]. The yeast Cvt pathway receptor Atg19 and its homolog Atg34 can be regarded as homologs of the mammalian SAR NBR1 [102]. After we had discovered the LIR-Atg8-family protein interaction by studying SQSTM1 [98], it was also shown that Atg19 contains a LIR or AIM which it uses to interact with yeast Atg8 [89].

As mentioned above, there is a dependency of multivalent LIR-Atg8-family protein interactions for efficient docking of the cargo to the phagophore membrane. This is solved either by oligomerization of the SARs, like that seen for SQSTM1 [17, 63]. Alternatively, SARs may have more than one LIR/AIM motif, like yeast Atg19 [108], or RTN3L [33]. Membrane-bound SARs may form a cluster on the membrane, like the reticulophagy receptor RETREG1/FAM134B which gets ubiquitinated to promote receptor clustering and binding to lipidated LC3B, thereby promoting selective autophagy of the ER, so called reticulophagy [30].

An essential function of a SAR, in addition to its role in cargo docking, is to connect with and activate the core autophagy machinery so that autophagosome formation is initiated. In selective autophagy in yeast, the selected cargo is transported to the PAS where the core components are localized and autophagosome formation initiated. In mammals, the core machinery is instead recruited to the cargo, and this defines the location where the new PAS is established. The same core machinery is used in selective autophagy as in starvation-induced bulk autophagy and the autophagosomes are basically formed in a similar way. However, unlike starvation-induced autophagy, selective autophagy is active also under nutrient-rich conditions and responsible for a continuous removal of damaged or superfluous cellular components. MTOR complex 1 (MTORC1) is a nutrient-sensitive kinase which phosphorylates and inactivates ULK1 and ULK2 and thereby inhibits bulk autophagy when the cell has enough nutrients. Upon amino acid starvation MTORC1 is inactivated and bulk autophagy induced [31]. Starvation-induced macroautophagy was recently reported to prioritize degradation and recycling of the membrane-bound organelles, mostly ER and Golgi [37]. The consensus now is that bulk autophagy is switched on when cells starve, and that selective autophagy is MTORC1 independent and can initiate without starvation. What is important to keep in mind is that the ULK1 complex is the gatekeeper of both forms of autophagy. During starvation, the basal inhibition of ULK1 by MTORC1 is relieved, allowing autophagy to initiate. How then can selective autophagy bypass ULK1 inhibition by MTORC1? The first clue to this came from studies of the budding yeast biosynthetic Cvt pathway. Here, the cargo receptor Atg19 delivers the amino-peptidase precursor prApe1 to the vacuole. Studies of the Cvt pathway showed that the adapter protein Atg11 binds to Atg19 after it has engaged multimerized cargo molecules. The binding of Atg11 results in a transport of the SAR-cargo complex to the PAS, and Atg11 also helps recruit core autophagy components, like Atg1 (ULK1/2 homolog) and Atg9 [18, 48, 121]. Further studies revealed that Atg11 acts as a general scaffold protein for selective autophagy in yeast. Atg11 combines with the SARs Atg32 in mitophagy [49, 95] and Atg36 in pexophagy [82], respectively. All yeast SAR-cargo complexes use Atg11 to activate Atg1, and the direct interaction of Atg11 with Atg1 results in an activation of Atg1 even under nutrient-rich conditions [48, 121].

## 4.1    What Deserves Attention

It was for a long time not understood how the core autophagy machinery was initially recruited to a SAR-cargo complex in mammals. No distinct yeast Atg11 homolog was characterized in mammalian cells. It was however noted that the C-terminal domain in RB1CC1 has similarity with Atg11. Smith et al. [115] discovered the ER membrane-localized mammalian reticulophagy receptor CCPG1 and identified a short motif they named RB1CC1/FIP200-interacting region (FIR) that binds to this domain in RB1CC1. CCPG1 has both a LIR that binds to Atg8-family proteins and a FIR that binds to RB1CC1. Indeed, several SARs use a FIR motif to interact directly with the so-called Claw domain in the C-terminal region of RB1CC1 to recruit and activate the core machinery. Although LIR and FIR are two distinct non-overlapping motifs in CCPG1, other autophagy receptors have overlapping LIR and FIR motifs [149]. As previously noted, the LDS in Atg8-family proteins has two hydrophobic pockets where the aromatic (W/F/Y) and hydrophobic (L/I/V) side chains of amino acids at positions 0 and 3 of LIR motifs can dock. The Claw domain of RB1CC1 differs from the LDS by having a single hydrophobic pocket. But the LIR and FIR sequences show similarity. Hence, Turco et al. [123] identified a FIR motif in SQSTM1 that overlapped with the LIR in SQSTM1 and interacted with the C-terminal Claw domain in RB1CC1. They could show that the Claw domain is recruited to SQSTM1/p62 bodies in a reconstituted system *in vitro*. The interactions of SQSTM1 with RB1CC1 and LC3B are mutually exclusive, and they proposed a model where an interaction with LC3B displaces RB1CC1 at the phagophore membrane to enable the docking of the SQSTM1-cargo complex to the membrane. FIR motifs have later been identified in several other LIR-containing proteins, including the SAR OPTN that has a dual LIR/FIR motif activated by phosphorylation [149].

It is not fully understood how the core machinery is induced by a recruitment of RB1CC1, but a key step in selective autophagy is activation of ULK1. RB1CC1 is a scaffold protein in the ULK1 complex, and its presence may therefore be sufficient to activate the ULK1 complex locally at the selected cargo. However, the exact mechanism is context dependent and different mammalian SARs appear to recruit different core autophagy proteins [12, 103, 128, 139]. Many mammalian SARs do not interact directly with RB1CC1, but instead interact with other components of the core machinery, or they recruit the core machinery indirectly via other proteins. The first evidence that a mammalian SAR can bind directly to components of the core machinery came from studies of TRIM family E3 ligases like TRIM5/TRIM5α and TRIM20 that were shown to bind to ULK1 and BECN1 [54, 72]. TRIM5 acts as a SAR for the HIV-1 capsid protein, and its simultaneous interaction with Atg8-family proteins, ULK1 and BECN1 suggested a role in recruiting and activating the core machinery [72]. Two articles were later published back-to-back [123, 128], describing how SQSTM1 and CALCOCO2 use different approaches to activate the core machinery by recruiting RB1CC1. CALCOCO2 has no FIR and uses its SKICH domain to interact with RB1CC1 [128].

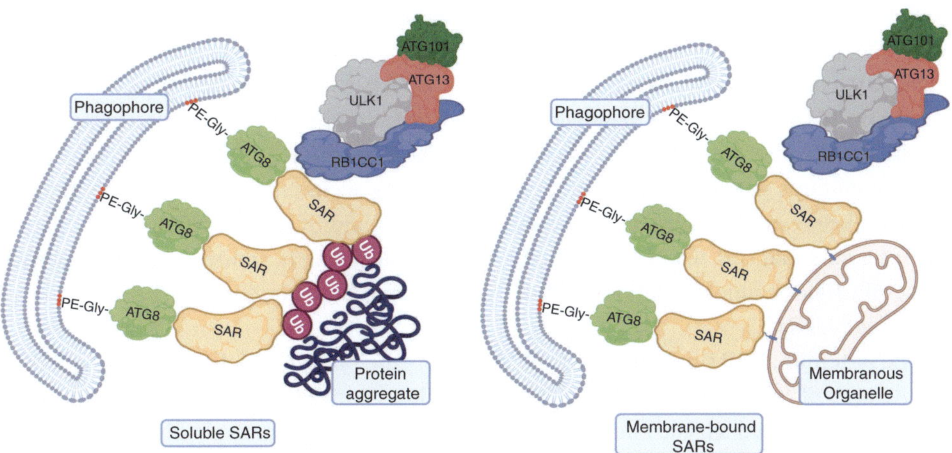


**Fig. 3** Soluble and membrane-bound SARs. SARs recruit cargo into the growing phagophore by LIR-mediated interaction with PE-conjugated Atg8-family proteins and interactions with the autophagy core machinery, here exemplified by the ULK1 complex. Soluble SARs oligomerize and bind ubiquitinated protein aggregates or other ubiquitinated cargo (left). Membrane-bound SARs are constitutively bound to organelle membranes (right). Both SAR types can recruit core autophagy machinery components. (Illustration created with BioRender.com). ATG8: mammalian Atg8-family protein

## 4.2    Soluble SARs

In the category of soluble SARs, we find the six SQSTM1-like receptors (SLRs) SQSTM1, NBR1, OPTN, CALCOCO2, TAX1BP1 and CALCOCO1 (Fig. 3 and Table 1). They all contain LIR motifs binding to Atg8-family proteins, PB1 or coiled-coil domains for oligomerization and, except for CALCOCO1 [91], ubiquitin-binding domains for binding to cargo [46, 105].

We will here only describe SLRs since they are, together with NCOA4, the most important group of soluble SARs in mammals. However, other soluble SARs are listed in Table 1. SLRs are distinguished from most other identified SARs by the fact that each SLR can be recruited to many different types of cargo, including intracellular bacteria, misfolded proteins, protein aggregates, or organelles like mitochondria or peroxisomes. Other soluble SARs and membrane bound SARs normally degrade only a single type of cargo (see Table 1). Another important feature of SLRs is that their substrates are in most cases ubiquitinated, and selective autophagy is induced when the SLRs are recruited by binding to ubiquitin attached to the selected cargo. Yeast lack SLRs, but yeast Atg19 is considered evolutionarily related to NBR1 [59, 102].

The degradation of misfolded or damaged proteins is an important part of normal homeostasis in mammals and selected substrates are either degraded by the ubiquitin-proteasome system/UPS or by selective autophagy via the formation of SQSTM1/p62 bodies. SQSTM1/p62 bodies are ubiquitin-positive condensates formed by the SLRs SQSTM1 and NBR1. Autophagy of SQSTM1/p62 bodies was recently named fluidophagy because the degraded structure is a fluid protein condensate [1]. The delivery of substrates into SQSTM1/p62 bodies is regulated by chaperones and recruited substrates are often ubiquitinated. The formation of SQSTM1/p62 bodies is transient but increases under stress conditions when misfolded or damaged proteins accumulate. There is a continuous crosstalk between the different degradation pathways, and SQSTM1/p62 body formation increases in situations when the capacity of the proteasome is limited or overwhelmed. Some substrates cannot be degraded by the proteasome and therefore must be degraded by fluidophagy [62].

The N-terminal PB1 domain in SQSTM1 allows the protein to self-interact into a helical filament of SQSTM1 molecules. Polymerization is achieved by repeated head-to-tail interactions between a basic cluster in one PB1 domain and a negatively charged so-called OPCA motif in another domain. Polymerization of SQSTM1 is essential for the formation of SQSTM1/p62 bodies. In addition, the interaction of the UBA domain of SQSTM1 with polyubiquitin is essential for the formation of SQSTM1/p62 bodies [17, 63]. The UBA domain in SQSTM1 has low intrinsic binding affinity, but ubiquitin binding is activated by phosphorylation of the UBA domain [75, 99]. It is the specific binding of polymeric SQSTM1 with polyubiquitin that induces SQSTM1/p62 body formation, but phase separation into an autophagy-competent structure that can be docked into a phagophore membrane presumably relies on other weaker interactions as well [117, 143].

NBR1 is required for efficient formation of SQSTM1/p62 bodies, and NBR1 is always present in SQSTM1/p62 bodies. NBR1 is recruited into SQSTM1/p62 bodies via its PB1 domain. The PB1 domain in NBR1 differs from the one in SQSTM1 in that it cannot self-interact into a polymer. The reason is that it has no negatively charged OPCA motif, but the PB1 domain in NBR1 uses its basic binding surface to interact with the acidic OPCA motif in the PB1 domain of SQSTM1, and NBR1 is recruited into SQSTM1/p62 bodies via this interaction [56]. It is not fully understood why NBR1 facilitates SQSTM1/p62 body formation but binding of NBR1 to SQSTM1 competes with polymerization of SQSTM1 and may regulate the length of SQSTM1 polymers in an SQSTM1/p62 body [42]. NBR1 is also suggested to facilitate ubiquitin binding [124], or it may recruit other proteins promoting SQSTM1/p62 body formation. One protein recruited by NBR1 is the SLR TAX1BP1, that facilitates degradation of SQSTM1/p62 bodies by recruiting RB1CC1 [107].

The evolution of mammalian SQSTM1 and NBR1 is interesting [102]. Non-metazoan eukaryotes like plants only express a single SQSTM1-NBR1 ortholog. We named the plant ortholog Nbr1 because it contains the FW domain and has strongest similarity to mammalian NBR1. Hence, NBR1 preceded SQSTM1 in evolution, and NBR1 orthologs are found throughout eukaryotic evolution. However, non-metazoan Nbr1 has a polymeric PB1 domain that is functionally very similar to the domain in mammalian SQSTM1 [42, 118]. This means that NBR1 orthologs in non-metazoan species forms ubiquitin-positive condensates resembling those formed by SQSTM1 in mammals, and these structures are similarly degraded by autophagy. The conclusion is therefore that SQSTM1-NBR1 ortholog-mediated fluidophagy is active in most eukaryotes making it a very ancient type of selective autophagy.

Another important autophagy process involving SLRs is ubiquitin-dependent degradation of depolarized mitochondria by mitophagy. The PINK1-PRKN pathway is the best understood mitophagy pathway, although other mitophagy mechanisms may be equally important under physiological conditions. A loss of membrane potential causes a stabilization of the protein kinase PINK1 at the outer mitochondria membrane (OMM) resulting in phosphorylation of PRKN and ubiquitin. PRKN is then activated to ubiquitinate OMM proteins leading to their proteasomal degradation. Ubiquitination also recruits SLRs to the mitochondria. The choice of SLR may be context dependent, but some studies indicate that CALCOCO2 and OPTN are the essential SLRs recruited in PINK1-PRKN induced mitophagy [125]. An important function of the SLRs is to recruit the core autophagy machinery. CALCOCO2 and OPTN are likely to cooperate since CALCOCO2 binds to RB1CC1 [103, 128] and OPTN can recruit ATG9 [139]. CALCOCO2 recruits the protein kinase TBK1 that is required for phosphorylating and thereby activating the LIR motif in OPTN. The LIR interaction of OPTN can then be used to dock the selected mitochondria to the phagophore membrane.

As a final example of SLR-mediated selective autophagy, we will mention the degradation of intracellular bacteria by xenophagy. Specialized intracellular pathogens have ways to escape the autophagy machinery, but nonpathogenic bacteria are rapidly ubiquitinated and degraded upon entering a cell. *Salmonella typhimurium* proliferate in so-called *Salmonella*-containing vacuoles (SCV) and are not detected unless the vacuole is damaged. Upon damage of the vacuole membrane, cytosolic lectins of the galectin family are recruited and bind to intralumenal sugars. Recruited galectins serve as "eat me" signals, and CALCOCO2 binds to LGALS8/GAL8. Ubiquitin-independent xenophagy of the SCV is then initiated [120], and docking of the SCV to the phagophore membrane is here shown to depend on an interaction of the noncanonical LIR of CALCOCO2 with LC3C [130]. However, if the bacteria escapes from the vacuole, it is ubiquitinated and all the different ubiquitin binding SLRs are then rapidly recruited to induce xenophagy of the bacteria [119, 147].

## 4.3    Membrane-Bound SARs

With all membrane-delimited organelles of the cell in principle being targets for autophagic degradation under certain conditions it is not surprising that a number of membrane-bound SARs are known (Table 1). Most SARs in yeast are membrane-bound, but also in mammals the list of identified membrane-bound SARs is increasing and has outnumbered the list of soluble SARs. Selective autophagy induction by membrane-bound SARs is ubiquitin independent. Most attention has so far been given to the autophagy of mitochondria (mitophagy), ER (reticulophagy) and lysosomes with membrane damage (lysophagy) [62, 129]. However, there is also autophagy of peroxisomes (pexophagy) [134], and of the Golgi apparatus (Golgiphagy) [37], being mediated by membrane-bound SARs in a ubiquitin-independent manner.

Membrane-bound SARs are constitutively attached to the cargo and selective autophagy induced by a membrane-bound SARs is restricted to this organelle. This means that the SAR function of a membrane-bound SAR must be tightly regulated. For some membrane-bound SARs including the mitophagy receptors BNIP3L/NIX, BNIP3 and FUNDC1, selective autophagy induction depends on elevated expression, but also the activity is normally regulated [96]. The LIR interaction is essential, and the function of the LIR motif of a membrane-resident SAR is typically regulated by phosphorylation. The above-mentioned mitophagy receptors BNIP3L, BNIP3 and FUNDC1 all have LIR motifs that are regulated by phosphorylation [96]. Another essential level of regulation is the recruitment and activation of the core autophagy machinery. In yeast, the mitophagy receptor Atg32 is phosphorylated by CK2 before it can bind to Atg11 and thereby activate the core machinery [96]. It is less understood how mammalian membrane-bound SARs recruit the core machinery, and direct interactions with core components are so far only shown for a minority of these proteins including the FUNDC1 interaction with ULK1 [136], the CCPG1 interaction with RB1CC1 [115], and the BNIP3L interaction with WIPI2 [12]. Most membrane-bound SARs have multiple functions and are associated with many other proteins that potentially participate in autophagy induction. Induction of autophagy may depend on oligomerization or clustering of the SAR, and degradation of some substrates like mitochondria or ER depends on fragmentation of the organelle. Some reticulophagy receptors have an intrinsic ability to induce membrane curvature, tubulation and fragmentation. The reticulon-homology domain (RHD) in yeast Atg40 creates highly curved cortical ER regions that can be efficiently packed into autophagosomes [81]. RHD is also found in some mammalian reticulophagy receptors including RETREG1 [8] and RTN3L [33]. But fragmentation may also involve other membrane proteins like RTN (reticulon) proteins or ATL (atlastin) GTPases that modulate the ER structure.

Some organelles like ER and mitochondria contain many organelle-resident SARs and we are only beginning to understand the interplay that exists between the different SARs that are involved in selective autophagy of a specific organelle. In addition, soluble SARs are also engaged in degradation of organelles and the interplay between membrane-bound and soluble SARs is poorly understood. An important reason for the high number of reticulophagy receptors in mammals is that the ER is composed of regional subdomains like sheets, tubules, and ER exit sites. The choice of SAR also depends on the purpose of the autophagy process. This can be non-selective volume reduction or a selective degradation of specific components that are damaged or in excess. Both mitochondria and the ER are membrane networks that are degraded as fragments. We here discuss macroautophagy, but these organelles are also degraded by microautophagy, and selected cargos in ER or mitochondria lumen can be delivered to the lysosome via small vesicles that are formed from the organelle (Chino and Mizushima 2020).

**Question**
4. What distinguishes selective autophagy from nonselective, bulk autophagy and how do we usually classify the different forms of selective autophagy?
5. What are the main features of SQSTM1/p62 relevant to its function as a selective autophagy receptor?
6. What are the two main groups of SARs, and how do they differ?

## 5    Introduction to Cardinal Articles

**Cardinal article 1**: Bjørkøy G, Lamark T, Brech A, Outzen H, Perander M, Øvervatn A, Stenmark H, Johansen T. p62/SQSTM1 forms protein aggregates degraded by autophagy and has a protective effect on huntingtin-induced cell death. J Cell Biol. 2005;171:603–14.

**Cardinal article 2**: Pankiv S, Høyvarde Clausen T, Lamark T, Brech A, Bruun J-A, Outzen H, Øvervatn A, Bjørkøy G, Johansen T. p62/SQSTM1 binds directly to Atg8/LC3 to facilitate degradation of ubiquitinated protein aggregates by autophagy. J Biol Chem. 2007;282:24131–45.

**Cardinal article 3**: Yorimitsu T, Klionsky DJ. Atg11 links cargo to the vesicle-forming machinery in the cytoplasm to vacuole targeting pathway. Mol Biol Cell. 2005;16:1593–605.

**Cardinal article 4**: Novak I, Kirkin V, McEwan DG, Zhang J, Wild P, Rozenknop A, Rogov V, Löhr F, Popovic D, Occhipinti A, Reichert AS, Terzic J, Dötsch V, Ney PA, Dikic I. Nix is a selective autophagy receptor for mitochondrial clearance. EMBO Rep. 2010;11:45–51.

## 5.1    Take Note Of

In our first paper on autophagy, we showed that SQSTM1/p62 is continuously degraded by autophagy, and we presented data to suggest that SQSTM1 in fact acted as an autophagy receptor. We suggested that SQSTM1 could, via LC3, link polyubiquitinated protein aggregates to the autophagy machinery [11]. We studied the scaffolding protein SQSTM1 because it interacted with the two atypical PRKCs/protein kinase Cs PRKCZ/ζ and PRKCI/ι through interactions between the N-terminal PB1 domains of SQSTM1 and these two proteins [63]. At that time, it was also clear that SQSTM1 was found in polyubiquitinated protein aggregates in protein aggregation diseases of both the brain and the liver [60, 61, 83, 144]. As described in the chapter by Barbosa et al., there was already evidence for a role for autophagy in the clearance of aggregation-prone mutant proteins.

In response to stress, SQSTM1 transiently forms so-called SQSTM1/p62 bodies that contain polyubiquitinated proteins. Bjørkøy et al. showed that LC3B is recruited to SQSTM1/p62 bodies [11]. Correlative light electron microscopy of cells containing GFP-tagged SQSTM1/p62 bodies revealed that SQSTM1/p62 bodies exist both as membrane-free protein aggregates and inside membrane-embedded autophagosomes or lysosomes. This indicated that SQSTM1, via LC3, could link polyubiquitinated protein aggregates to the autophagy machinery. Before this study was published, autophagy was mainly considered to be nonselective. This study is important because it showed that autophagy of ubiquitinated protein aggregates can be highly selective, and SQSTM1 itself was the first mammalian protein shown to be selectively degraded by autophagy. This initiated further studies to identify other proteins participating in selective autophagy, leading to a subsequent discovery of other ubiquitin binding SARs including NBR1 [56], CALCOCO2 [119], OPTN [133] and TAX1BP1 [86].

In a follow-up paper we showed that SQSTM1 interacts directly with six mammalian Atg8-family proteins and we identified the LIR motif used for this interaction [98]. The LIR motif included the essential HP1 binding Trp (W) 338 residue in the core motif and three adjacent Asp (D) residues that are essential for binding. Subsequent structural studies solved the SQSTM1 LIR-LC3B structure, and the last essential residue in the W-x-x-L core LIR motif of SQSTM1 (Leu (L) 341) was found [40, 89]. Since then, many more SARs have been identified, and it became clear that the LIR motif is evolutionarily conserved and found in many proteins, including core autophagy proteins, and it has become a signature motif for SARs.

Pankiv et al. [98] used the mCherry-EGFP tandem tag to verify the importance of the LIR motif for autophagy of transiently transfected SQSTM1, and along with the use of mRFP-EGFP tag in another simultaneous study [53], this established a new method to visualize autophagy by fluorescence imaging of cells. The identification of the LIR in SQSTM1 enabled us to propose a molecular model for selective autophagy proved to be valid also for other SARs. The SAR interacts directly with the selected cargo and with Atg8-family proteins on the phagophore. The SAR-cargo complex is thereby encapsulated by the phagophore.

Early studies of the Cvt pathway were extremely useful in characterizing how selectivity is achieved in selective autophagy, and these studies were the first to describe how a SAR acts in bringing the cargo to the core autophagy machinery. The function of the Cvt pathway is to transport vacuolar hydrolyses into the vacuole, but mechanistically it uses the core autophagy machinery. The Cvt pathway therefore became a perfect model for describing selective autophagy in yeast. Early studies by the group of Dan Klionsky revealed the importance of adaptor proteins in connecting selected substrates to the core autophagy machinery.

Yorimitsu and Klionsky [142] is an important study investigating how the yeast SAR Atg19 and the adaptor protein Atg11 cooperate in connecting the cargo (Cvt complex) to the PAS. Atg11 had already been shown to be essential for the Cvt pathway and pexophagy, but not for bulk autophagy [51]. In this new study they mapped the interaction between Atg19 and Atg11 and found that Atg19 interacted with the C terminus of Atg11. It appeared that Atg19 interacts both with the cargo (Cvt complex) and Atg11, before the cargo could be transported to the PAS. This study is important because it increased our knowledge in how a cargo is selected for selective autophagy, and the specific recruitment of the cargo-Atg19-Atg11 complex to the PAS was shown to play a direct role in organizing the PAS and inducing formation of a phagophore membrane. Atg11 was already known to bind to Atg1 [114], and the authors suggested that this interaction could play a role. Shortly after, a LIR motif was identified in yeast Atg19 [89]. Hence, Atg19 uses an Atg11 interaction to connect with core autophagy machinery, and an Atg8 interaction to dock the cargo to the phagophore membrane.

The identification of BNIP3L as a membrane-bound SAR represented a new type of selective autophagy in mammals [90]. This study followed two yeast studies where the stress induced mitochondria membrane protein Atg32 was identified as a SAR for stressed mitochondria [49, 95]. Mitophagy induced by Atg32 was shown to depend both on a LIR motif binding to Atg8 and on its interaction with the selective autophagy adapter Atg11. Like yeast Atg32, BNIP3L sits in the outer mitochondrial membrane. Previous studies had implicated BNIP3L in the programmed clearance of mitochondria during reticulocyte maturation [106, 110]. The protein was known to interact with GABARAP [109]. Novak et al. [90] identified a LIR motif in BNIP3L, and they demonstrated that BNIP3L is a SAR responsible for mitophagy during reticulocyte maturation. The LIR in BNIP3L had highest affinity for GABARAPL1 and was shown to recruit ectopically expressed GABARAPL1 to the mitochondria. The LIR in BNIP3L was essential for efficient clearance of mitochondria in reticulocytes isolated from mice transplanted with *bnip3l*$^{-/-}$ bone-marrow, and then transduced with either wild-type (WT) or LIR-mutated BNIP3L.

The paper by Novak et al. [90] is important because it, together with the studies of yeast Atg32, revealed the existence of organelle-resident SARs. The identification of BNIP3L as a SAR led to a search for other membrane-bound mammalian SARs. Multiple organelle-resident SARs are now known for organelles including the ER and mitochondria (Table 1). Typical characteristics that distinguish membrane-bound SARs from soluble SARs is that the autophagy is normally ubiquitin independent and restricted to the organelle containing the SAR [90]. LIRs of membrane-bound SARs are seemingly often more tightly regulated by post-translational modifications than those of soluble SARs. Subsequent studies of BNIP3L revealed that its binding affinity is increased by phosphorylation of serines S17 and S24 flanking the LIR motif [150].

**Take Home Message**

- Selective autophagy is the targeted lysosomal degradation of cellular components mediated by a set of autophagy receptors.
- Selective autophagy differs from bulk autophagy by the involvement of selective autophagy receptors (SARs) that are responsible for the docking of selected cargos to the phagophore membrane.
- Cells express different types of SARs, and the choice of SAR depends on the stress condition and type of substrate that is selected for degradation.
- It is common to distinguish between organelle-resident SARs that are stably attached to their cargo and soluble SARs that must be recruited to the cargo before they can perform selective autophagy.
- Recruitment of a soluble SAR often depends on ubiquitination of the selected cargo, but also other types of "eat-me" signals can be involved.
- All SARs contain a LIR/AIM motif. This motif interacts directly with Atg8-family proteins that are covalently conjugated to phosphatidylethanolamine in the inner membrane surface of the phagophore.
- The LIR/AIM motif is needed for efficient docking of the SAR-cargo complex to the phagophore, and the selected cargo can then be encapsulated by the growing phagophore.
- Induction of selective autophagy depends on the core autophagy machinery. There is now strong evidence that SARs play a direct role in recruiting the core autophagy machinery to the SAR-cargo complex. Many SARs interact directly with components of the core machinery, but the interaction can also be indirect via associated proteins.
- A crucial role for selective autophagy is in cellular quality control where it is responsible for the continuous removal of damaged or excess organelles, and it collaborates with the ubiquitin-proteasome system in degrading excess, mis-folded, or damaged proteins.
- Selective autophagy is also essential for preventing the occurrence of pathogenic protein aggregates, and a decline in selective autophagy during aging contributes to age-related diseases including neurodegenerative diseases, cardiovascular dis-eases and cancer.

**Answer Guide to Questions**

1. The two main purposes served by autophagy are (1) removal of toxic and surplus components, and (2) supply of energy and building blocks upon nutrient deprivation.
2. The main function of the Atg8-family proteins anchored in the phagophore mem-brane is to act as adaptors that recruit other proteins, like autophagy receptors, core machinery proteins (i.e. ULK1/ULK2, ATG13, ATG14, ATG4B etc.), and transport adapters like FYCO1 and MAPK8IP1/JIP1.

(continued)

(continued)

3. Atg8-family proteins recruit selective autophagy receptors (SARs) and other interacting proteins mainly through the LIR-LDS interaction. The interacting proteins presents one or more LIR motif consisting of a core sequence W/F/YxxL/I/V often flanked by acidic or phosphorylatable residues presented on a beta strand where the aromatic residue (W, F or Y) docks into one hydrophobic pocket, and the hydrophobic residue (L, I or V) into another hydrophobic pocket, in the LDS (LIR docking site) of the Atg8-family proteins of the LC3 or GABARAP subfamily.

4. Selective autophagy employs selective autophagy receptors (SARs), sometimes referred to as selective autophagy adaptors, to engage the cargo to be degraded and connect it to the forming autophagosomes by interacting with Atg8-family proteins and by recruiting core machinery components. Nonselective, bulk autophagy does not employ SARs and there is a random degradation of cytoplasmic components. This form of autophagy is inhibited in nutrient-rich conditions because MTORC1 kinase is active and inhibits ULK1/ULK2 kinases needed to initiate autophagy. Upon amino acid starvation MTORC1 is inactivated and ULK1/ULK2 become active and nonselective, bulk autophagy is initiated. The selective autophagy can occur under nutrient-rich conditions because autophagy receptors recruit the core machinery, so ULK1/ULK2 is activated at the selected cargo where the autophagosome forms.

We usually classify the different forms of selective autophagy after the cargo or substrate that is degraded so that degradation of protein aggregates is called aggrephagy, autophagy of mitochondria is called mitophagy and autophagy of the ER is named reticulophagy.

5. The main molecular features of SQSTM1 relevant to its function as a selective autophagy receptor are the following: its ability to bind to ubiquitin-labeled cargo via its UBA domain, its ability to polymerize via its PB1 domain, its binding to Atg8-family proteins in the forming autophagosome membrane via its LIR motif, and ability to bind to the core machinery component RB1CC1 via its FIR motif. The latter overlaps with the LIR motif.

6. The two main groups of SARs are the soluble SARs and the membrane bound SARS. They differ from each other obviously in that the latter group are made up of transmembrane proteins located in the membranes of organelles like mitochondria, ER, peroxisomes and the Golgi apparatus. The autophagy involving membrane-bound SARs is usually ubiquitin independent and restricted to the organelle containing the SARs, whereas soluble SARs (with some exceptions) most often rely on ubiquitin tagging of the cargo to be degraded.

## References


1. Agudo-Canalejo J, Schultz SW, Chino H, Migliano SM, Saito C, Koyama-Honda I, Stenmark H, Brech A, May AI, Mizushima N, Knorr RL. Wetting regulates autophagy of phase-separated compartments and the cytosol. Nature. 2021;591:142–6.

2. Alemu EA, Lamark T, Torgersen KM, Birgisdottir AB, Larsen KB, Jain A, Olsvik H, Overvatn A, Kirkin V, Johansen T. ATG8 family proteins act as scaffolds for assembly of the ULK complex: sequence requirements for LC3-interacting region (LIR) motifs. J Biol Chem. 2012;287:39275–90.

3. Ames J, Yadavalli T, Suryawanshi R, Hopkins J, Agelidis A, Patil C, Fredericks B, Tseng H, Valyi-Nagy T, Shukla D. OPTN is a host intrinsic restriction factor against neuroinvasive HSV-1 infection. Nat Commun. 2021;12:5401.

4. An H, Ordureau A, Paulo JA, Shoemaker CJ, Denic V, Harper JW. TEX264 is an endoplasmic reticulum-resident ATG8-interacting protein critical for ER remodeling during nutrient stress. Mol Cell. 2019;74(5):891–908.e10.

5. Ashford TP, Porter KR. Cytoplasmic components in hepatic cell lysosomes. J Cell Biol. 1962;12:198–202.

6. Axe EL, Walker SA, Manifava M, Chandra P, Roderick HL, Habermann A, Griffiths G, Ktistakis NT. Autophagosome formation from membrane compartments enriched in phosphatidylinositol 3-phosphate and dynamically connected to the endoplasmic reticulum. J Cell Biol. 2008;182:685–701.

7. Baskaran S, Carlson LA, Stjepanovic G, Young LN, Kim DJ, Grob P, Stanley RE, Nogales E, Hurley JH. Architecture and dynamics of the autophagic phosphatidylinositol 3-kinase complex. elife. 2014;3:e05115.

8. Bhaskara RM, Grumati P, Garcia-Pardo J, Kalayil S, Covarrubias-Pinto A, Chen W, Kudryashev M, Dikic I, Hummer G. Curvature induction and membrane remodeling by FAM134B reticulon homology domain assist selective ER-phagy. Nat Commun. 2019;10:2370.

9. Bhujabal Z, Birgisdottir AB, Sjottem E, Brenne HB, Overvatn A, Habisov S, Kirkin V, Lamark T, Johansen T. FKBP8 recruits LC3A to mediate Parkin-independent mitophagy. EMBO Rep. 2017;18:947–61.

10. Birgisdottir AB, Mouilleron S, Bhujabal Z, Wirth M, Sjottem E, Evjen G, Zhang W, Lee R, O'Reilly N, Tooze SA, Lamark T, Johansen T. Members of the autophagy class III phosphatidylinositol 3-kinase complex I interact with GABARAP and GABARAPL1 via LIR motifs. Autophagy. 2019;15:1333–55.

11. Bjørkøy G, Lamark T, Brech A, Outzen H, Perander M, Øvervatn A, Stenmark H, Johansen T. p62/SQSTM1 forms protein aggregates degraded by autophagy and has a protective effect on huntingtin-induced cell death. J Cell Biol. 2005;171:603–14.

12. Bunker EN, Le Guerroue F, Wang C, Strub MP, Werner A, Tjandra N, Youle RJ. Nix interacts with WIPI2 to induce mitophagy. EMBO J. 2023;42:e113491.

13. Chauhan S, Kumar S, Jain A, Ponpuak M, Mudd MH, Kimura T, Choi SW, Peters R, Mandell M, Bruun JA, Johansen T, Deretic V. TRIMs and galectins globally cooperate and TRIM16 and galectin-3 co-direct autophagy in endomembrane damage homeostasis. Dev Cell. 2016;39:13–27.

14. Chen Q, Xiao Y, Chai P, Zheng P, Teng J, Chen J. ATL3 is a tubular ER-phagy receptor for GABARAP-mediated selective autophagy. Curr Biol. 2019;29:846–855.e846.

15. Chino H, Hatta T, Natsume T, Mizushima N. Intrinsically disordered protein TEX264 mediates ER-phagy. Mol Cell. 2019;74(5):909–921.e6.

16. Chu CT, Ji J, Dagda RK, Jiang JF, Tyurina YY, Kapralov AA, Tyurin VA, Yanamala N, Shrivastava IH, Mohammadyani D, Wang KZQ, Zhu J, Klein-Seetharaman J, Balasubramanian

K, Amoscato AA, Borisenko G, Huang Z, Gusdon AM, Cheikhi A, Steer EK, Wang R, Baty C, Watkins S, Bahar I, Bayir H, Kagan VE. Cardiolipin externalization to the outer mitochondrial membrane acts as an elimination signal for mitophagy in neuronal cells. Nat Cell Biol. 2013;15:1197–205.

17. Ciuffa R, Lamark T, Tarafder AK, Guesdon A, Rybina S, Hagen WJ, Johansen T, Sachse C. The selective autophagy receptor p62 forms a flexible filamentous helical scaffold. Cell Rep. 2015;11:748–58.

18. Coudevylle N, Banas B, Baumann V, Schuschnig M, Zawadzka-Kazimierczuk A, Kozminski W, Martens S. Mechanism of Atg9 recruitment by Atg11 in the cytoplasm-to-vacuole targeting pathway. J Biol Chem. 2022;298:101573.

19. Crooks GE, Hon G, Chandonia JM, Brenner SE. WebLogo: a sequence logo generator. Genome Res. 2004;14:1188–90.

20. De Duve C, Wattiaux R. Functions of lysosomes. Annu Rev Physiol. 1966;28:435–92.

21. Deosaran E, Larsen KB, Hua R, Sargent G, Wang Y, Kim S, Lamark T, Jauregui M, Law K, Lippincott-Schwartz J, Brech A, Johansen T, Kim PK. NBR1 acts as an autophagy receptor for peroxisomes. J Cell Sci. 2013;126:939–52.

22. Deretic V, Duque T, Trosdal E, Paddar M, Javed R, Akepati P. Membrane atg8ylation in canonical and noncanonical autophagy. J Mol Biol. 2024;436:168532.

23. Dooley HC, Razi M, Polson HE, Girardin SE, Wilson MI, Tooze SA. WIPI2 links LC3 conjugation with PI3P, autophagosome formation, and pathogen clearance by recruiting Atg12-5-16L1. Mol Cell. 2014;55:238–52.

24. Dou Z, Xu C, Donahue G, Shimi T, Pan JA, Zhu J, Ivanov A, Capell BC, Drake AM, Shah PP, Catanzaro JM, Ricketts MD, Lamark T, Adam SA, Marmorstein R, Zong WX, Johansen T, Goldman RD, Adams PD, Berger SL. Autophagy mediates degradation of nuclear lamina. Nature. 2015;527:105–9.

25. Dowdle WE, Nyfeler B, Nagel J, Elling RA, Liu S, Triantafellow E, Menon S, Wang Z, Honda A, Pardee G, Cantwell J, Luu C, Cornella-Taracido I, Harrington E, Fekkes P, Lei H, Fang Q, Digan ME, Burdick D, Powers AF, Helliwell SB, D'Aquin S, Bastien J, Wang H, Wiederschain D, Kuerth J, Bergman P, Schwalb D, Thomas J, Ugwonali S, Harbinski F, Tallarico J, Wilson CJ, Myer VE, Porter JA, Bussiere DE, Finan PM, Labow MA, Mao X, Hamann LG, Manning BD, Valdez RA, Nicholson T, Schirle M, Knapp MS, Keaney EP, Murphy LO. Selective VPS34 inhibitor blocks autophagy and uncovers a role for NCOA4 in ferritin degradation and iron homeostasis in vivo. Nat Cell Biol. 2014;16:1069–79.

26. Durgan J, Lystad AH, Sloan K, Carlsson SR, Wilson MI, Marcassa E, Ulferts R, Webster J, Lopez-Clavijo AF, Wakelam MJ, Beale R, Simonsen A, Oxley D, Florey O. Non-canonical autophagy drives alternative ATG8 conjugation to phosphatidylserine. Mol Cell. 2021;81:2031–2040.e2038.

27. Fujioka Y, Alam JM, Noshiro D, Mouri K, Ando T, Okada Y, May AI, Knorr RL, Suzuki K, Ohsumi Y, Noda NN. Phase separation organizes the site of autophagosome formation. Nature. 2020;578:301–5.

28. Fumagalli F, Noack J, Bergmann TJ, Cebollero E, Pisoni GB, Fasana E, Fregno I, Galli C, Loi M, Solda T, D'Antuono R, Raimondi A, Jung M, Melnyk A, Schorr S, Schreiber A, Simonelli L, Varani L, Wilson-Zbinden C, Zerbe O, Hofmann K, Peter M, Quadroni M, Zimmermann R, Molinari M. Translocon component Sec62 acts in endoplasmic reticulum turnover during stress recovery. Nat Cell Biol. 2016;18:1173–84.

29. Ghanbarpour A, Valverde DP, Melia TJ, Reinisch KM. A model for a partnership of lipid transfer proteins and scramblases in membrane expansion and organelle biogenesis. Proc Natl Acad Sci USA. 2021;118:e2101562118.

30. Gonzalez A, Covarrubias-Pinto A, Bhaskara RM, Glogger M, Kuncha SK, Xavier A, Seemann E, Misra M, Hoffmann ME, Brauning B, Balakrishnan A, Qualmann B, Dotsch V, Schulman BA, Kessels MM, Hubner CA, Heilemann M, Hummer G, Dikic I. Ubiquitination regulates ER-phagy and remodelling of endoplasmic reticulum. Nature. 2023;618:394–401.

31. Goul C, Peruzzo R, Zoncu R. The molecular basis of nutrient sensing and signalling by mTORC1 in metabolism regulation and disease. Nat Rev Mol Cell Biol. 2023;24:857–75.

32. Grasso D, Ropolo A, Lo Re A, Boggio V, Molejon MI, Iovanna JL, Gonzalez CD, Urrutia R, Vaccaro MI. Zymophagy, a novel selective autophagy pathway mediated by VMP1-USP9x-p62, prevents pancreatic cell death. J Biol Chem. 2011;286:8308–24.

33. Grumati P, Morozzi G, Holper S, Mari M, Harwardt MI, Yan R, Muller S, Reggiori F, Heilemann M, Dikic I. Full length RTN3 regulates turnover of tubular endoplasmic reticulum via selective autophagy. elife. 2017;6:e25555.

34. Hanada T, Noda NN, Satomi Y, Ichimura Y, Fujioka Y, Takao T, Inagaki F, Ohsumi Y. The Atg12-Atg5 conjugate has a novel E3-like activity for protein lipidation in autophagy. J Biol Chem. 2007;282:37298–302.

35. Hanna RA, Quinsay MN, Orogo AM, Giang K, Rikka S, Gustafsson AB. Microtubule-associated protein 1 light chain 3 (LC3) interacts with Bnip3 protein to selectively remove endoplasmic reticulum and mitochondria via autophagy. J Biol Chem. 2012;287:19094–104.

36. Heo JM, Ordureau A, Paulo JA, Rinehart J, Harper JW. The PINK1-PARKIN mitochondrial ubiquitylation pathway drives a program of OPTN/NDP52 recruitment and TBK1 activation to promote mitophagy. Mol Cell. 2015;60:7–20.

37. Hickey KL, Swarup S, Smith IR, Paoli JC, Miguel Whelan E, Paulo JA, Harper JW. Proteome census upon nutrient stress reveals Golgiphagy membrane receptors. Nature. 2023;623:167–74.

38. Holzer E, Martens S, Tulli S. The role of ATG9 vesicles in autophagosome biogenesis. J Mol Biol. 2024;436:168489.

39. Hurley JH, Young LN. Mechanisms of autophagy initiation. Annu Rev Biochem. 2017;86:225–44.

40. Ichimura Y, Kumanomidou T, Sou YS, Mizushima T, Ezaki J, Ueno T, Kominami E, Yamane T, Tanaka K, Komatsu M. Structural basis for sorting mechanism of p62 in selective autophagy. J Biol Chem. 2008;283:22847–57.

41. Isakson P, Lystad AH, Breen K, Koster G, Stenmark H, Simonsen A. TRAF6 mediates ubiquitination of KIF23/MKLP1 and is required for midbody ring degradation by selective autophagy. Autophagy. 2013;9:1955–64.

42. Jakobi AJ, Huber ST, Mortensen SA, Schultz SW, Palara A, Kuhm T, Shrestha BK, Lamark T, Hagen WJH, Wilmanns M, Johansen T, Brech A, Sachse C. Structural basis of p62/SQSTM1 helical filaments and their role in cellular cargo uptake. Nat Commun. 2020;11:440.

43. Jatana N, Ascher DB, Pires DEV, Gokhale RS, Thukral L. Human LC3 and GABARAP subfamily members achieve functional specificity via specific structural modulations. Autophagy. 2019;16:239–55.

44. Jiang S, Wells CD, Roach PJ. Starch-binding domain-containing protein 1 (Stbd1) and glycogen metabolism: identification of the Atg8 family interacting motif (AIM) in Stbd1 required for interaction with GABARAPL1. Biochem Biophys Res Commun. 2011;413:420–5.

45. Johansen T, Lamark T. Selective autophagy mediated by autophagic adapter proteins. Autophagy. 2011;7:279–96.

46. Johansen T, Lamark T. Selective autophagy: ATG8 family proteins, LIR motifs and cargo receptors. J Mol Biol. 2020;432:80–103.

47. Kabeya Y, Mizushima N, Ueno T, Yamamoto A, Kirisako T, Noda T, Kominami E, Ohsumi Y, Yoshimori T. LC3, a mammalian homologue of yeast Apg8p, is localized in autophagosome membranes after processing. EMBO J. 2000;19:5720–8.

48. Kamber RA, Shoemaker CJ, Denic V. Receptor-bound targets of selective autophagy use a scaffold protein to activate the Atg1 kinase. Mol Cell. 2015;59:372–81.

49. Kanki T, Wang K, Cao Y, Baba M, Klionsky DJ. Atg32 is a mitochondrial protein that confers selectivity during mitophagy. Dev Cell. 2009;17:98–109.

50. Khaminets A, Heinrich T, Mari M, Grumati P, Huebner AK, Akutsu M, Liebmann L, Stolz A, Nietzsche S, Koch N, Mauthe M, Katona I, Qualmann B, Weis J, Reggiori F, Kurth I, Hubner CA, Dikic I. Regulation of endoplasmic reticulum turnover by selective autophagy. Nature. 2015;522:354–8.

51. Kim J, Kamada Y, Stromhaug PE, Guan J, Hefner-Gravink A, Baba M, Scott SV, Ohsumi Y, Dunn WA Jr, Klionsky DJ. Cvt9/Gsa9 functions in sequestering selective cytosolic cargo destined for the vacuole. J Cell Biol. 2001;153:381–96.

52. Kim I, Rodriguez-Enriquez S, Mizushima N, Ohsumi Y, Lemasters J. Autophagic degradation of mitochondria in GFP-LC3 transgenic mouse hepatocytes after nutrient deprivation. Hepatology. 2004;40. John Wiley & Sons Inc 111 River St, Hoboken, NJ 07030 USA. 291A.

53. Kimura S, Noda T, Yoshimori T. Dissection of the autophagosome maturation process by a novel reporter protein, tandem fluorescent-tagged LC3. Autophagy. 2007;3:452–60.

54. Kimura T, Jain A, Choi SW, Mandell MA, Schroder K, Johansen T, Deretic V. TRIM-mediated precision autophagy targets cytoplasmic regulators of innate immunity. J Cell Biol. 2015;210:973–89.

55. Kirisako T, Ichimura Y, Okada H, Kabeya Y, Mizushima N, Yoshimori T, Ohsumi M, Takao T, Noda T, Ohsumi Y. The reversible modification regulates the membrane-binding state of Apg8/Aut7 essential for autophagy and the cytoplasm to vacuole targeting pathway. J Cell Biol. 2000;151:263–76.

56. Kirkin V, Lamark T, Sou YS, Bjorkoy G, Nunn JL, Bruun JA, Shvets E, McEwan DG, Clausen TH, Wild P, Bilusic I, Theurillat JP, Overvatn A, Ishii T, Elazar Z, Komatsu M, Dikic I, Johansen T. A role for NBR1 in autophagosomal degradation of ubiquitinated substrates. Mol Cell. 2009;33:505–16.

57. Koerver L, Papadopoulos C, Liu B, Kravic B, Rota G, Brecht L, Veenendaal T, Polajnar M, Bluemke A, Ehrmann M, Klumperman J, Jaattela M, Behrends C, Meyer H. The ubiquitin-conjugating enzyme UBE2QL1 coordinates lysophagy in response to endolysosomal damage. EMBO Rep. 2019;20:e48014.

58. Korac J, Schaeffer V, Kovacevic I, Clement AM, Jungblut B, Behl C, Terzic J, Dikic I. Ubiquitin-independent function of optineurin in autophagic clearance of protein aggregates. J Cell Sci. 2013;126:580–92.

59. Kraft C, Peter M, Hofmann K. Selective autophagy: ubiquitin-mediated recognition and beyond. Nat Cell Biol. 2010;12:836–41.

60. Kuusisto E, Salminen A, Alafuzoff I. Ubiquitin-binding protein p62 is present in neuronal and glial inclusions in human tauopathies and synucleinopathies. Neuroreport. 2001;12:2085–90.

61. Kuusisto E, Salminen A, Alafuzoff I. Early accumulation of p62 in neurofibrillary tangles in Alzheimer's disease: possible role in tangle formation. Neuropathol Appl Neurobiol. 2002;28:228–37.

62. Lamark T, Johansen T. Mechanisms of selective autophagy. Annu Rev Cell Dev Biol. 2021;37:143–69.

63. Lamark T, Perander M, Outzen H, Kristiansen K, Øvervatn A, Michaelsen E, Bjørkøy G, Johansen T. Interaction codes within the family of mammalian Phox and Bem1p domain-containing proteins. J Biol Chem. 2003;278:34568–81.

64. Lazarou M, Sliter DA, Kane LA, Sarraf SA, Wang C, Burman JL, Sideris DP, Fogel AI, Youle RJ. The ubiquitin kinase PINK1 recruits autophagy receptors to induce mitophagy. Nature. 2015;524:309–14.

65. Leber R, Silles E, Sandoval IV, Mazon MJ. Yol082p, a novel CVT protein involved in the selective targeting of aminopeptidase I to the yeast vacuole. J Biol Chem. 2001;276:29210–7.

66. Liu L, Feng D, Chen G, Chen M, Zheng Q, Song P, Ma Q, Zhu C, Wang R, Qi W, Huang L, Xue P, Li B, Wang X, Jin H, Wang J, Yang F, Liu P, Zhu Y, Sui S, Chen Q. Mitochondrial outer-membrane protein FUNDC1 mediates hypoxia-induced mitophagy in mammalian cells. Nat Cell Biol. 2012;14:177–85.

67. Lopez AR, Jorgensen MH, Havelund JF, Arendrup FS, Kolapalli SP, Nielsen TM, Pais E, Beese CJ, Abdul-Al A, Vind AC, Bartek J, Bekker-Jensen S, Montes M, Galanos P, Faergeman N, Happonen L, Frankel LB. Autophagy-mediated control of ribosome homeostasis in oncogene-induced senescence. Cell Rep. 2023;42:113381.

68. Lynch-Day MA, Klionsky DJ. The Cvt pathway as a model for selective autophagy. FEBS Lett. 2010;584:1359–66.

69. Ma X, Lu C, Chen Y, Li S, Ma N, Tao X, Li Y, Wang J, Zhou M, Yan YB, Li P, Heydari K, Deng H, Zhang M, Yi C, Ge L. CCT2 is an aggrephagy receptor for clearance of solid protein aggregates. Cell. 2022;185:1325–1345.e1322.

70. Maeda S, Yamamoto H, Kinch LN, Garza CM, Takahashi S, Otomo C, Grishin NV, Forli S, Mizushima N, Otomo T. Structure, lipid scrambling activity and role in autophagosome formation of ATG9A. Nat Struct Mol Biol. 2020;27:1194–201.

71. Mancias JD, Wang X, Gygi SP, Harper JW, Kimmelman AC. Quantitative proteomics identifies NCOA4 as the cargo receptor mediating ferritinophagy. Nature. 2014;509:105–9.

72. Mandell MA, Jain A, Arko-Mensah J, Chauhan S, Kimura T, Dinkins C, Silvestri G, Munch J, Kirchhoff F, Simonsen A, Wei Y, Levine B, Johansen T, Deretic V. TRIM proteins regulate autophagy and can target autophagic substrates by direct recognition. Dev Cell. 2014;30:394–409.

73. Mandell MA, Jain A, Kumar S, Castleman MJ, Anwar T, Eskelinen EL, Johansen T, Prekeris R, Deretic V. TRIM17 contributes to autophagy of midbodies while actively sparing other targets from degradation. J Cell Sci. 2016;129:3562–73.

74. Matoba K, Kotani T, Tsutsumi A, Tsuji T, Mori T, Noshiro D, Sugita Y, Nomura N, Iwata S, Ohsumi Y, Fujimoto T, Nakatogawa H, Kikkawa M, Noda NN. Atg9 is a lipid scramblase that mediates autophagosomal membrane expansion. Nat Struct Mol Biol. 2020;27:1185–93.

75. Matsumoto G, Wada K, Okuno M, Kurosawa M, Nukina N. Serine 403 phosphorylation of p62/SQSTM1 regulates selective autophagic clearance of ubiquitinated proteins. Mol Cell. 2011;44:279–89.

76. Mattera R, Park SY, De Pace R, Guardia CM, Bonifacino JS. AP-4 mediates export of ATG9A from the trans-Golgi network to promote autophagosome formation. Proc Natl Acad Sci USA. 2017;114:E10697–706.

77. Mirdita M, Schutze K, Moriwaki Y, Heo L, Ovchinnikov S, Steinegger M. ColabFold: making protein folding accessible to all. Nat Methods. 2022;19:679–82.

78. Mizushima N, Noda T, Yoshimori T, Tanaka Y, Ishii T, George MD, Klionsky DJ, Ohsumi M, Ohsumi Y. A protein conjugation system essential for autophagy. Nature. 1998;395:395–8.

79. Mizushima N, Noda T, Ohsumi Y. Apg16p is required for the function of the Apg12p-Apg5p conjugate in the yeast autophagy pathway. EMBO J. 1999;18:3888–96.

80. Mizushima N, Yoshimori T, Ohsumi Y. The role of Atg proteins in autophagosome formation. Annu Rev Cell Dev Biol. 2011;27:107–32.

81. Mochida K, Yamasaki A, Matoba K, Kirisako H, Noda NN, Nakatogawa H. Super-assembly of ER-phagy receptor Atg40 induces local ER remodeling at contacts with forming autophago-somal membranes. Nat Commun. 2020;11:3306.

82. Motley AM, Nuttall JM, Hettema EH. Pex3-anchored Atg36 tags peroxisomes for degradation in Saccharomyces cerevisiae. EMBO J. 2012;31:2852–68.

83. Nagaoka U, Kim K, Jana NR, Doi H, Maruyama M, Mitsui K, Oyama F, Nukina N. Increased expression of p62 in expanded polyglutamine-expressing cells and its association with polyglutamine inclusions. J Neurochem. 2004;91:57–68.

84. Nakatogawa H. Mechanisms governing autophagosome biogenesis. Nat Rev Mol Cell Biol. 2020;21:439–58.

85. Nakatogawa H, Suzuki K, Kamada Y, Ohsumi Y. Dynamics and diversity in autophagy mechanisms: lessons from yeast. Nat Rev Mol Cell Biol. 2009;10:458–67.

86. Newman AC, Scholefield CL, Kemp AJ, Newman M, McIver EG, Kamal A, Wilkinson S. TBK1 kinase addiction in lung cancer cells is mediated via autophagy of Tax1bp1/Ndp52 and non-canonical NF-kappaB signalling. PLoS One. 2012;7:e50672.

87. Nguyen A, Lugarini F, David C, Hosnani P, Alagoz C, Friedrich A, Schlutermann D, Knotkova B, Patel A, Parfentev I, Urlaub H, Meinecke M, Stork B, Faesen AC. Metamorphic proteins at the basis of human autophagy initiation and lipid transfer. Mol Cell. 2023;83:2077–2090.e2012.

88. Noda NN. Structural view on autophagosome formation. FEBS Lett. 2024;598:84–106.

89. Noda NN, Kumeta H, Nakatogawa H, Satoo K, Adachi W, Ishii J, Fujioka Y, Ohsumi Y, Inagaki F. Structural basis of target recognition by Atg8/LC3 during selective autophagy. Genes Cells. 2008;13:1211–8.

90. Novak I, Kirkin V, McEwan DG, Zhang J, Wild P, Rozenknop A, Rogov V, Lohr F, Popovic D, Occhipinti A, Reichert AS, Terzic J, Dotsch V, Ney PA, Dikic I. Nix is a selective autophagy receptor for mitochondrial clearance. EMBO Rep. 2010;11:45–51.

91. Nthiga TM, Kumar Shrestha B, Sjottem E, Bruun JA, Bowitz Larsen K, Bhujabal Z, Lamark T, Johansen T. CALCOCO1 acts with VAMP-associated proteins to mediate ER-phagy. EMBO J. 2020;39:e103649.

92. Nthiga TM, Shrestha BK, Bruun JA, Larsen KB, Lamark T, Johansen T. Regulation of Golgi turnover by CALCOCO1-mediated selective autophagy. J Cell Biol. 2021;220:e202006128.

93. Ohsumi Y. Molecular mechanism of autophagy in yeast, Saccharomyces cerevisiae. Philos Trans R Soc Lond Ser B Biol Sci. 1999;354:1577–80; discussion 1580–1571.

94. Ohsumi Y. Historical landmarks of autophagy research. Cell Res. 2014;24:9–23.

95. Okamoto K, Kondo-Okamoto N, Ohsumi Y. Mitochondria-anchored receptor Atg32 mediates degradation of mitochondria via selective autophagy. Dev Cell. 2009;17:87–97.

96. Onishi M, Yamano K, Sato M, Matsuda N, Okamoto K. Molecular mechanisms and physiological functions of mitophagy. EMBO J. 2021;40:e104705.

97. Orvedahl A, MacPherson S, Sumpter R Jr, Talloczy Z, Zou Z, Levine B. Autophagy protects against Sindbis virus infection of the central nervous system. Cell Host Microbe. 2010;7:115–27.

98. Pankiv S, Clausen TH, Lamark T, Brech A, Bruun JA, Outzen H, Overvatn A, Bjorkoy G, Johansen T. p62/SQSTM1 binds directly to Atg8/LC3 to facilitate degradation of ubiquitinated protein aggregates by autophagy. J Biol Chem. 2007;282:24131–45.

99. Pilli M, Arko-Mensah J, Ponpuak M, Roberts E, Master S, Mandell MA, Dupont N, Ornatowski W, Jiang S, Bradfute SB, Bruun JA, Hansen TE, Johansen T, Deretic V. TBK-1 promotes autophagy-mediated antimicrobial defense by controlling autophagosome maturation. Immunity. 2012;37:223–34.

100. Pohl C, Jentsch S. Midbody ring disposal by autophagy is a post-abscission event of cytokinesis. Nat Cell Biol. 2009;11:65–70.

101. Princely Abudu Y, Pankiv S, Mathai BJ, Hakon Lystad A, Bindesboll C, Brenne HB, Ng MYW, Thiede B, Yamamoto A, Nthiga TM, Lamark T, Esguerra CV, Johansen T, Simonsen A. NIPSNAP1 and NIPSNAP2 act as "eat me" signals for mitophagy. Dev Cell. 2019;49:509–525.e512.

102. Rasmussen NL, Kournoutis A, Lamark T, Johansen T. NBR1: the archetypal selective autophagy receptor. J Cell Biol. 2022;221:e202208092.

103. Ravenhill BJ, Boyle KB, von Muhlinen N, Ellison CJ, Masson GR, Otten EG, Foeglein A, Williams R, Randow F. The cargo receptor NDP52 initiates selective autophagy by recruiting the ULK complex to cytosol-invading bacteria. Mol Cell. 2019;74(2):320–329.e6.

104. Ren X, Nguyen TN, Lam WK, Buffalo CZ, Lazarou M, Yokom AL, Hurley JH. Structural basis for ATG9A recruitment to the ULK1 complex in mitophagy initiation. Sci Adv. 2023;9:eadg2997.

105. Rogov VV, Nezis IP, Tsapras P, Zhang H, Dagdas Y, Noda NN, Nakatogawa H, Wirth M, Mouilleron S, McEwan DG, Behrends C, Deretic V, Elazar Z, Tooze SA, Dikic I, Lamark T, Johansen T. Atg8 family proteins, LIR/AIM motifs and other interaction modes. Autophagy Rep. 2023;2:27694127.2023.2188523.

106. Sandoval H, Thiagarajan P, Dasgupta SK, Schumacher A, Prchal JT, Chen M, Wang J. Essential role for Nix in autophagic maturation of erythroid cells. Nature. 2008;454:232–5.

107. Sarraf SA, Shah HV, Kanfer G, Pickrell AM, Holtzclaw LA, Ward ME, Youle RJ. Loss of TAX1BP1-directed autophagy results in protein aggregate accumulation in the brain. Mol Cell. 2022;82:1383–5.

108. Sawa-Makarska J, Abert C, Romanov J, Zens B, Ibiricu I, Martens S. Cargo binding to Atg19 unmasks further Atg8-binding sites to mediate membrane-cargo apposition during selective autophagy. Nat Cell Biol. 2014;16:425–33.

109. Schwarten M, Mohrluder J, Ma P, Stoldt M, Thielmann Y, Stangler T, Hersch N, Hoffmann B, Merkel R, Willbold D. Nix directly binds to GABARAP: a possible crosstalk between apoptosis and autophagy. Autophagy. 2009;5:690–8.

110. Schweers RL, Zhang J, Randall MS, Loyd MR, Li W, Dorsey FC, Kundu M, Opferman JT, Cleveland JL, Miller JL, Ney PA. NIX is required for programmed mitochondrial clearance during reticulocyte maturation. Proc Natl Acad Sci USA. 2007;104:19500–5.

111. Scott SV, Guan J, Hutchins MU, Kim J, Klionsky DJ. Cvt19 is a receptor for the cytoplasm-to-vacuole targeting pathway. Mol Cell. 2001;7:1131–41.

112. Sentelle RD, Senkal CE, Jiang W, Ponnusamy S, Gencer S, Selvam SP, Ramshesh VK, Peterson YK, Lemasters JJ, Szulc ZM, Bielawski J, Ogretmen B. Ceramide targets autophagosomes to mitochondria and induces lethal mitophagy. Nat Chem Biol. 2012;8:831–8.

113. Shimizu T, Tamura N, Nishimura T, Saito C, Yamamoto H, Mizushima N. Comprehensive analysis of autophagic functions of WIPI family proteins and their implications for the pathogenesis of beta-propeller associated neurodegeneration. Hum Mol Genet. 2023;32:2623–37.

114. Shintani T, Huang WP, Stromhaug PE, Klionsky DJ. Mechanism of cargo selection in the cytoplasm to vacuole targeting pathway. Dev Cell. 2002;3:825–37.

115. Smith MD, Harley ME, Kemp AJ, Wills J, Lee M, Arends M, von Kriegsheim A, Behrends C, Wilkinson S. CCPG1 is a non-canonical autophagy cargo receptor essential for ER-phagy and pancreatic ER proteostasis. Dev Cell. 2018;44:217–232.e211.

116. Strappazzon F, Nazio F, Corrado M, Cianfanelli V, Romagnoli A, Fimia GM, Campello S, Nardacci R, Piacentini M, Campanella M, Cecconi F. AMBRA1 is able to induce mitophagy via LC3 binding, regardless of PARKIN and p62/SQSTM1. Cell Death Differ. 2015;22:419–32.

117. Sun D, Wu R, Zheng J, Li P, Yu L. Polyubiquitin chain-induced p62 phase separation drives autophagic cargo segregation. Cell Res. 2018;28:405–15.

118. Svenning S, Lamark T, Krause K, Johansen T. Plant NBR1 is a selective autophagy substrate and a functional hybrid of the mammalian autophagic adapters NBR1 and p62/SQSTM1. Autophagy. 2011;7:993–1010.

119. Thurston TL, Ryzhakov G, Bloor S, von Muhlinen N, Randow F. The TBK1 adaptor and autophagy receptor NDP52 restricts the proliferation of ubiquitin-coated bacteria. Nat Immunol. 2009;10:1215–21.

120. Thurston TL, Wandel MP, von Muhlinen N, Foeglein A, Randow F. Galectin 8 targets damaged vesicles for autophagy to defend cells against bacterial invasion. Nature. 2012;482:414–8.

121. Torggler R, Papinski D, Brach T, Bas L, Schuschnig M, Pfaffenwimmer T, Rohringer S, Matzhold T, Schweida D, Brezovich A, Kraft C. Two independent pathways within selective autophagy converge to activate Atg1 kinase at the vacuole. Mol Cell. 2016;64:221–35.

122. Tumbarello DA, Manna PT, Allen M, Bycroft M, Arden SD, Kendrick-Jones J, Buss F. The autophagy receptor TAX1BP1 and the molecular motor myosin VI are required for clearance of Salmonella typhimurium by autophagy. PLoS Pathog. 2015;11:e1005174.

123. Turco E, Witt M, Abert C, Bock-Bierbaum T, Su MY, Trapannone R, Sztacho M, Danieli A, Shi X, Zaffagnini G, Gamper A, Schuschnig M, Fracchiolla D, Bernklau D, Romanov J, Hartl M, Hurley JH, Daumke O, Martens S. FIP200 claw domain binding to p62 promotes autophagosome formation at ubiquitin condensates. Mol Cell. 2019;74(2):330–346.e11.

124. Turco E, Savova A, Gere F, Ferrari L, Romanov J, Schuschnig M, Martens S. Reconstitution defines the roles of p62, NBR1 and TAX1BP1 in ubiquitin condensate formation and autophagy initiation. Nat Commun. 2021;12:5212.

125. Uoselis L, Nguyen TN, Lazarou M. Mitochondrial degradation: mitophagy and beyond. Mol Cell. 2023;83:3404–20.

126. Van Humbeeck C, Cornelissen T, Hofkens H, Mandemakers W, Gevaert K, De Strooper B, Vandenberghe W. Parkin interacts with Ambra1 to induce mitophagy. J Neurosci. 2011;31:10249–61.

127. van Vliet AR, Chiduza GN, Maslen SL, Pye VE, Joshi D, De Tito S, Jefferies HBJ, Christodoulou E, Roustan C, Punch E, Hervas JH, O'Reilly N, Skehel JM, Cherepanov P, Tooze SA. ATG9A and ATG2A form a heteromeric complex essential for autophagosome formation. Mol Cell. 2022;82:4324–4339.e4328.

128. Vargas JNS, Wang C, Bunker E, Hao L, Maric D, Schiavo G, Randow F, Youle RJ. Spatiotemporal control of ULK1 activation by NDP52 and TBK1 during selective autophagy. Mol Cell. 2019;74(2):347–362.e6.

129. Vargas JNS, Hamasaki M, Kawabata T, Youle RJ, Yoshimori T. The mechanisms and roles of selective autophagy in mammals. Nat Rev Mol Cell Biol. 2023;24:167–85.

130. von Muhlinen N, Akutsu M, Ravenhill BJ, Foeglein A, Bloor S, Rutherford TJ, Freund SM, Komander D, Randow F. LC3C, bound selectively by a noncanonical LIR motif in NDP52, is required for antibacterial autophagy. Mol Cell. 2012;48:329–42.

131. Wang L, Zhou J, Yan S, Lei G, Lee CH, Yin XM. Ethanol-triggered lipophagy requires SQSTM1 in AML12 hepatic cells. Sci Rep. 2017;7:12307.

132. Wei Y, Chiang WC, Sumpter R Jr, Mishra P, Levine B. Prohibitin 2 is an inner mitochondrial membrane mitophagy receptor. Cell. 2017;168:224–238.e210.

133. Wild P, Farhan H, McEwan DG, Wagner S, Rogov VV, Brady NR, Richter B, Korac J, Waidmann O, Choudhary C, Dotsch V, Bumann D, Dikic I. Phosphorylation of the autophagy receptor optineurin restricts salmonella growth. Science. 2011;333:228–33.

134. Wilhelm LP, Zapata-Munoz J, Villarejo-Zori B, Pellegrin S, Freire CM, Toye AM, Boya P, Ganley IG. BNIP3L/NIX regulates both mitophagy and pexophagy. EMBO J. 2022;41:e111115.

135. Wong YC, Holzbaur EL. Optineurin is an autophagy receptor for damaged mitochondria in parkin-mediated mitophagy that is disrupted by an ALS-linked mutation. Proc Natl Acad Sci USA. 2014;111:E4439–48.

136. Wu W, Tian W, Hu Z, Chen G, Huang L, Li W, Zhang X, Xue P, Zhou C, Liu L, Zhu Y, Zhang X, Li L, Zhang L, Sui S, Zhao B, Feng D. ULK1 translocates to mitochondria and phosphorylates FUNDC1 to regulate mitophagy. EMBO Rep. 2014;15:566–75.

137. Wyant GA, Abu-Remaileh M, Frenkel EM, Laqtom NN, Dharamdasani V, Lewis CA, Chan SH, Heinze I, Ori A, Sabatini DM. NUFIP1 is a ribosome receptor for starvation-induced ribophagy. Science. 2018;360:751–8.
138. Yamamoto H, Fujioka Y, Suzuki SW, Noshiro D, Suzuki H, Kondo-Kakuta C, Kimura Y, Hirano H, Ando T, Noda NN, Ohsumi Y. The intrinsically disordered protein Atg13 mediates supramolecular assembly of autophagy initiation complexes. Dev Cell. 2016;38:86–99.
139. Yamano K, Kikuchi R, Kojima W, Hayashida R, Koyano F, Kawawaki J, Shoda T, Demizu Y, Naito M, Tanaka K, Matsuda N. Critical role of mitochondrial ubiquitination and the OPTN-ATG9A axis in mitophagy. J Cell Biol. 2020;219:e201912144.
140. Yamasaki A, Noda NN. Structural biology of the Cvt pathway. J Mol Biol. 2017;429:531–42.
141. Yang M, Chen P, Liu J, Zhu S, Kroemer G, Klionsky DJ, Lotze MT, Zeh HJ, Kang R, Tang D. Clockophagy is a novel selective autophagy process favoring ferroptosis. Sci Adv. 2019;5:eaaw2238.
142. Yorimitsu T, Klionsky DJ. Atg11 links cargo to the vesicle-forming machinery in the cytoplasm to vacuole targeting pathway. Mol Biol Cell. 2005;16:1593–605.
143. Zaffagnini G, Savova A, Danieli A, Romanov J, Tremel S, Ebner M, Peterbauer T, Sztacho M, Trapannone R, Tarafder AK, Sachse C, Martens S. p62 filaments capture and present ubiquitinated cargos for autophagy. EMBO J. 2018;37:e98308.
144. Zatloukal K, Stumptner C, Fuchsbichler A, Heid H, Schnoelzer M, Kenner L, Kleinert R, Prinz M, Aguzzi A, Denk H. p62 is a common component of cytoplasmic inclusions in protein aggregation diseases. Am J Pathol. 2002;160:255–63.
145. Zhang Y, Yao Y, Qiu X, Wang G, Hu Z, Chen S, Wu Z, Yuan N, Gao H, Wang J, Song H, Girardin SE, Qian Y. Listeria hijacks host mitophagy through a novel mitophagy receptor to evade killing. Nat Immunol. 2019;20:433–46.
146. Zhao YG, Zhang H. Formation and maturation of autophagosomes in higher eukaryotes: a social network. Curr Opin Cell Biol. 2018;53:29–36.
147. Zheng YT, Shahnazari S, Brech A, Lamark T, Johansen T, Brumell JH. The adaptor protein p62/SQSTM1 targets invading bacteria to the autophagy pathway. J Immunol. 2009;183:5909–16.
148. Zheng Q, Chen Y, Chen D, Zhao H, Feng Y, Meng Q, Zhao Y, Zhang H. Calcium transients on the ER surface trigger liquid-liquid phase separation of FIP200 to specify autophagosome initiation sites. Cell. 2022;185:4082–4098.e4022.
149. Zhou Z, Liu J, Fu T, Wu P, Peng C, Gong X, Wang Y, Zhang M, Li Y, Wang Y, Xu X, Li M, Pan L. Phosphorylation regulates the binding of autophagy receptors to FIP200 Claw domain for selective autophagy initiation. Nat Commun. 2021;12:1570.
150. Zhu Y, Massen S, Terenzio M, Lang V, Chen-Lindner S, Eils R, Novak I, Dikic I, Hamacher-Brady A, Brady NR. Modulation of serines 17 and 24 in the LC3-interacting region of Bnip3 determines pro-survival mitophagy versus apoptosis. J Biol Chem. 2013;288:1099–113.

## Further Reading

Johansen T, Lamark T. Selective autophagy: ATG8 family proteins, LIR motifs and cargo receptors. J Mol Biol. 2020;432:80–103.
Kirkin V. History of the selective autophagy research: how did it begin and where does it stand today? J Mol Biol. 2020;432:3–27.
Lamark T, Johansen T. Mechanisms of selective autophagy. Ann Rev Cell Dev Biol. 2021;37:143–69.
Lynch-Day MA, Klionsky DJ. The Cvt pathway as a model for selective autophagy. FEBS Lett. 2010;584:1359–66.

Rogov VV, Nezis IP, Tsapras P, Zhang H, Dagdas Y, Noda NN, Nakatogawa H, Wirth M, Mouilleron S, McEwan DG, Behrends C, Deretic V, Elazar Z, Tooze SA, Dikic I, Lamark T, Johansen T. Atg8 family proteins, LIR/AIM motifs and other interaction modes. Autophagy Rep. 2023;2:2188523.

Vargas JNS, Hamasaki M, Kawabata T, Youle RJ, Yoshimori T. The mechanisms and roles of selective autophagy in mammals. Nat Rev Mol Cell Biol. 2023;24:167–85.

Yamamoto H, Zhang S, Mizushima N. Autophagy genes in biology and disease. Nat Rev Genet. 2023;24:382–400.

Chino H, Mizushima N. ER-Phagy: Quality and quantity control of the endoplasmic reticulum by autophagy. Cold Spring Harb Perspect Biol. 2023;15:a041256.

# The Source of Membrane During Autophagy and the Early Steps in Autophagosome Formation

Nicholas T. Ktistakis

**What You Will Learn in This Chapter**

This chapter will describe work aiming to determine the origins of the autophagosomal membrane, which is a topic as old as the study of autophagy itself. Several potential ideas have been proposed over the years with many cellular membranes being considered as sites of autophagosome formation, and including the possibility that autophagosomes form *de novo* in the cytoplasm without interaction with any other membrane. A consensus view that has emerged and will be discussed is that autophagosomes form in primary association with the endoplasmic reticulum (ER) but also in secondary contact with other membrane compartments. Contact with the ER allows transfer of lipids to the forming autophagosomal membrane, and it also provides a cradle that shapes the formation of autophagosomes. To arrive at this conclusion, a large number of laboratories used morphological, biochemical and structural studies to define and describe each step of the process and to understand the protein machineries involved in functional and structural detail. These approaches will be described in this chapter as a way to introduce you to a variety of experimental work used in the study of many cell biological problems, including autophagy. Finally, we will dedicate a brief section to the comparison of autophagosome formation mechanisms in selective and non-selective autophagy.

**Outline**

1. Origin of the Autophagosomal Membrane: The Early Years.
2. Current Understanding of Autophagosome Formation: Morphological Studies.
3. Biochemical Reactions and Structural Arrangements Leading to Autophagosome Formation.

---

N. T. Ktistakis (✉)
Signalling Programme, Babraham Institute, Cambridge, UK
e-mail: nicholas.ktistakis@babraham.ac.uk

    101
B. Loos, D. J. Klionsky (eds.), *Autophagy - From Molecular Mechanisms to Flux Control in Health and Disease*, Learning Materials in Biosciences,
https://doi.org/10.1007/978-3-031-88121-3_5

4. A pathway of Autophagosome Formation.

5. Differences Between Non-selective and Selective Autophagy at the Initiation Stage.

# 1　　Origin of the Autophagosomal Membrane: The Early Years

*Thus, one might picture two different mechanisms of autophagy. In the first, the enveloping elements actually belong to the vacuolar apparatus (VA) and possess membranes of the VA type. In the other, they are originally of ER type and suffer subsequent conversion to the VA form in the GERL region* [1].

*In 1964 we suggested that in hepatocytes the ER is involved in forming autophagic vacuoles… It is even possible that in life the tortuous or serpentine areas of ER are constantly undulating, while creating micro-isolates through internalization of adjacent cytosol.* [2].

## 1.1　　Early Morphological Studies

The early (circa 1960s) discovery by electron microscopy (EM) of membranous vacuoles containing cytoplasm (including mitochondria and ER) suggested that there must be a way for the cell to trap and engulf some cytosolic material for eventual degradation. Remarkably, the presence of these vacuoles was from the beginning linked to starvation conditions or to treatments with glucagon, i.e. it was thought to be a regulated phenomenon (reviewed in [3]).

A little later, Christian de Duve and colleagues coined the term autophagy in order to describe the process by which cellular material undergoes digestion in the lysosomal system [4]. In fact, de Duve differentiated between autophagy and heterophagy, the latter having been defined as cellular digestion of exogenous material. Of some historical interest is the fact that the term autophagy was in use from the middle of the nineteenth century to signify self-digestion at an organismal level, but without its modern interpretation as a cell-based response [5].

The EM morphology of autophagosomes as double-membrane vesicles did not immediately suggest a pathway for their formation [3]. Some visual clues were provided later by the accumulation of curving membrane structures in cells undergoing an autophagic response. Could those be the precursors of autophagosomes? Coupled with significant technical advances in sample preparation which allowed better preservation of membranous structures for EM analysis, it was subsequently realized that mature autophagosomes formed via osmiophilic pre-autophagosomal structures which were lipid-rich but protein-poor. These precursors are termed phagophores [6, 7]. The simplest diagram depicting this process is shown in Fig. 1 top panel.

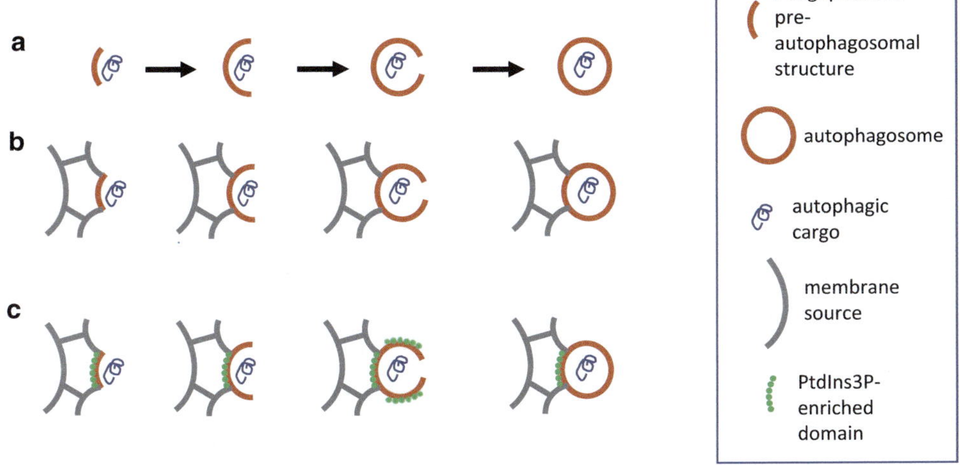


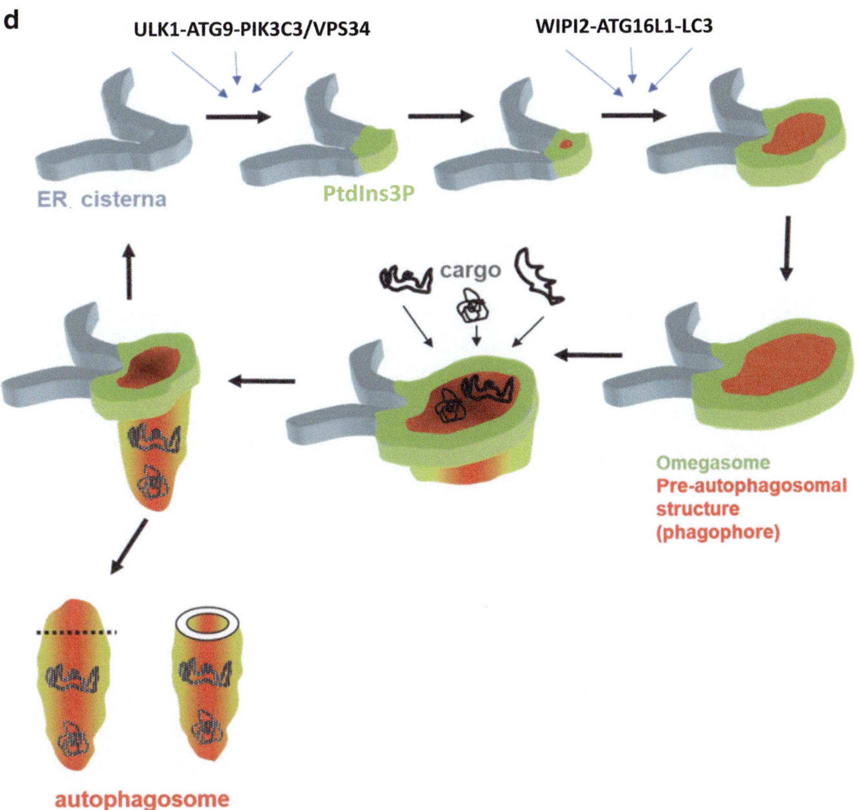


**Fig. 1** An evolving view of the process of autophagosome formation in mammalian cells. (**a**) Based on older EM observations, autophagosome formation was thought to involve the gradual expansion of a flat membrane sheet (phagophore) into a vesicle engulfing cargo. (**b**) Later work showed that this process occurs in association with a cellular membrane, e.g. the ER, mitochondria or endosomes. (**c**) For at the least ER-associated omegasomes, the process also depends on a PtdIns3P-enriched region of the ER. (**d**) A more nuanced view of the process involving the PtdIns3P-enriched omegasomes forming on an ER cisterna which then enable formation of autophagosomal membranes in an ER cradle, eventually exiting the omegasomes to become fully formed autophagosomes

## 1.2 How Is the Phagophore Formed—Early Discussions

The appearance of a novel membranous structure in the cellular interior after starvation or glucagon treatment which gave rise to a mature autophagosome initiated a long discussion as to its origin and provenance. The most important early question was whether the phagophores arose from a pre-existing membrane/organelle, or did it form *de novo*. Evidence for both possibilities was discussed, but neither option was supported by unassailable data [3]. Arguing for an origin via pre-existing membrane (most notably the ER, the Golgi complex and less so endolysosomal vesicles were candidate membranes) were EM observations showing that the phagophore was frequently seen between ER strands, and its staining pattern resembled vesicles emanating from the Golgi complex [3]. Arguing for *de novo* formation was the absence of any evidence that (a) the phagophore membranes made connections to other cellular membranes, and (b) autophagosomes and their precursors stained positive for pre-existing membrane compartments [3].

## 1.3 Early Biochemical Studies Failed to Resolve the Provenance of the Autophagosomal Membrane

In addition to the morphological studies described above, several laboratories attacked the problem on the origin of the autophagosomal membrane using biochemical techniques. It should be recalled that this is the era of cell fractionation using centrifugation: for example, de Duve and collaborators succeeded in isolating lysosomes via rate zonal or equilibrium sedimentation and characterization of these fractions allowed them to identify several acid hydrolases as constituents of lysosomes [8]. In related work, de Duve and colleagues also isolated a lysosomal fraction from livers of glucagon-treated rats which differed from untreated tissues. Given that glucagon was known to induce robust autophagy in those animals, the biochemical properties of that novel fraction were correlated to the presence of autophagic vacuoles [9].

One of the earliest attempts to isolate autophagosomes and their precursors was by Hans Glaumann and colleagues [10]. They used vinblastine treatment to accumulate autophagic structures in rat liver, and a discontinuous metrizamide gradient to purify a fraction composed primarily of autophagic vacuoles. This fraction contained mitochondria, ER and cytosol as well as lysosomal enzymes, but such a composition—which reflected the cargo of autophagosomes and their eventual destination—did not illuminate the question on the origin of the autophagosomal membrane. In fact, many subsequent attempts to isolate pure autophagic fractions have run into a similar problem in that these fractions reflect the cargo of autophagosomes without showing an enrichment of a membrane source [11].

## 1.4    Early Biochemical Studies Defined Important Requirements for Autophagosome Formation

The mechanism of autophagosome formation was a subject of many investigations in the early years of autophagy research, which however did not yield definitive answers in the absence of a genetic understanding of the pathway. Nevertheless, some important results were obtained which allowed subsequent work to proceed more rationally. A very important early result was that inhibition of lysosomal function by either lysosomotropic amines (ammonium chloride, methylamine) or by lysosomal protease inhibitors (leupeptin) had a strong inhibitory effect on the degradative function of autophagy and led to accumulation of autophagosomal structures [12, 13]. Another important early observation was that the chemical 3-methyladenine was a strong inhibitor of autophagy [14]. Subsequent work showing that this chemical inhibits the PIK3C3/VPS34 lipid kinase responsible for synthesising the lipid phosphatidylinositol 3-phosphate (PtdIns3P) first implicated this lipid in the regulation of autophagosome formation [15]. Finally, a large number of historical studies using kinase and phosphatase inhibitors revealed other tool compounds that could be used to modulate autophagic responses [16, 17].

Special mention must be made here to early work showing that the protein kinase MTOR (mechanistic target of rapamycin) is a master regulator of autophagy: when MTOR is active autophagy is suppressed whereas inactivation of MTOR—in the majority of cases—leads to rapid induction of autophagy. First identified in yeast by genetic means, this discovery was facilitated and greatly exploited by the availability of the drug rapamycin, a specific inhibitor of MTOR [18]. Rapamycin, and subsequent inhibitors of MTOR, have allowed the precise induction of autophagy independently of nutrient availability and hormonal stimulation(reviewed in [19]). This has allowed, in turn, the investigation of the dynamics and molecular requirements of the autophagic response in most eukaryotic cells.

1. How are "phagophores" defined, and what is their relationship to autophagosomes?
2. What were the two competing theories on the origin of autophagosomes?
3. Who coined the term "autophagy" in the modern era and what other important contributions to cell biology did he make?

## 2    Current Understanding of Autophagosome Formation: Morphological Studies

The identification of yeast genes involved in the autophagic response, and the realisation that these genes were largely conserved in higher eukaryotes, heralded a new era in morphological studies of autophagy [20]. By attaching fluorescent reporters to autophagy genes, it was possible to follow the dynamics of this process by wide field microscopy in great detail. One of the earliest studies used a fluorescent version of LC3 (a mammalian homolog of yeast Atg8), to show in fixed cells that autophagosomes could be clearly separated from lysosomes, and the overlap between the two organelles was minimal [21]. Subsequent work using a fluorescent version of ATG5 (a protein involved in the conjugation of LC3 with phosphatidylethanolamine), revealed for the first time by live imaging the intricate details of autophagosome formation: ATG5 puncta formed non-synchronously throughout the cell, which expanded into cup shaped structures before becoming spherical in morphology [22]. These studies verified in a dynamic way the pathway of autophagosome formation via a phagophore intermediate as deduced by earlier EM studies. Much additional work followed using an ever-increasing panoply of microscopical techniques that has provided us with a very clear view of the pathway of autophagosome formation [23] (Fig. 1 lower panels). Some important conclusions from these studies are that (i) the process depends on a hierarchical sequence of intermediates that transiently associate with the forming structure, (ii) autophagosomes always form in association with an internal membrane and never by themselves in the cytosol and (iii) both non-selective and selective autophagy utilise the same mechanism of autophagosome formation with some small differences. In the following sections we will describe in some detail the 4 stages involved in the biogenesis of autophagosomes, initiation, nucleation, expansion, and sealing off, from a morphological perspective.

## 2.1 Initiation of Autophagosome Formation

The first step in starvation-induced autophagy is the inactivation of the protein kinase MTOR complex 1 (MTORC1) and the subsequent activation of the ULK protein kinase complex (composed of the kinase ULK1, and the adaptor proteins RB1CC1, ATG13 and ATG101). The activated ULK complex translocates to a cellular membrane which is in most cases the ER where it initiates a cascade of events leading to autophagosome formation [24] Super-resolution microscopy using both SIM and STORM suggests a quasi-spherical morphology of ULK complex components coalescing on tubulovesicular elements of the ER, and this appears to be the earliest discernible assembly which gives rise to autophagosomes [25] (Fig. 2 and also see below). One of the important contributors to this early phagophore assembly site are vesicles containing the protein ATG9, the only known transmembrane protein among the autophagy machinery that is required for autophagosome formation. In this early stage, the ATG9 vesicles appear to mark the places in the ER where the ULK machinery translocate, but exact details of this are not understood [25, 26]. The final step during initiation is the ULK-mediated activation of the PIK3C3/VPS34 complex (composed of the lipid kinase PIK3C3/VPS34 and the adaptor proteins PIK3R4/VPS15, BECN1 and ATG14) and its translocation to the same punctate structures that contain the ULK complex proteins. Although it has not been possible to follow the dynamics of the PIK3C3/VPS34 complex by live imaging, it has been straight-

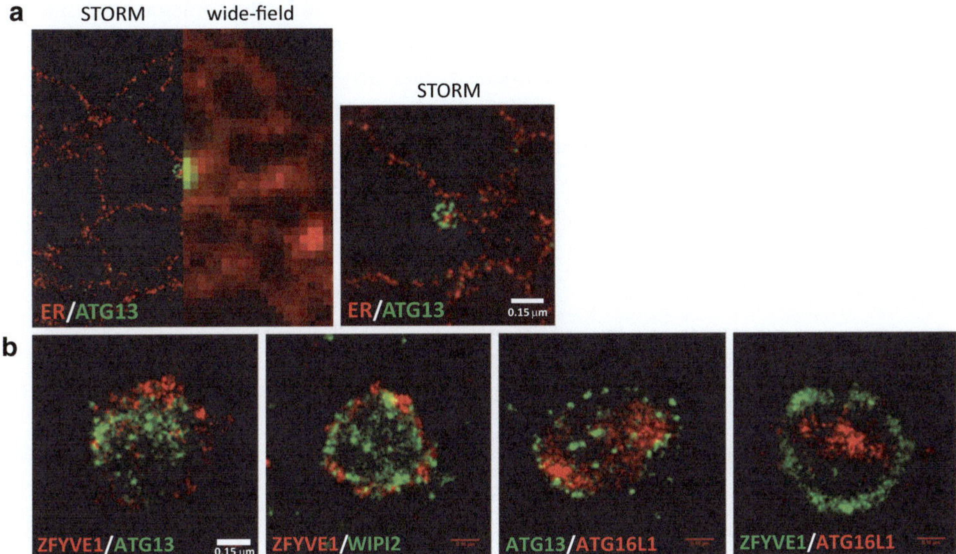



**Fig. 2** Super-resolution microscopy can reveal the spatial relationship between autophagy proteins. (**a**) The relationship between ATG13 and the ER is clearly shown by STORM in comparison to regular wide-field (compare in left panel clear vs blurry image). The whole STORM image is shown in the panel on the right. (**b**) STORM images of various autophagy protein during starvation. Note that, although they all decorate a spherical autophagic structure, their spatial relationships are very distinct

forward to image the formation of its lipid product PtdIns3P by its effector proteins ZFYVE1 and WIPI2 and in this way deduce the behavior of the complex itself (see below).

## 2.2 Nucleation of Autophagosome Formation

The activated PIK3C3/VPS34 complex synthesizes PtdIns3P in special ER-associated regions that have been termed "omegasomes", and which represent a major assembly site for autophagosome formation. The presence of this PtdIns3P-enriched compartment was initially discovered by following the dynamics of the protein ZFYVE1 during autophagy induction [27]. It was observed that early in the process ZFYVE1 puncta associated with tubular extensions of the ER where eventually autophagy proteins such as LC3 began to accumulate. Once omegasome and LC3-containing mixed membranes expanded to their maximum diameter, the LC3-associated membranes exited the structure giving rise to a complete autophagosome. Remarkably, this whole sequence of membrane expansions and extrusions was situated on the underlying ER, which served as a cradle for these events. In the diagram of the bottom part of Fig. 1, this is shown schematically.

Live imaging of the ULK complex component ATG13, the omegasome marker ZFYVE1 and the autophagosomal protein LC3 under the same conditions allowed an appreciation of the dynamics of this pathway at the early stages [28]. The appearance of the PtdIns3P-enriched structures was fractionally delayed in comparison to the appearance of ULK complex proteins in those punctate structures, demonstrating that PIK3C3/VPS34 activity is downstream of ULK activation and translocation (Fig. 3a). At the same time, formation of the LC3-positive structures was delayed in comparison to the ATG13-positive structures (and to the omegasomes), again demonstrating that autophagosome formation

**Fig. 3** Dynamics of early autophagy responses. Cell expressing the ULK complex protein ATG13, the omegasome marker protein ZFYVE1 or the autophagosomal marker protein LC3 were imaged during starvation and the intensity of the various punctate structures was quantified. (**a**) It can be seen that ATG13 puncta occur slightly earlier than omegasomes, and they disappear faster than omegasomes. (**b**) ATG13 puncta appear well in advance of LC3 puncta, and their duration is shorter than LC3 puncta. (**c**) Stylized graph of the dynamic relationship between ULK complex (ATG13), PtdIns3P effector (ZFYVE1) and LC3 during autophagosome formation. This is derived from averages of the quantification of the three components as shown in **a** and **b**

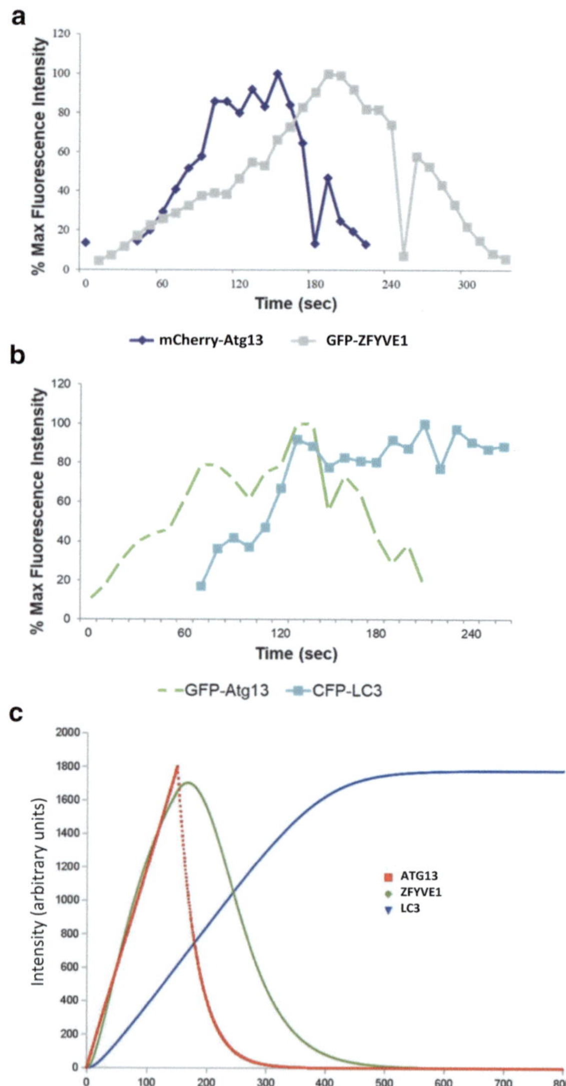

follows the translocation of the ULK complex to phagophore assembly sites (Fig. 3b). Deducing a sequence of events involving all three of those components (ATG13, omegasomes, LC3) shows that formation of autophagosomes involves sequential translocation and disengagement of the two early components (ULK complex and omegasomes) whereas the LC3-containing fully formed autophagosomes are longer lived (Fig. 3c).

In addition to the temporal regulation of those components, the distribution of early autophagy intermediates is spatially distinct, as revealed by super-resolution multi-color STORM, a technique well-suited to investigations of such relatively large and complex structures. In comparison to regular microscopy, STORM offers a much cleaner view of the relationship between ATG13-containing early structures and the ER where it can be seen that the ATG13-positive puncta localize to ER extensions (Fig. 2a). The relationship between ZFYVE1, ATG13, WIPI2 and ATG16L1 is also very illuminating as to the way an early autophagic structure assembles. Early machinery is for the most part on the surface of such structures (ZFYVE1, ATG13 and WIPI2 in Fig. 2b) whereas ATG16L1 which is involved in LC3 lipidation appears to be present on the interior of such structures. This agrees with the live imaging data showing LC3-containing membranes in the interior of omegasome-marked structures.

## 2.3    Expansion of Phagophores and Sealing Off of the Double Membranes

It has been estimated that a 500-nm autophagosome requires 3 million lipids during its formation and supply of these lipids has a significant role in the expansion of the phagophore membrane [29]. A second important component of phagophore growth is formation of the lipidated Atg8-family proteins, briefly discussed for the case of LC3 above, which decorate the expanding phagophores [30].

In terms of lipid supply, none of the essential autophagy proteins have been shown to synthesize lipids, whereas the protein ATG2 in combination with ATG9 have been shown to synergistically transport lipids to the forming autophagosomes and to redistribute them to the bilayers of the vesicle. Lipid transport is mediated by ATG2, which has also been shown by live imaging to reside on ER membranes during this process [31], whereas lipid redistribution has been shown to occur by the scramblase activity of ATG9 in an interaction with ATG2 [32]. Interestingly, ATG9 containing vesicles were shown by live imaging to transiently interact with growing phagophores [33], perhaps as a reflection of the lipid scrambling activity during this process. Given that ATG2 is localized on the ER membranes a very plausible model is that during the process of autophagosome formation the ER via ATG2 and ATG9 provides the lipids for the membrane expansion [34].

Formation of the lipidated Atg8-family proteins and their coincidental incorporation in the autophagosomal membranes occurs via two ubiquitin-like conjugation systems in a reaction that starts with the proteolytic removal of the last 4 C-terminal amino acids of LC3 by the protease ATG4 to expose a glycine residue which is the acceptor for the covalent conjugation with phosphatidylethanolamine [30]. In terms of the dynamics of this stage, it appears that growth of the autophagosomal membrane is a smooth process which commences from a single small punctate structure which expands into a semi-spherical morphology before the sphere is completed and a new vesicle formed (Fig. 4a). Such a smooth growth of the membrane marked by LC3 was also shown for selective autophagy of Salmonella [35](Fig. 4b).

The final early step in autophagosome formation is the sealing off of the double membrane [36]. Perhaps counter-intuitively, this is a fission (or scission) and not a fusion reaction somewhat akin to formation of an invaginated inner vesicle during the process of multivesicular body formation [37]. The dynamics of this process have not been reported, but it appears to depend on the protein CHMP2A, a component of the ESCRT-III machinery that is also involved in multivesicular body formation [38].

## 2.4 How Are the ER and Other Membranes Implicated in the Process of Autophagosome Formation?

Major advancements in EM sample preparation and analysis, especially electron tomography, have provided for the first time evidence that the forming autophagosomes are connected with the ER [39, 40]. Two independent groups showed that the autophagosomal membrane is sandwiched between two ER strands, and the ER connects to this membrane via one or multiple thin tubules which were likely to have been missed in earlier work due to fixation problems. A proposal from such work is that the ER, in addition to providing a lipid source, also cradles the forming autophagosome for reasons not well understood. This model has verified previous work — ushered in with the identification of the molecu-

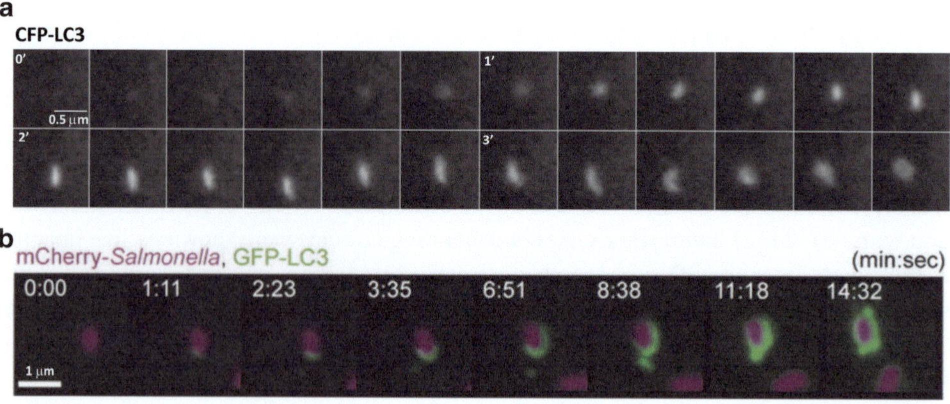



**Fig. 4** Dynamics of LC3 during autophagy. (**a**) Live imaging of CFP-LC3 during starvation, showing the bending of the phagophore before assuming a spherical morphology. (**b**) Engulfment of Salmonella by LC3 during infection-note the bending of the LC3-conatining phagophore around the bacterium. (Image in **b** is used with permission from reference [30])

lar machinery of autophagy — which showed that early phagophore assembly sites as well as LC3-containing autophagosomes were frequently seen to colocalize with ER proteins by immunofluorescent microscopy (Fig. 5).

Although the ER has a prominent role in autophagosome formation, various other cellular organelles such as the mitochondria, the ER to Golgi intermediate compartment, the Golgi complex, the early and recycling endosomes, and the plasma membrane have also been implicated in this process [41–43]. In most cases, evidence at the light microscopy level or some fractionation studies have led to such proposals. Significantly, when electron tomography was used to map autophagosomes to other cellular membranes it was also shown that, in addition to a prominent role for the ER, several other membranes could connect to forming autophagosomes [44]. A consensus opinion is that autophagosomes have a ubiquitous connection with the ER during their biogenesis, and a secondary connection/interaction with various other membrane compartments depending on conditions [45, 46]. Interestingly, the function of the ER as a cradle for autophagosome formation seems to be maintained for at least one type of selective autophagy against mitochondria targets [47].

4. How did identification of autophagy genes help in determining the dynamics of this process?
5. What could be advantages and disadvantages of using a pre-existing membrane for lipid supply during autophagosome formation?
6. If you consider the sequence ULK complex activation → PIK3C3/VPS34 activation → WIPI2 puncta formation → lipidation → LC3 puncta formation, try to predict what intermediates may accumulate if you block (a) ULK activity, (b) PIK3C3/VPS34 activity, (c) lipidation.

## 3    Biochemical Reactions and Structural Arrangements Leading to Autophagosome Formation

Four types of biochemical reactions regulate autophagosome formation: (1) signalling cascades involving the ULK kinase and the master regulator MTOR, a kinase that integrates inputs related to nutrient and growth factor availability with outputs including cell growth, autophagy and metabolism; (2) formation of bioactive lipids, most notably PtdIns3P, via activation of the lipid kinase PIK3C3/VPS34; (3) ubiquitin-like conjugation reactions that covalently modify Atg8-family proteins with phosphatidylethanolamine and (4) lipid transfer and scrambling via the ATG2 and ATG9 proteins which allow autophagosome to incorporate phospholipids as they expand.

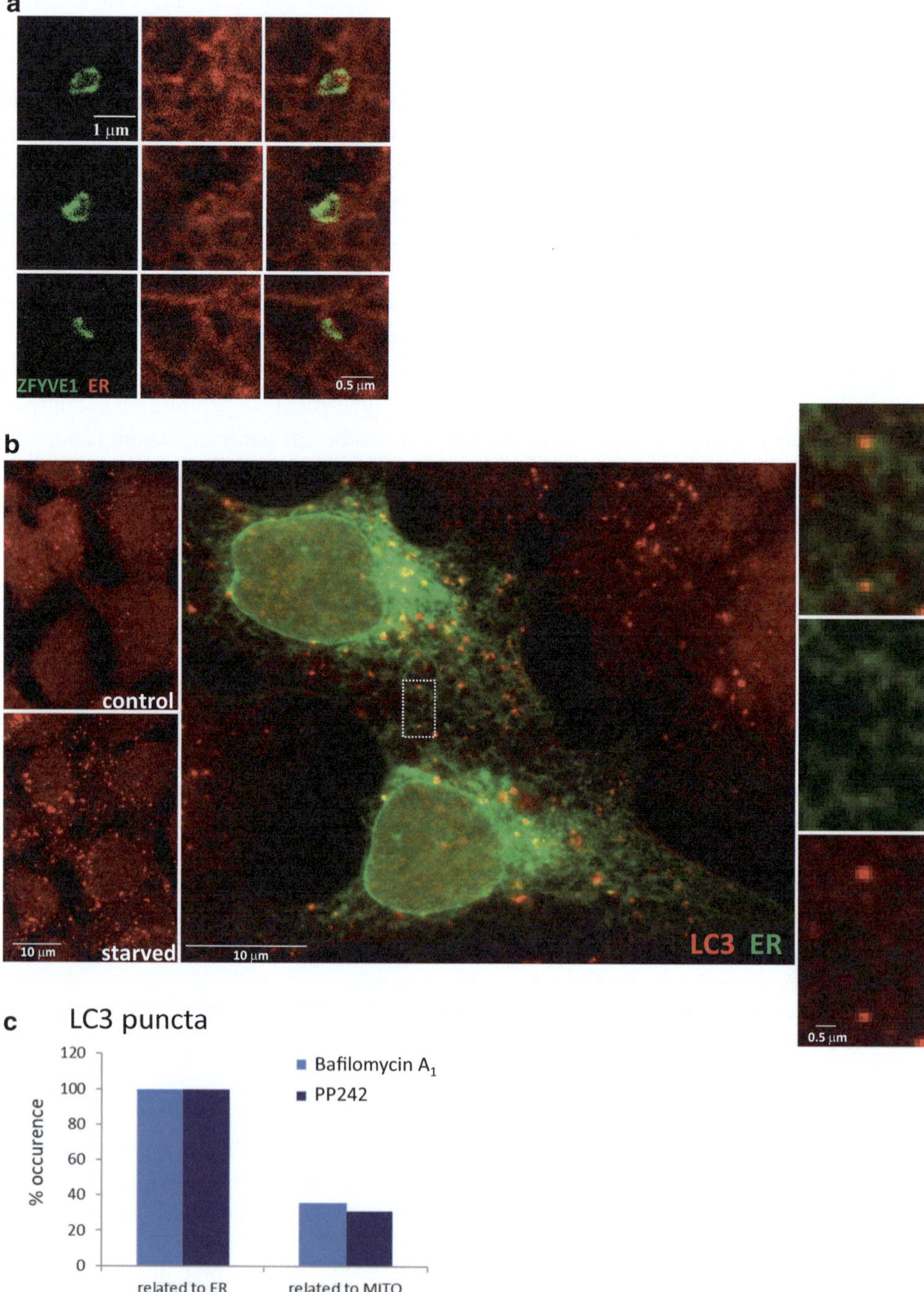



**Fig. 5** (**a**) Tight relationship between autophagosomal structures and the ER is evident by live imaging and immunofuorescence microscopy. (**a**) Selected frames from a live imaging experiment

## 3.1    Signalling via MTOR and ULK, and Scaffolding Function of the ULK Complex

Under conditions of nutrient availability MTORC1 remains active and this results in phosphorylation of ULK1 and inactivation. Upon nutrient limitation, MTORC1 becomes inactivated leading to dephosphorylation of ULK1 and its activation [48, 49]. Activated ULK1 phosphorylates a plethora of autophagy proteins, ATG13, RB1CC1, ATG101, BECN1, PIK3C3/VPS34, ATG14, BECN1 and ATG9 and in this way it positively regulates both the ULK complex as well as the PIK3C3/VPS34 complex. Most of those ULK1-dependent phosphorylations have been shown to take place *in vivo*, and for the vast majority of them phosphor-specific antibodies which can follow the dynamics of the response are available. What is less clear is the precise effect of these phosphorylations on the activity of the target proteins, but such work is proceeding rapidly. One general concept that has emerged is that the ULK complex regulates autophagy both via phosphorylation reactions but also by providing a scaffold on which additional autophagy proteins can coalesce and induce autophagosome nucleation and growth [50, 51]. Although the entire ULK complex has not been crystallized or analyzed by cryo-EM, a subcomplex consisting of an N-terminal fragment of RB1CC1 together with ULK1, ATG13 and ATG101 has allowed some important structural insights on how assembly of the complex may initiate autophagy [52, 53]. The organising hub is a dimer of the RB1CC1 fragment to which monomers of ATG13, ULK1 and ATG101 assemble. In the fully assembled subcomplex, the extended conformation of RB1CC1 becomes semi-spherical in a C shape, and this would allow formation of higher order oligomers which could scaffold autophagy initiation. For selective autophagy where a role for MTORC1 is not clear, this scaffolding function of the ULK complex may be the main regulator of the induction as will be discussed later.

A minor signalling role in the early steps of autophagy is played by the AMPK kinase, a protein that senses and responds to energy status in the cell. AMPK mediates a number of positive phosphorylations on subunits of the ULK (ULK1, ATG13) and the PIK3C3/VPS34 (BECN1, PIK3C3/VPS34) complex, as well as a negative phosphorylation on MTOR [54]. However, because autophagy can proceed without any AMPK proteins, the role of this kinase is at most auxiliary and perhaps context-dependent.

---

**Fig. 5** (continued) with cells under starvation and expressing the omegasome marker ZFYVE1 and an ER protein. In all three frames, it can be seen that omegasomes form aligned to the underlying ER. (**b**) Cells in fed or starved conditions were fixed and stained for LC3, an Atg8-family protein, together with an ER marker protein. All LC3-containing autophagosomes co-localize with the ER. (**c**) Cells treated with PP242, an MTOR inhibitor, to reveal a starvation-induced autophagic response, or with bafilomycin A$_1$, an inhibitor of lysosomal function, to reveal a basal autophagy response, were fixed and stained with ER or mitochondrial markers. The percentage of LC3 puncta colocalizing with ER or with mitochondria is shown in the graph

## 3.2 Activation of PIK3C3/VPS34 and Formation of PtdIns3P

Two main PIK3C3/VPS34 complexes exist in mammalian cells: PIK3C3/VPS34 complex I which is composed as discussed before of PIK3C3/VPS34, PIK3R4/VPS15, BECN1 and ATG14, and PIK3C3/VPS34 complex II which exchanges ATG14 for UVRAG [55] (Fig. 6a). Complex I is involved in autophagy whereas complex II regulates endosomal traffic at the level of early endosomes and the formation of multivesicular bodies [56]. The four subunits of complex I are found assembled under all growth conditions, but activation of the lipid kinase activity is enhanced by ULK1-mediated phosphorylation of ATG14 and BECN1 during starvation. ATG14 is also involved in the translocation of complex I to pre-autophagosomal ER-associated regions for the subsequent generation of PtdIns3P during omegasome formation [57]. PIK3C3/VPS34 activity that enhances the lipidation reaction has also been shown to reside in the ER-to-Golgi intermediate compartment, and its transport to pre-autophagosomal regions may be mediated by coated vesicles [58, 59].

The structures of both PIK3C3/VPS34 complex I and II have been determined by a combination of X-ray diffraction and cryo-EM studies [55] (Fig. 6b). Both complexes display a similar Y shape, with the position occupied by ATG14 in complex I being substituted almost precisely by UVRAG in complex II. Catalysis occurs at the far end of the Y arm, whereas membrane binding involves both arms of the Y structure. In addition to ULK1-mediated phosphorylations, the PIK3C3/VPS34 complex I (and complex II) are subjected to a multitude of other phosphorylations (both activating and inhibitory) by MTOR, AMPK and additional kinases (Fig. 6a). How all of these post translational modifications regulate PIK3C3/VPS34 activity are aims of current research.

A very useful recent development with respect to PIK3C3/VPS34 is the discovery of specific inhibitors of the kinase activity [55] (Fig. 6c). Much previous work examining the function of PIK3C3/VPS34 in autophagy utilised either 3-methyladenine which works at millimolar concentrations and is likely to have non-specific effects, or wortmannin which inhibits all PtdIns and PI 3-kinases including the enzymes that synthesize PtdIns(3,4,5)P$_3$. The availability of cleaner and more specific PIK3C3/VPS34 inhibitors has been a positive development, although these compounds work equally well for both autophagic and endocytic PIK3C3/VPS34 complexes.

In addition to PIK3C3/VPS34 activity, levels of PtdIns3P during autophagy are also regulated by a family of lipid phosphatases termed myotubularin-related 3-phosphatases (MTMRs) which dephosphorylate PtdIns3P to generate PtdIns [60]. Three of the MTMRs,

**Fig. 6** The PIK3C3/VPS34 complex and its inhibitors. (**a**) Domain representation of the four subunits of PIK3C3/VPS34 complex I (PIK3C3/VPS34, PIK3R4/VPS15, BECN1 and ATG14) together with UVRAG which replaces ATG14 in PIK3C3/VPS34 complex II are shown, along with various mutations and post-translational modifications that may impinge on function. (**b**) Space-filling model of the structure of PIK3C3/VPS34 complex I and complex II are shown oriented on the membranes on which catalytic activity occurs. (**c**) List of PIK3C3/VPS34-specific inhibitors that have recently been developed. (Modified from reference [55])

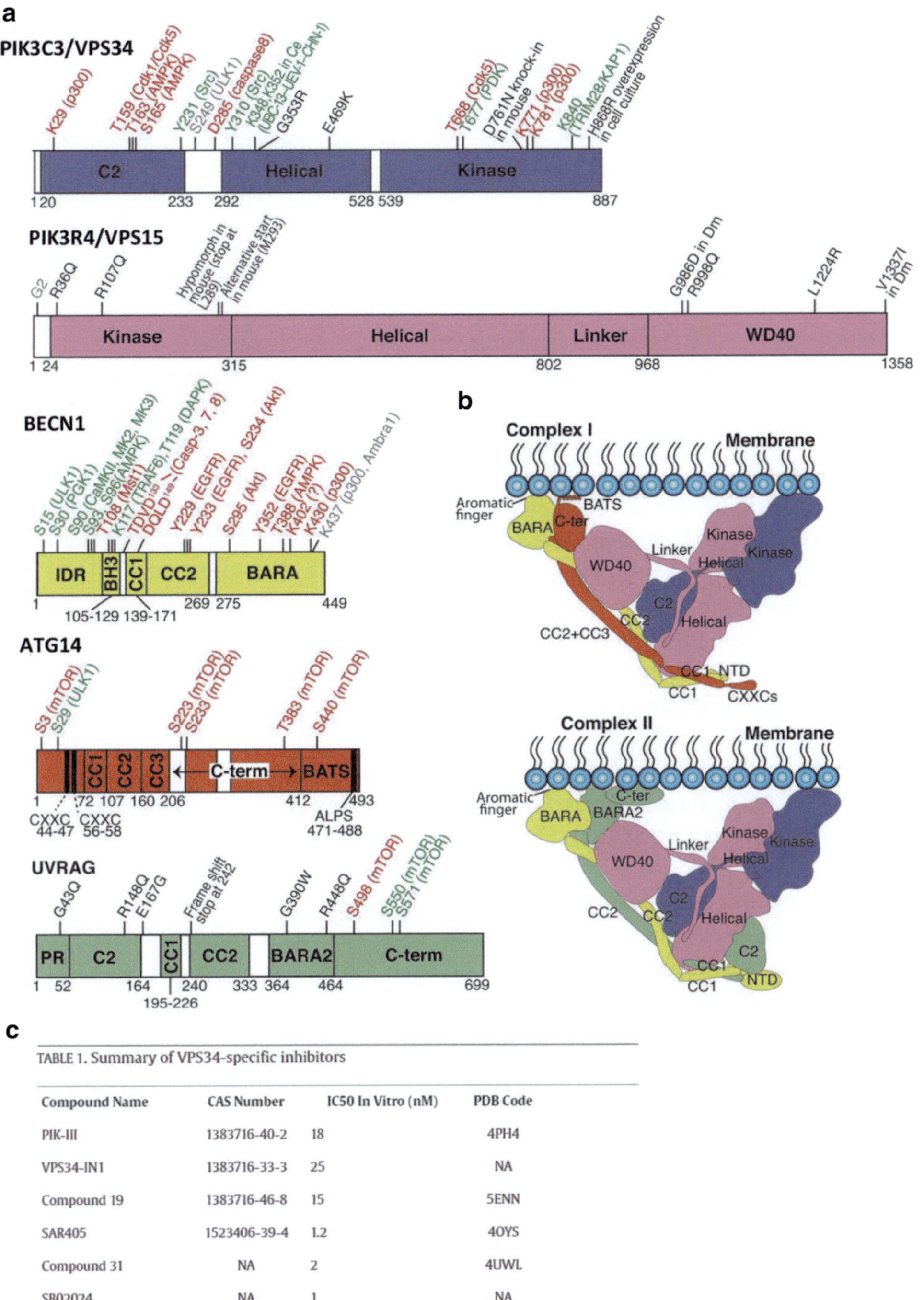


c

TABLE 1. Summary of VPS34-specific inhibitors

| Compound Name | CAS Number | IC50 In Vitro (nM) | PDB Code |
| --- | --- | --- | --- |
| PIK-III | 1383716-40-2 | 18 | 4PH4 |
| VPS34-IN1 | 1383716-33-3 | 25 | NA |
| Compound 19 | 1383716-46-8 | 15 | 5ENN |
| SAR405 | 1523406-39-4 | 1.2 | 4OYS |
| Compound 31 | NA | 2 | 4UWL |
| SB02024 | NA | 1 | NA |

CAS, Chemical Abstracts Service; NA, not available.

MTMR3, MTMR6 and MTMR14 have been shown to affect levels of PtdIns3P during autophagy and in this way influence the dynamics of the response [61–63]. Interestingly, the PtdIns3P signal as revealed by live imaging of the ZFYVE1 reporter diminishes everywhere but continues to be present at the region where the autophagosomal membrane closes upon itself and exits the omegasome, indicating an additional important role for this lipid in the last stage of autophagosome formation [27].

## 3.3  Feedback Regulation of ULK and PIK3C3/VPS34 Complexes

Inhibition of PIK3C3/VPS34 activity has a strong negative effect on the translocation of ULK complex proteins to phagophore assembly sites [64]. This has led to the hypothesis that the presence of PtdIns3P has a stabilising and an enhancing role for the translocation of the ULK complex during autophagy via a positive feedback loop. Supporting this idea are recent results using single particle imaging where it was shown that formation of ULK1-containing clusters involved in autophagy initiation were reduced upon PIK3C3/VPS34 inhibition [65] and that the ULK1 complex and PIK3C3/VPS34 cooperate in a positive feedback loop for autophagosome formation [34]. The structural basis for this may be related to the finding of a supercomplex composed of the ULK complex and the PIK3C3/VPS34 complex bound together [53]. If this is the functional unit during autophagy, it would suggest the possibility of tight coordination between the two activities as the process of autophagosome formation gets underway.

## 3.4  The Atg8-Family Protein Conjugation Reactions During Autophagy Require a PtdIns3P Platform

The conjugation of Atg8-family proteins with lipid is a signature biochemical reaction during autophagy induction and it involves the co-ordinated activities of 8 different proteins [30]. One reaction creates an ATG12–ATG5 conjugate in complex with an ATG16L1 dimer which is important for the localization of the lipid conjugation in the appropriate membrane. The second reaction uses Atg8-family proteins as the ubiquitin-like molecule (after it has been proteolytically clipped at a C terminal glycine) and three E1-E2-E3-like activities (ATG7, ATG10, ATG12–ATG5-ATG16L1) for conjugation of Atg8-family proteins to either phosphatidylethanolamine (for autophagic responses) of phosphatidylserine (for non-canonical autophagic functions). Our understanding of these complex biochemical reactions has been greatly enabled by the discovery of conditions allowing lipidation *in vitro* with purified proteins and phospholipids. The minimum requirement for lipidation *in vitro* is a mixture of Atg8-family proteins truncated at the C terminal glycine residue, ATG7, ATG3, ATP and liposomes [66]. At the same time, crystal structures for the majority of the proteins of the lipidation machinery are available, allowing us a deep understanding of the structural characteristics of this process [67, 68].

The discovery that the PtdIns3P effector WIPI2 directly interacts with ATG16L1 has provided a rationale for the fact that formation of PtdIns3P is important during phagophore expansion [69]. Of note, the ability of ATG16L1 to direct Atg8-family protein lipidation independently of PtdIns3P has also provided an explanation for the punctate distribution of Atg8-family proteins on single membranes of endocytic origin in a pathway originally termed non-canonical autophagy and more recently CASM (conjugation of ATG8s to single membranes) [70]. The function of CASM is still not clear.

How important is the lipidation reaction for autophagosome formation? Somewhat surprisingly, it has been reported than in the absence of all Atg8-family proteins or when the lipidation reaction is inhibited, formation of autophagosomes with seemingly normal morphology still takes place. What seems to be affected under these conditions is the fusion of autophagosomes with lysosomes and the degradation of the inner autophagosomal membrane [71, 72]. Therefore, although lipid conjugation is a signature reaction of this pathway, it is not a strict requirement for autophagosome formation.

## 3.5  Lipid Transfer and Scrambling Feed the Forming Autophagosome

If Atg8-family proteins lipid conjugation is not essential for phagophore growth, what reaction(s) ensure that this expansion occurs? As mentioned above, two long-sought activities were recently identified among the essential autophagy proteins which account for membrane expansion: ATG2 which transports lipids from the ER to the growing phagophore and ATG9 which randomizes the distribution of these lipids by directly interacting with ATG2 [73–75]. At the same time, two additional lipid scrambling proteins involved in autophagy are localized on the ER (VMP1 and TMEM41B) and interact with ATG2 [76, 77]. These proteins are not exclusively involved in autophagy, since they also have roles in lipid droplet formation, but they are thought to regulate the supply of lipids from the ER to the growing phagophore.

The structure of ATG9A in complex with ATG2A has provided us with a very detailed view of how lipids are transported during autophagosome formation [78]. The complex contains many interacting regions between the two proteins, and significant local rearrangements occur upon binding of the two subunits. Lipids are directly delivered to the perpendicular region of ATG9A by ATG2A owing to the physical interaction of the two proteins.

## 3.6  Closing of the Phagophore Membrane

It has been challenging to identify the machinery responsible for sealing off of the autophagosome since determination of the exact closure status of phagophores can only be done via laborious EM techniques not easily adaptable for screening protocols. Recently, a biochemical assay designed to distinguish between partially open and closed early autophagosomal structures has shown that the ESCRT-III protein CHMP2A has an essential role in autophagosome closure [38]. Other ESCRT-III components, such as CHMP3 and CHMP7, are also involved but not in an essential way [38]. In addition to CHMP2A, the AAA ATPase VPS4 is involved in this process as well as the ESCRT-I subunit VPS37A [79]. Interestingly, although both ESCRT-I and ESRCT-III provide functionality for autophagosome closure, their involvement is distinct in comparison to their other functions in membrane sealing during endocytosis [80], i.e. the autophagic pathway usurps part of their canonical activity. In addition to the ESCRT components, Atg8-family proteins proteins appear to have a role in autophagosome closure and integrity, by binding and recruiting ESCRT-I to closed autophagosomes to maintain their sealed status [81].

7. Discuss if a selective inhibitor of MTORC1 is a good activator of autophagy *in vivo*.
8. Given the structural and biochemical principles discussed here, how would you design a specific autophagy inhibitor?
9. Why is a scramblase activity required at the donor ER membrane during phagophore expansion?

## 4     A Pathway of Autophagosome Formation

In this section we will summarize the current understanding of the pathway of autophagosome formation in one continuous narrative-some repetition with previous sections is inevitable but we aim to highlight the most salient points. Figure 7 accompanies this narrative.

The process of autophagosome formation during starvation-induced autophagy starts with the inactivation of MTOR due to limitations in nutrient and growth factor levels. Inactivation of MTOR relieves an inactivating phosphorylation on ULK1 making this kinase active for the autophagic response. Activated ULK1 phosphorylates a number of downstream autophagy proteins, leading to their activation. Concomitantly to this, the ULK complex translocates to peri-nuclear, ER-associated sites where its subunits serve as scaffolds for the assembly of phagophore assembly sites. Partially in response to activating phosphorylations, and perhaps as a result of direct protein-protein interactions with the ULK complex, the PIK3C3/VPS34 complex translocates to the same site as the ULK complex where it initiates formation of PtdIns3P on ER extensions called omegasomes. A positive feedback loop between ULK and PIK3C3/VPS34 complexes ensures that formation of PIK3C3/VPS34 will be fast and efficient at omegasomes. A third protein likely present in those phagophore assembly sites is ATG9 which arrives there in the form of small vesicles and with an accompanying phosphorylation by the ULK1 kinase. Three important events take place at the omegasome-ER junction: (i) the lipidation machinery

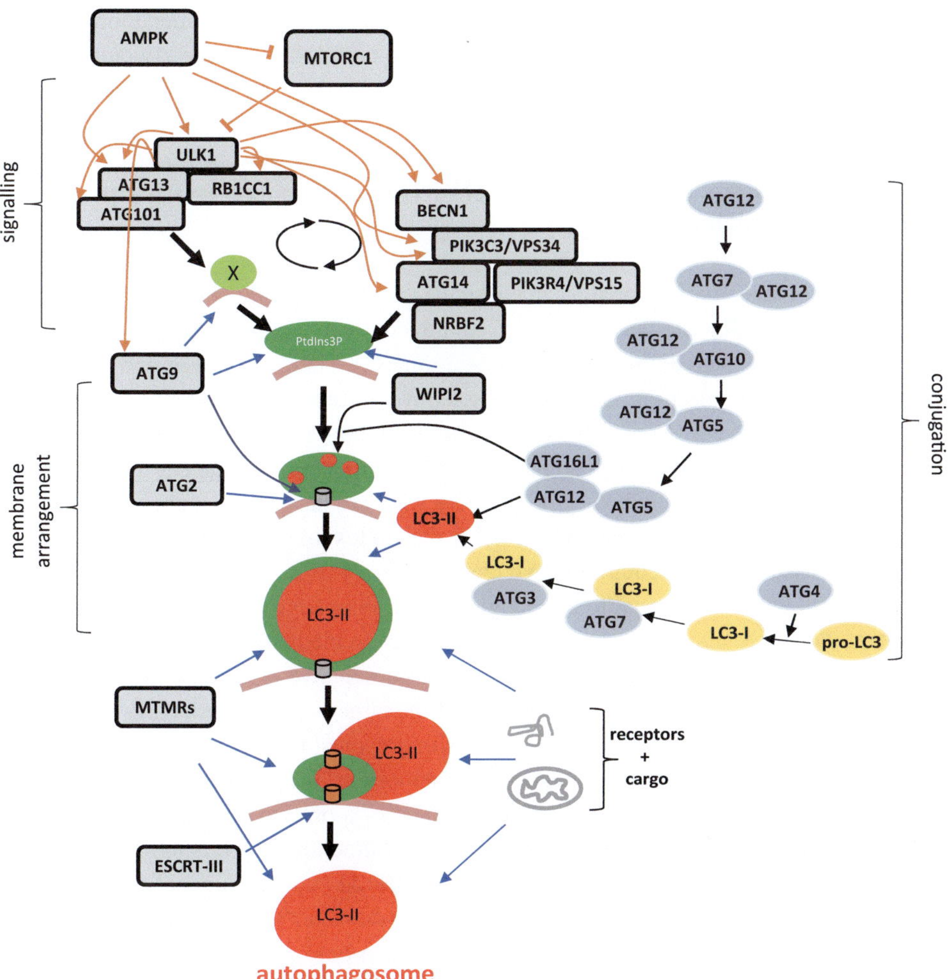



**Fig. 7** A pathway of autophagosome formation during non-selective autophagy. The scheme incorporates current knowledge of this process, and is explained throughout the text and in Sect. 4. Orange arrows and lines signify activating or inhibiting phosphorylations respectively. Black arrows indicate stages of autophagosome formation, whereas blue arrows indicate translocation of relevant proteins to autophagosomal sites

via the interaction of ATG16L1 with the PtdIns3P effector WIPI2 starts to covalently conjugate Atg8-family proteins with phosphatidylethanolamine on forming autophagosomal membranes; (ii) a lipid supply chain from the ER to autophagosomal membranes is set up with a complex between the lipid transfer protein ATG2 and the scramblase ATG9; (iii) cargo to be engulfed and transported to lysosomes is incorporated within the forming autophagosomal membrane. As the process of autophagosome formation reaches its completion, levels of PtdIns3P are reduced via the action of the MTMRs, the autophagosomal membrane disengages from the omegasomes in a mechanism not understood, and the edges of the phagophore are closed to generate the double membrane structure via the action of ESCRT-III.

## 5    Differences Between Non-selective and Selective Autophagy at the Initiation Stage

The pathway of selective autophagy can eliminate virtually all cellular components, differing in size from small protein aggregates of 10 nm length up to mitochondria, peroxisomes or invading bacteria with diameters exceeding 2 μm [82]. There are two important items to consider when comparing non-selective to selective autophagy: how is the process triggered and how is the cargo recognized. The answer to both relates to the adaptor proteins that determine how the autophagosomal machinery is nucleated adjacent to the targeted material. We will describe here very briefly how what we have learned about initiation of autophagosome formation can be understood in the context of selective autophagy. A more detailed discussion would require its own chapter.

In order for cellular material to be recognized for autophagosomal degradation it must exhibit an "eat me signal" [83]. Those signals are divided into ubiquitin-positive and ubiquitin-negative [84, 85]. In the case of ubiquitin negative signals, some pre-existing cellular protein or lipid appears on the surface of the targeted structure having as a characteristic the ability to bind to the autophagic machinery, most frequently to the Atg8-family proteins which decorate the autophagosomal membrane. For the case of mitochondrial autophagy not involving ubiquitination, appearance of a protein or lipid "eat me signal" takes place in response to the depolarization of the mitochondrial membrane upon damage. In the case of ubiquitin-positive signals, the initial "eat me signal" is the specific ubiquitination of the targeted structure in response to damage. This ubiquitin signal is then recognized by a family of adaptor proteins which also bind Atg8-family proteins and in this way mediate the bridging of the targeted cargo to the forming autophagosomal membrane [86].

Although both pathways of ubiquitin-dependent and -independent selective autophagy provide a mechanism for the binding of the cargo to phagophores via the Atg8-family proteins [86], they do not explain how the upstream machinery (involving the ULK and PIK3C3/VPS34 complexes as well as ATG9 vesicles) is recruited. This is especially puzzling because most pathways of selective autophagy lack a signal from MTOR inactivation which in non-selective autophagy is the trigger for the process. It appears that the answer to this conundrum lies within the scaffolding function of the ULK complex as it interacts with the selective autophagy adaptors [87, 88]. Recent work from several laboratories has mapped specific interactions between the adaptors and RB1CC1, a nucleating subunit of the ULK complex. In this sense, the missing signal from MTOR to ULK1 that initiates non-selective autophagy is supplied during selective autophagy by an interaction between adaptors and RB1CC1 which ultimately results in the local activation of the ULK complex driving the rest of the pathway as described before [84].

**Cardinal Articles**

- Dynamics of autophagosome formation marked with GFP-LC3 and GFP-ATG5 were first described by the labs of Yoshimori and Mizushima:

  Kabeya Y, Mizushima N, Ueno T, Yamamoto A, Kirisako T, Noda T, et al. LC3, a mammalian homologue of yeast Apg8p, is localized in autophagosome membranes after processing. EMBO J. 2000;19(21):5720–8.

  Mizushima N, Yamamoto A, Hatano M, Kobayashi Y, Kabeya Y, Suzuki K, et al. Dissection of autophagosome formation using Apg5-deficient mouse embryonic stem cells. J Cell Biol. 2001;152(4):657–68.

- Dynamics of autophagosome formation from the ER involving omegasomes were described for the first time in the author's laboratory.

  Axe EL, Walker SA, Manifava M, Chandra P, Roderick HL, Habermann A, et al. Autophagosome formation from membrane compartments enriched in phosphatidylinositol 3-phosphate and dynamically connected to the endoplasmic reticulum. J Cell Biol. 2008;182(4):685–701.

- Two publication independently established that autophagosomes are connected to the ER via narrow tubules.

  Yla-Anttila P, Vihinen H, Jokitalo E, Eskelinen EL. 3D tomography reveals connections between the phagophore and endoplasmic reticulum. Autophagy. 2009;5(8):1180–5.

  Hayashi-Nishino M, Fujita N, Noda T, Yamaguchi A, Yoshimori T, Yamamoto A. A subdomain of the endoplasmic reticulum forms a cradle for autophagosome formation. Nat Cell Biol. 2009;11(12):1433–7.

**Take Home Messages**
- Autophagosome formation involves a complicated set of morphological and biochemical steps mediated by a set of autophagy proteins.
- The process can be divided into 4 stages, initiation, nucleation, expansion, and sealing off
- Initiation of regulated by two protein kinases, MTOR and ULK1.
- Nucleation involves a lipid kinase, PIK3C3/VPS34, and formation of PtdIns3P-enriched membrane domains that give rise to autophagosomes.
- Elongation involves lipid supply by ATG2 and ATG9, and two ubiquitin-like conjugation systems that covalently attach phosphatidylethanolamine on the C terminus of Atg8-family proteins.
- Sealing off is a fission reaction and involves the ESCRT complex and the Atg8-family proteins.
- Autophagosomes form in association with a cellular compartment which is the ER, with contributions from mitochondria, Golgi and endosomes.

(continued)

- Non-selective and selective autophagy differ at the mechanism of initiation, with selective autophagy depending on nucleation of the early machinery by adaptor proteins.

**Answers to Questions**

1. Phagophores are straight or curved membranes that eventually expand into a spherical shape and become double-membrane autophagosomes (the double membrane occurs because two a membrane of single bilayers fuses to make a vesicle).
2. Formation *de novo* vs from a pre-existing membrane.
3. Christian de Duve, he also discovered the lysosomes for which he received the Nobel Prize. A very nice interview on his work was conducted by Professor Daniel Klionsky [89].
4. having the genes allowed knowing the proteins so that imaging and biochemical tools could be developed to study their function. Best example is the GFP-LC3 marker which was developed after the LC3 coding gene was cloned from its homology to yeast Atg8.
5. Advantage would be that a novel route of lipid synthesis is not required, a disadvantage is that all autophagy events must be connected to a cellular lipid supplying membrane.
6. For (a), cargo marked for selective autophagy by ubiquitination, but not eliminated; for (b) ULK complex-containing precursors; for (c) PtdIns3P- and ULK complex-positive intermediates, such as multilamellar membranes related to ER.
7. A strong MTOR inhibitor would be a good autophagy activator, but it would also cause an inhibition in many other cellular pathways regulated by MTOR, for example protein synthesis, lipid synthesis etc.
8. Open question, perhaps aiming for a PIK3C3/VPS34 complex I-specific activator, or an activator of lipid conjugation. Other more imaginative possibilities would be desirable.
9. The extensive flow of lipids from a donor ER membrane would create an imbalance in lipid content between the two ER bilayers, and perhaps an imbalance in lipid composition. Under these conditions, an activity that "normalises" lipid content and composition at the donor membrane would be desirable.

**Acknowledgments** I want to thank present and past members of my laboratory for their contributions to our understanding of the autophagic pathway, and the Biotechnology and Biological Sciences Research Council for financial support.

## References


1. De Duve C, Wattiaux R. Functions of lysosomes. Annu Rev Physiol. 1966;28:435–92.
2. Novikoff AB, Shin WY. Endoplasmic reticulum and autophagy in rat hepatocytes. Proc Natl Acad Sci USA. 1978;75(10):5039–42.
3. Eskelinen E-L, Reggiori F, Baba M, Kovács AL, Seglen PO. Seeing is believing: the impact of electron microscopy on autophagy research. Autophagy. 2011;7(9):935–56.
4. De Duve C. The lysosome. Sci Am. 1963;208:64–72.
5. Ktistakis NT. In praise of M. Anselmier who first used the term "autophagie" in 1859. Autophagy. 2017;13(12):2015–7.
6. Seglen P. Regulation of autophagic degradation in isolated liver cells. In: Lysosomes: their role in protein breakdown; 1987. p. 369–414.
7. Fengsrud M, Erichsen ES, Berg TO, Raiborg C, Seglen PO. Ultrastructural characterization of the delimiting membranes of isolated autophagosomes and amphisomes by freeze-fracture electron microscopy. Eur J Cell Biol. 2000;79(12):871–82.
8. de Duve C. The lysosome turns fifty. Nat Cell Biol. 2005;7(9):847–9.
9. Deter RL, Baudhuin P, De Duve C. Participation of lysosomes in cellular autophagy induced in rat liver by glucagon. J Cell Biol. 1967;35(2):C11–6.
10. Marzella L, Ahlberg J, Glaumann H. Isolation of autophagic vacuoles from rat liver: morphological and biochemical characterization. J Cell Biol. 1982;93(1):144–54.
11. Øverbye A, Fengsrud M, Seglen PO. Proteomic analysis of membrane-associated proteins from rat liver autophagosomes. Autophagy. 2007;3(4):300–22.
12. Kovacs AL, Reith A, Seglen PO. Accumulation of autophagosomes after inhibition of hepatocytic protein degradation by vinblastine, leupeptin or a lysosomotropic amine. Exp Cell Res. 1982;137(1):191–201.
13. Furuno K, Ishikawa T, Kato K. Appearance of autolysosomes in rat liver after leupeptin treatment. J Biochem. 1982;91(5):1485–94.
14. Seglen PO, Gordon PB. 3-Methyladenine: specific inhibitor of autophagic/lysosomal protein degradation in isolated rat hepatocytes. Proc Natl Acad Sci USA. 1982;79(6):1889–92.
15. Petiot A, Ogier-Denis E, Blommaart EF, Meijer AJ, Codogno P. Distinct classes of phosphatidylinositol 3′-kinases are involved in signaling pathways that control macroautophagy in HT-29 cells. J Biol Chem. 2000;275(2):992–8.
16. Holen I, Gordon PB, Seglen PO. Inhibition of hepatocytic autophagy by okadaic acid and other protein phosphatase inhibitors. Eur J Biochem. 1993;215(1):113–22.
17. Klionsky DJ, Seglen PO. The Norse god of autophagy. Interviewed by Daniel J Klionsky. Autophagy. 2010;6(8):1017–31.
18. Noda T, Ohsumi Y. Tor, a phosphatidylinositol kinase homologue, controls autophagy in yeast. J Biol Chem. 1998;273(7):3963–6.
19. Jung CH, Ro SH, Cao J, Otto NM, Kim DH. mTOR regulation of autophagy. FEBS Lett. 2010;584(7):1287–95.
20. Ohsumi Y. Historical landmarks of autophagy research. Cell Res. 2014;24(1):9–23.
21. Kabeya Y, Mizushima N, Ueno T, Yamamoto A, Kirisako T, Noda T, et al. LC3, a mammalian homologue of yeast Apg8p, is localized in autophagosome membranes after processing. EMBO J. 2000;19(21):5720–8.
22. Mizushima N, Yamamoto A, Hatano M, Kobayashi Y, Kabeya Y, Suzuki K, et al. Dissection of autophagosome formation using Apg5-deficient mouse embryonic stem cells. J Cell Biol. 2001;152(4):657–68.
23. Karanasios E, Ktistakis NT. Live-cell imaging for the assessment of the dynamics of autophagosome formation: focus on early steps. Methods. 2015;75:54–60.

24. Itakura E, Mizushima N. Characterization of autophagosome formation site by a hierarchical analysis of mammalian Atg proteins. Autophagy. 2010;6(6):764–76.

25. Karanasios E, Walker SA, Okkenhaug H, Manifava M, Hummel E, Zimmermann H, et al. Autophagy initiation by ULK complex assembly on ER tubulovesicular regions marked by ATG9 vesicles. Nat Commun. 2016;7:12420.

26. Judith D, Jefferies HBJ, Boeing S, Frith D, Snijders AP, Tooze SA. ATG9A shapes the forming autophagosome through Arfaptin 2 and phosphatidylinositol 4-kinase IIIβ. J Cell Biol. 2019;218(5):1634–52.

27. Axe EL, Walker SA, Manifava M, Chandra P, Roderick HL, Habermann A, et al. Autophagosome formation from membrane compartments enriched in phosphatidylinositol 3-phosphate and dynamically connected to the endoplasmic reticulum. J Cell Biol. 2008;182(4):685–701.

28. Karanasios E, Stapleton E, Walker SA, Manifava M, Ktistakis NT. Live cell imaging of early autophagy events: omegasomes and beyond. J Vis Exp. 2013;77:50484.

29. Melia TJ, Lystad AH, Simonsen A. Autophagosome biogenesis: From membrane growth to closure. J Cell Biol. 2020;219(6):e202002085.

30. Mizushima N. The ATG conjugation systems in autophagy. Curr Opin Cell Biol. 2020;63:1–10.

31. Valverde DP, Yu S, Boggavarapu V, Kumar N, Lees JA, Walz T, et al. ATG2 transports lipids to promote autophagosome biogenesis. J Cell Biol. 2019;218(6):1787–98.

32. van Vliet AR, Chiduza GN, Maslen SL, Pye VE, Joshi D, De Tito S, et al. ATG9A and ATG2A form a heteromeric complex essential for autophagosome formation. Mol Cell. 2022;82(22):4324–39.e8.

33. Orsi A, Razi M, Dooley HC, Robinson D, Weston AE, Collinson LM, et al. Dynamic and transient interactions of Atg9 with autophagosomes, but not membrane integration, are required for autophagy. Mol Biol Cell. 2012;23(10):1860–73.

34. Broadbent DG, Barnaba C, Perez GI, Schmidt JC. Quantitative analysis of autophagy reveals the role of ATG9 and ATG2 in autophagosome formation. J Cell Biol. 2023;222(7):e202210078.

35. Kageyama S, Omori H, Saitoh T, Sone T, Guan JL, Akira S, et al. The LC3 recruitment mechanism is separate from Atg9L1-dependent membrane formation in the autophagic response against Salmonella. Mol Biol Cell. 2011;22(13):2290–300.

36. Yu S, Melia TJ. The coordination of membrane fission and fusion at the end of autophagosome maturation. Curr Opin Cell Biol. 2017;47:92–8.

37. Knorr RL, Lipowsky R, Dimova R. Autophagosome closure requires membrane scission. Autophagy. 2015;11(11):2134–7.

38. Takahashi Y, He H, Tang Z, Hattori T, Liu Y, Young MM, et al. An autophagy assay reveals the ESCRT-III component CHMP2A as a regulator of phagophore closure. Nat Commun. 2018;9(1):2855.

39. Yla-Anttila P, Vihinen H, Jokitalo E, Eskelinen EL. 3D tomography reveals connections between the phagophore and endoplasmic reticulum. Autophagy. 2009;5(8):1180–5.

40. Hayashi-Nishino M, Fujita N, Noda T, Yamaguchi A, Yoshimori T, Yamamoto A. A subdomain of the endoplasmic reticulum forms a cradle for autophagosome formation. Nat Cell Biol. 2009;11(12):1433–7.

41. Dikic I, Elazar Z. Mechanism and medical implications of mammalian autophagy. Nat Rev Mol Cell Biol. 2018;19(6):349–64.

42. Lamb CA, Yoshimori T, Tooze SA. The autophagosome: origins unknown, biogenesis complex. Nat Rev Mol Cell Biol. 2013;14(12):759–74.

43. Bento CF, Renna M, Ghislat G, Puri C, Ashkenazi A, Vicinanza M, et al. Mammalian Autophagy: how does it work? Annu Rev Biochem. 2016;85:685–713.

44. Biazik J, Ylä-Anttila P, Vihinen H, Jokitalo E, Eskelinen E-L. Ultrastructural relationship of the phagophore with surrounding organelles. Autophagy. 2015;11(3):439–51.

45. Ktistakis NT, Tooze SA. Digesting the expanding mechanisms of Autophagy. Trends Cell Biol. 2016;26(8):624–35.

46. Ktistakis NT. ER platforms mediating autophagosome generation. Biochim Biophys Acta Mol Cell Biol Lipids. 2020;1865(1):158433.

47. Zachari M, Gudmundsson SR, Li Z, Manifava M, Cugliandolo F, Shah R, et al. Selective Autophagy of Mitochondria on a ubiquitin-endoplasmic-reticulum platform. Dev Cell. 2020;55(2):251.

48. Mizushima N, Komatsu M. Autophagy: renovation of cells and tissues. Cell. 2011;147(4):728–41.

49. Zachari M, Ganley IG. The mammalian ULK1 complex and autophagy initiation. Essays Biochem. 2017;61(6):585–96.

50. Stanley RE, Ragusa MJ, Hurley JH. The beginning of the end: how scaffolds nucleate autophagosome biogenesis. Trends Cell Biol. 2014;24(1):73–81.

51. Hurley JH, Young LN. Mechanisms of Autophagy Initiation. Annu Rev Biochem. 2017;86:225–44.

52. Shi X, Yokom AL, Wang C, Young LN, Youle RJ, Hurley JH. ULK complex organization in autophagy by a C-shaped FIP200 N-terminal domain dimer. J Cell Biol. 2020;219(7):e201911047.

53. Chen M, Ren X, Cook ASI, Hurley JH. Structure and activation of the human autophagy-initiating ULK1C:PI3KC3-C1 supercomplex. bioRxiv. 2023:2023.06.01.543278.

54. Li Y, Chen Y. AMPK and autophagy. Adv Exp Med Biol. 2019;1206:85–108.

55. Ohashi Y, Tremel S, Williams RL. VPS34 complexes from a structural perspective. J Lipid Res. 2019;60(2):229–41.

56. Schink KO, Raiborg C, Stenmark H. Phosphatidylinositol 3-phosphate, a lipid that regulates membrane dynamics, protein sorting and cell signalling. BioEssays. 2013;35(10):900–12.

57. Matsunaga K, Morita E, Saitoh T, Akira S, Ktistakis NT, Izumi T, et al. Autophagy requires endoplasmic reticulum targeting of the PI3-kinase complex via Atg14L. J Cell Biol. 2010;190(4):511–21.

58. Ge L, Baskaran S, Schekman R, Hurley JH. The protein-vesicle network of autophagy. Curr Opin Cell Biol. 2014;29C:18–24.

59. Ge L, Melville D, Zhang M, Schekman R. The ER-Golgi intermediate compartment is a key membrane source for the LC3 lipidation step of autophagosome biogenesis. elife. 2013;2:e00947.

60. Vergne I, Deretic V. The role of PI3P phosphatases in the regulation of autophagy. FEBS Lett. 2010;584(7):1313–8.

61. Taguchi-Atarashi N, Hamasaki M, Matsunaga K, Omori H, Ktistakis NT, Yoshimori T, et al. Modulation of local PtdIns3P levels by the PI phosphatase MTMR3 regulates constitutive autophagy. Traffic. 2010;11(4):468–78.

62. Vergne I, Roberts E, Elmaoued RA, Tosch V, Delgado MA, Proikas-Cezanne T, et al. Control of autophagy initiation by phosphoinositide 3-phosphatase Jumpy. EMBO J. 2009;28(15):2244–58.

63. Allen EA, Amato C, Fortier TM, Velentzas P, Wood W, Baehrecke EH. A conserved myotubularin-related phosphatase regulates autophagy by maintaining autophagic flux. J Cell Biol. 2020;219(11):e201909073.

64. Karanasios E, Stapleton E, Manifava M, Kaizuka T, Mizushima N, Walker SA, et al. Dynamic association of the ULK1 complex with omegasomes during autophagy induction. J Cell Sci. 2013;126(Pt 22):5224–38.

65. Banerjee C, Mehra D, Song D, Mancebo A, Park JM, Kim DH, et al. ULK1 forms distinct oligomeric states and nanoscopic structures during autophagy initiation. Sci Adv. 2023;9(39):eadh4094.

66. Sou YS, Tanida I, Komatsu M, Ueno T, Kominami E. Phosphatidylserine in addition to phosphatidylethanolamine is an in vitro target of the mammalian Atg8 modifiers, LC3, GABARAP, and GATE-16. J Biol Chem. 2006;281(6):3017–24.

67. Hurley JH, Schulman BA. Atomistic autophagy: the structures of cellular self-digestion. Cell. 2014;157(2):300–11.

68. Noda NN, Inagaki F. Mechanisms of Autophagy. Annu Rev Biophys. 2015;44:101–22.

69. Dooley HC, Razi M, Polson HE, Girardin SE, Wilson MI, Tooze SA. WIPI2 Links LC3 Conjugation with PI3P, Autophagosome Formation, and Pathogen Clearance by Recruiting Atg12-5-16L1. Mol Cell. 2014;55:238–52.

70. Durgan J, Florey O. Many roads lead to CASM: Diverse stimuli of noncanonical autophagy share a unifying molecular mechanism. Sci Adv. 2022;8(43):eabo1274.

71. Tsuboyama K, Koyama-Honda I, Sakamaki Y, Koike M, Morishita H, Mizushima N. The ATG conjugation systems are important for degradation of the inner autophagosomal membrane. Science. 2016;354(6315):1036–41.

72. Nguyen TN, Padman BS, Usher J, Oorschot V, Ramm G, Lazarou M. Atg8 family LC3/ GABARAP proteins are crucial for autophagosome–lysosome fusion but not autophagosome formation during PINK1/Parkin mitophagy and starvation. J Cell Biol. 2016;215(6):857–74.

73. Tang Z, Takahashi Y, Wang HG. ATG2 regulation of phagophore expansion at mitochondria-associated ER membranes. Autophagy. 2019;15(12):2165–6.

74. Matoba K, Kotani T, Tsutsumi A, Tsuji T, Mori T, Noshiro D, et al. Atg9 is a lipid scramblase that mediates autophagosomal membrane expansion. Nat Struct Mol Biol. 2020;27(12):1185–93.

75. Ghanbarpour A, Valverde DP, Melia TJ, Reinisch KM. A model for a partnership of lipid transfer proteins and scramblases in membrane expansion and organelle biogenesis. Proc Natl Acad Sci USA. 2021;118(16):e2101562118.

76. Morita K, Hama Y, Mizushima N. TMEM41B functions with VMP1 in autophagosome formation. Autophagy. 2019;15(5):922–3.

77. Hama Y, Morishita H, Mizushima N. Regulation of ER-derived membrane dynamics by the DedA domain-containing proteins VMP1 and TMEM41B. EMBO Rep. 2022;23(2):e53894.

78. van Vliet AR, Chiduza GN, Maslen SL, Pye VE, Joshi D, De Tito S, et al. ATG9A and ATG2A form a heteromeric complex essential for autophagosome formation. Mol Cell. 2022;82:4324–4339.e8.

79. Takahashi Y, Liang X, Hattori T, Tang Z, He H, Chen H, et al. VPS37A directs ESCRT recruitment for phagophore closure. J Cell Biol. 2019;218(10):3336–54.

80. Vietri M, Radulovic M, Stenmark H. The many functions of ESCRTs. Nat Rev Mol Cell Biol. 2020;21(1):25–42.

81. Javed R, Jain A, Duque T, Hendrix E, Paddar MA, Khan S, et al. Mammalian ATG8 proteins maintain autophagosomal membrane integrity through ESCRTs. EMBO J. 2023;42(14):e112845.

82. Kirkin V, Rogov VV. A Diversity of Selective Autophagy Receptors Determines the Specificity of the Autophagy Pathway. Mol Cell. 2019;76(2):268–85.

83. Randow F, Youle RJ. Self and nonself: how autophagy targets mitochondria and bacteria. Cell Host Microbe. 2014;15(4):403–11.

84. Vargas JNS, Hamasaki M, Kawabata T, Youle RJ, Yoshimori T. The mechanisms and roles of selective autophagy in mammals. Nat Rev Mol Cell Biol. 2023;24(3):167–85.

85. Georgakopoulos ND, Wells G, Campanella M. The pharmacological regulation of cellular mitophagy. Nat Chem Biol. 2017;13(2):136–46.

86. Lamark T, Johansen T. Mechanisms of Selective Autophagy. Annu Rev Cell Dev Biol. 2021;37:143–69.

87. Ravenhill BJ, Boyle KB, von Muhlinen N, Ellison CJ, Masson GR, Otten EG, et al. The cargo receptor NDP52 initiates selective autophagy by recruiting the ULK complex to cytosol-invading bacteria. Mol Cell. 2019;74(2):320–9. e6

88. Vargas JNS, Wang C, Bunker E, Hao L, Maric D, Schiavo G, et al. Spatiotemporal control of ULK1 activation by NDP52 and TBK1 during selective autophagy. Mol Cell. 2019;74(2):347–62. e6

89. Klionsky DJ. Autophagy revisited: a conversation with Christian de Duve. Autophagy. 2008;4(6):740–3.

# Autophagy and Neurodegenerative Diseases


Antonio Daniel Barbosa, Jennifer E. Palmer, Xinyi Li, and David C. Rubinsztein


> **What Will You Learn in This Chapter**
> In this chapter, you will learn about the roles of autophagy in neurons and glia, how mutations in core autophagy genes or in regulators of the autophagic machinery cause neurodevelopmental and neurodegenerative diseases, and how autophagy-enhancing strategies can act as potential treatments for neurodegeneration. You will learn how the post-mitotic nature of neurons and their high metabolic demands and morphology render them particularly vulnerable to impaired autophagic function. You will be able to recognise that autophagy is not only important for neurons but also for glia. You will learn how mutations in autophagic genes impair autophagic function and cause neurodevelopmental and neurodegenerative disorders. Finally, you will learn about strategies to enhance autophagy and their potential applications to neurodegenerative diseases. Overall, you will gain an understating of why neurons are critically vulnerable to impaired autophagy, how disease-associated mutations affect autophagy at different stages, and how it can be exploited in disease treatment by inducing the pathway or by selectively targeting its substrates.


A. D. Barbosa
Department of Medical Genetics, Cambridge Institute for Medical Research (CIMR), Cambridge, UK

J. E. Palmer · X. Li · D. C. Rubinsztein (✉)
Department of Medical Genetics, Cambridge Institute for Medical Research (CIMR), Cambridge, UK

UK Dementia Research Institute, Cambridge Institute for Medical Research (CIMR), Cambridge, UK
e-mail: dcr1000@cam.ac.uk



© The Author(s), under exclusive license to Springer Nature Switzerland AG 2025
B. Loos, D. J. Klionsky (eds.), *Autophagy - From Molecular Mechanisms to Flux Control in Health and Disease*, Learning Materials in Biosciences,
https://doi.org/10.1007/978-3-031-88121-3_6

# 1 Neurons Are Selectively Vulnerable to Impaired Autophagic Function

Neurons are extremely long-lived, highly metabolically active cells that have a complex and elongated morphology. While efficient autophagy is important in all cells for mediating the clearance of substrates, such as aggregation-prone proteins and damaged or surplus organelles, the unique biology of neurons makes them especially dependent on this clearance pathway. Consequently, although most autophagy proteins are ubiquitously expressed, their mutation tends to have the greatest effect on the brain, resulting in neurodevelopmental conditions and neurodegeneration.

Ubiquitous loss of core autophagic genes causes neonatal death in mice due to a suckling defect and metabolic insufficiency [1]. Therefore, the use of conditional knockouts of key autophagy genes either in the nervous system or in adult mice is essential to address the effects of autophagy defects on the nervous system. Intriguingly, the lethality of *atg5* knockout mice can be rescued by expressing ATG5 only in neurons [2], highlighting the unique importance of neuronal autophagy. Mice with neuron-specific knockouts of the key autophagy genes *Atg5* or *Atg7* have early-onset neurodegeneration [3, 4]. These examples show that loss of neuronal autophagy is sufficient to cause neurodegeneration. Indeed, conditional loss of autophagy throughout all tissues in adult mice results in death in 2–3 months, with the most common cause of death being neurodegeneration [5]. This suggests that neurons are particularly vulnerable to autophagic loss.

The main reasons for the vulnerability of neurons to impaired autophagic function are: (a) the post-mitotic nature of neurons that prevents cell division-mediated dilution of autophagic substrates; (b) the high metabolic demand of neurons that necessitates a rapid turnover of mitochondria; (c) and the highly elongated and polarised morphology that exacerbates deficits in the trafficking of autophagosomes. These will be explored in the following paragraphs.

## 1.1 Neurons Are Post-mitotic and Cannot Dilute Autophagic Substrates Through Cell Division

Autophagy is the main degradative route for aggregation-prone proteins associated with neurodegenerative diseases. These include, for example, hyperphosphorylated MAPT/tau in Alzheimer disease (AD), frontotemporal dementia and other tauopathies; TARDBP/TDP-43 in other forms of frontotemporal dementia and in amyotrophic lateral sclerosis; SNCA/α-synuclein that forms Lewy bodies in PD; or polyglutamine expansions in several neurodegenerative disorders, including mutant HTT (huntingtin) in Huntington disease (HD) [6–9]. Almost all of the neurons in the adult brain are post-mitotic and cannot divide to dilute any accumulated autophagic substrates [10]. Over the decade- or even century-long lifespan of a neuron, this can lead to a detrimental and eventually overwhelming accumulation of toxic aggregation-prone proteins. Accumulation of these toxic aggregation-prone proteins is a shared tenet of all neurodegenerative diseases [11]. This suggests that impaired clearance by autophagy is a core hallmark of the neurodegeneration process.

## 1.2 The Highly Elongated Morphology of Neurons Exacerbates Cellular Trafficking Defects

The highly elongated and polarized morphology of neurons exacerbates minor defects in cellular trafficking, impairing autophagic function. Autophagosome biogenesis appears to be concentrated in the distal ends of axons and dendrites [12–15]. Autophagosomes must then travel back towards the cell body, where they can fuse with acidic lysosomes to degrade the autophagic cargo [14, 16]. Due to the long distances that autophagosomes must travel down the axons and dendrites, even a minor impairment in this trafficking can lead to an accumulation of autophagosomes in the axons and dendrites [17–21]. This results in a block in autophagic flux, preventing the degradation of autophagic substrates and exacerbating neurodegeneration [18, 20, 21].

## 1.3 Neurons Require Efficient Mitophagy to Enable Their High Metabolic Activity

Neurons require an efficient and rapid turnover of mitochondria by mitophagy to facilitate their high metabolic activity [22, 23]. Indeed, much of the autophagic cargo of neurons appears to be mitochondria [24, 25]. If autophagy is impaired, neurons cannot efficiently clear these damaged mitochondria, which reduces cellular energy production and increases the generation of reactive oxygen species, leading to impaired neuronal function and even cell death [22, 26, 27]. Furthermore, there is a strong genetic association between mitophagy and neurodegenerative diseases, as will be explored later in this chapter.

> **Question**
> 1. Why are neurons selectively vulnerable to impaired autophagy?

## 2 The Functions of Autophagy in Neurons and Glia

In addition to maintaining protein and organellar homeostasis, autophagy is also required for a multitude of other functions in the brain, including synaptic function and memory formation, neurodevelopmental pathways, and the maintenance of astrocyte, microglia, and oligodendrocyte functions (Fig. 1).

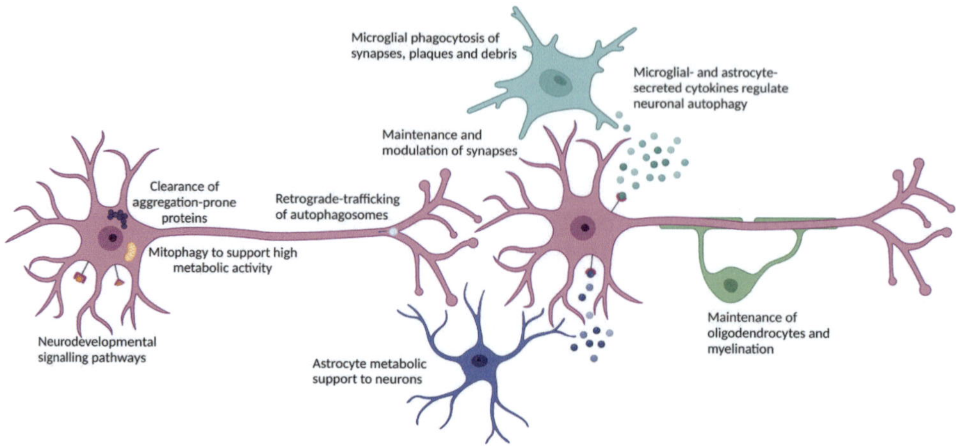


**Fig. 1** Roles of autophagy in neurons and glia

## 2.1 Synaptic Function, Learning and Memory Require Autophagy

Synapses are continuously remodelled over time in an autophagy-dependent manner by internalising and degrading neurotransmitter receptors, synaptic scaffolding proteins, entire synaptic spines, and synaptic vesicles [28–32]. Additionally, autophagy and specifically reticulophagy (autophagic degradation of ER) are important regulators of excitatory neurotransmission, by regulating the morphology of the ER and calcium release [33]. Furthermore, autophagy is required for the processes of long-term synaptic potentiation and long-term synaptic depression, which modulate the strength of a synapse over time [29, 34, 35]. Given the important roles for autophagy in maintaining and regulating synaptic function, it is unsurprising that autophagy is also required for learning and memory. Pharmacological inhibition of autophagy impairs the ability of mice to undertake learning and memory tasks, and induction of autophagy improves the performance of old mice on such activities [35]. It is not yet clear whether the impaired learning and memory is due to a direct effect of autophagy on the molecular processes required for memory formation, or whether this effect is indirect due to impaired neuronal health.

## 2.2 Neurodevelopmental Pathways Require Autophagy

Autophagy regulates neurogenesis, cell death, synaptic processes, and signalling pathways, by regulating the degradation of metabolites, signalling ligands and their receptors [36]. The expression of key autophagy genes is increased during neurogenesis, and loss of autophagic genes results in decreased neural progenitor cell differentiation, impaired neurogenesis, and decreased astrocyte differentiation [37–39]. Impaired autophagy also contributes to abnormal neurodevelopment due to altered synaptic remodelling, as described earlier. Given that microglial autophagic loss impairs synaptic phagocytosis [40], glial autophagic impairment may also contribute to abnormal neuronal circuit development and synaptic spine density [41, 42]. Furthermore, mutations in many autophagy genes are associated with neurodevelopmental diseases, as will be discussed later in this chapter.

## 2.3	Astrocyte Autophagy Modulates Neuronal Autophagy and Function

While most of the autophagy and neurodegeneration research has focused on neurons, glial autophagy is emerging as a critical component in neurodegeneration. Glia, which include astrocytes, microglia, oligodendrocytes and Schwann cells, regulate neuronal function in a non-cell autonomous manner [43]. Consequently, glial autophagy does not just regulate the health and function of the glia themselves but also affects the function and survival of neurons [44, 45]. Astrocytes provide metabolic support to neurons, regulate synapses by uptaking excess neurotransmitters or ions, modulate brain blood flow, can phagocytose neuronal debris, secrete cytokines to regulate neuronal function, and protect neurons during injury [46, 47]. Given the importance of autophagy for cellular health, and the non-cell-autonomous regulation between neurons and astrocytes, it is unsurprising that impaired astrocyte autophagy exacerbates neurodegeneration [45, 48]. Impaired autophagy in astrocytes also leads to a pro-inflammatory transition, increasing the secretion of cytokines that contribute to neurodegeneration [49]. Recovery of astrocyte mitochondrial dynamics after inflammation also requires autophagy [50]. Damaged neuronal mitochondria can be transferred to astrocytes to be degraded by mitophagy (autophagy-dependent mitochondrial degradation) [51, 52], which suggests that astrocytic autophagy might help to supplement neuronal autophagic function.

## 2.4	Microglial Autophagy Modulates Neuronal Health and Survival

Microglia are macrophage-like cells that phagocytose dead cells and other debris and secrete cytokines that regulate both the immune response and neuronal health [53, 54]. Under normal conditions, microglial autophagy is required for efficient synaptic pruning [40, 41] and for the efficient clearance of extracellular aggregates and cellular debris [55, 56]. A recent study also suggested that autophagy is required for microglia to transition from a homeostatic state into disease-associated microglia, a protective subtype of microglia, and that loss of autophagy in microglia impairs their ability to proliferate and to engage amyloid plaques in the context of AD [57]. Importantly, loss of autophagy specifically in microglia exacerbates neurodegeneration in multiple mouse models [57, 58]. Microglial-specific deletion of *Atg7* [57, 59] or *Atg5* [58] in mice results in a proinflammatory phenotype and increased neurodegeneration.

Microglia and astrocytes also regulate neuronal autophagy through the secretion of cytokines and other signalling molecules that bind to receptors on neurons to regulate neuronal autophagic signalling. For example, activated microglia secrete multiple cytokines that inhibit neuronal autophagy, and blocking this non-cell-autonomous inhibition of autophagy through pharmacological or genetic manipulation of the G-protein-coupled-receptor, CCR5, prevented the reduction in neuronal autophagy and ameliorated neurodegeneration in mouse models of tauopathy and HD [44, 60]. As microglia and astrocytes become increasingly activated during the ageing or disease process, it is possible that such cytokine-mediated, non-cell-autonomous inhibition of autophagy may further impair neuronal autophagic clearance.

## 2.5 Oligodendrocyte and Schwann Cell Autophagy Is Required for Myelination

Oligodendrocytes and Schwann cells insulate neuronal axons by producing a myelin sheath that facilitates the rapid conductance of electrical potentials down the axon [61]. Autophagy is required for the clearance of cytoplasmic organelles from oligodendrocytes to form the myelin sheath [62]. Mice with an oligodendrocyte-specific knockout of *Atg5* had impaired survival and maturation of oligodendrocytes, resulting in abnormal myelination [63]. It is currently unknown whether impairment of oligodendrocyte autophagy contributes to the demyelination that is often observed in neurodegenerative diseases.

> **Question**
> 2. What are the roles of autophagy in neurons and glia?

## 3 Defects in Autophagy Are Associated With Neurodevelopmental and Neurodegenerative Diseases

As we have seen above, the removal of long-lived, aggregate-prone proteins and damaged organelles is critical for the health and survival of neurons. It is therefore unsurprising that defects in autophagy are intrinsically linked with neurodevelopmental and neurodegenerative diseases. It is worth stressing that mutations in the autophagy core machinery are extremely rare, most likely due to the role of autophagy during development, with loss of autophagy proteins in mice being embryonic lethal or leading to death shortly after birth [64]. Yet, autophagy may be impaired by a wide range of proteins that control the expression, stability, interaction, trafficking, or recruitment of the core autophagy machinery, the maturation of the autophagosome or lysosomal homeostasis. Mutations in these proteins also lead to neurodevelopmental and neurodegenerative diseases by affecting autophagy at all stages of the pathway (Fig. 2), as we will see below.

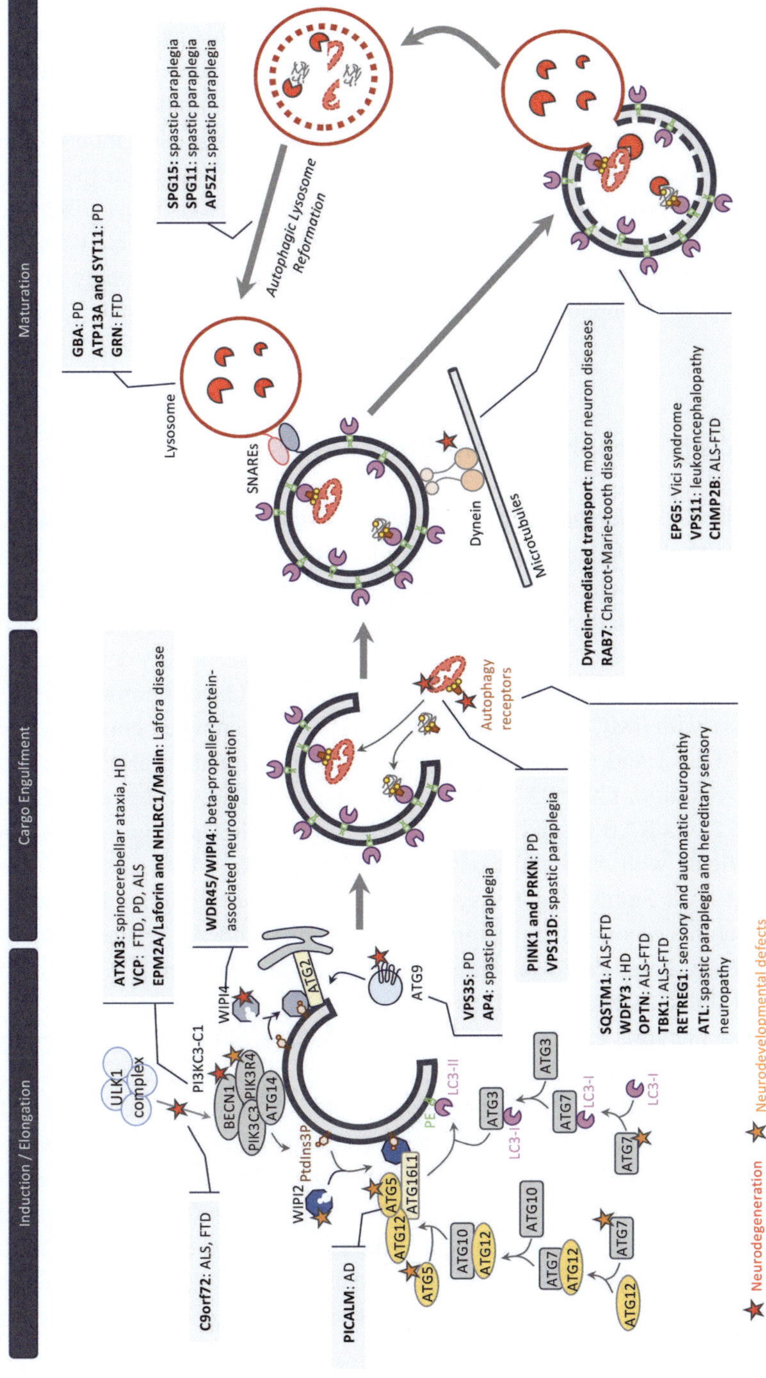


**Fig. 2** Impairment of autophagy at multiple stages cause neurodevelopmental and neurodegenerative diseases

The accumulation of aggregated proteins in pathological inclusions is a hallmark of several neurodegenerative diseases. The effects of these proteins in neurodegeneration often results from a toxic gain-of-function associated with this aggregation-propensity, which interferes with cellular processes, including autophagy, as described below, rather than a significant disruption of their normal biological function. Indeed, inducible depletion of HTT (the protein mutated in Huntington's disease) in the adult mouse brain does not induce neurodegeneration [65]. Selective degradation of these aggregate-prone proteins by aggrephagy prevents more severe outcomes in neurodegenerative diseases.

Here we will review how human mutations in genes regulating autophagy cause neurodegenerative or neurological diseases. Note, however, that the disease may also involve loss of non-autophagic functions of such proteins, particularly if they are not core autophagic machinery.

## 3.1    Defects in Autophagosome Biogenesis

Activation of the ULK1 complex initiates autophagosome biogenesis by promoting the translocation of the class III phosphatidylinositol 3-kinase complex I (PtdIns3K-C1) to the nascent phagophore, where it generates phosphatidylinositol-3-phosphate (PtdIns3P). C9orf72, a protein mutated in amyotrophic lateral sclerosis (ALS), mediates RAB1A-dependent targeting of the ULK1 complex to the phagophore, and its deficiency reduces autophagy in several cell types including in patient-derived induced neurons [66]. C9orf72 forms a complex with SMCR8, WDR41 and ATG101. This complex controls expression and activity of ULK1 and participates both in autophagy initiation and autophagosome maturation. In addition, C9orf72 interacts with the autophagy receptor SQSTM1/p62 (sequestosome 1) via RAB8A and RAB39B [67–70]. SMCR8 is downregulated in brain tissue from patients with ALS and frontotemporal dementia (FTD), and it is deleted in several patients with Smith-Magenis syndrome, a developmental disorder that leads to intellectual disability among other features [71]. Mutations in RAB39B cause intellectual disability and early-onset Parkinson disease (PD) [72]. Furthermore, C9orf72 deficiency also impairs MTORC1 signalling, which regulates ULK1 activity [73, 74].

The missense mutation L1224R in PIK3R4 (also known as VPS15), a core component of PtdIns3K-C1, is associated with a neurodevelopmental disease presenting as intellectual impairment, severe cortical and optic nerve atrophy, ataxia, localized cortical dysplasia, spasticity, psychomotor delay, muscle wasting, pseudobulbar palsy, late-onset epilepsy and mild hearing deficit. This mutation is a substitution in a highly conserved residue within the fourth strand of the WD40 domain (important for folding and stability) of PIK3R4, which may explain the impairment in the assembly of the PtdIns3K complex and the accumulation of autophagy substrates [75]. Because PIK3R4 is a core component of PtdIns3K complexes, its defects may contribute to neurodevelopmental diseases by also affecting other stages of autophagy.

BECN1 (beclin 1) degradation by the proteasome is prevented by the ATXN3 (ataxin 3) deubiquitinase. The polyQ domain of wild-type ATXN3 mediates its interaction with BECN1. Expansion of the polyQ domain of ATXN3 causes spinocerebellar ataxia type 3 (Machado-Joseph disease) and decreases its deubiquitinase activity toward BECN1. Either mutant ATXN3 and other disease proteins with polyQ domains like mutant HTT can compete with wild-type ATXN3, increasing BECN1 ubiquitination and degradation, which impairs autophagy [76]. ATXN3-mediated stabilization of BECN1 and PtdIns3K-C1 assembly also requires VCP/p97 (valosin containing protein) [77], which is also required for autophagosome maturation [78]. A loss-of-function VCP mutation causes dementia with MAPT/tau accumulation [79].

The four members of the WIPI/WDR family—WIPI1 (WD repeat domain, phosphoinositide interacting 1), WIPI2, WDR45B/WIPI3 (WD repeat domain 45B), and WDR45/WIPI4—bind to PtdIns3P generated by the PtdIns3K-C1 complex to recruit the downstream autophagic machinery. WIPI1 and WIPI2 recruit the ATG12–ATG5-ATG16L1 complex to promote LC3 lipidation. Missense variants in WIPI1 may contribute to the lethal neural tube defect anencephaly, and mutations in WIPI2 are associated with intellectual developmental disorders of variable severity [80]. These WIPI2 mutations were identified in highly conserved residues: R242 within the FRRG motif, which is essential for WIPI2 binding to phosphoinositides, and V249 is located near this lipid binding region; V184 is in a beta-sheet within blade 4, which may compromise folding, membrane association and protein-protein interactions. Interestingly, while the mutation V184 failed to rescue LC3 defects of *WIPI2* KO cells, the mutation in V242 (V242W) increased LC3 lipidation [81]. Conversely, the V249M mutation, which (like V242W) may also affect PtdIns3P binding, reduces the interaction with ATG16L1, and the autophagic flux [82]. In contrast to V249M that reduces the recruitment of WIPI2 to the site of phagophore formation, the V242W mutation may enhance the binding of WIPI2 to PtdIns3P and dysregulate autophagy by compromising the clearance of PtdIns3P [83], thus preventing the progression of the pathway. Therefore, further characterization is needed to assess whether the severity of the diseases associated with these mutations is due to defects in phosphoinositide binding and/or recruitment of the ATG12–ATG5-ATG16L1 complex. Interestingly, the dynamics of WIPI2 phosphorylation is impaired during aging in mice, contributing to age-dependent decline of autophagy due to an accumulation of stalled autophagosomes [84, 85]. PtdIns3K-C1 activity is enhanced by K63-linked polyubiquitination of its components mediated by a complex comprised of the glucan phosphatase EPM2A/Laforin and the RING-type E3-ubiquitin ligase NHLRC1/Malin [86]. Mutations in the genes encoding for EPM2A/Laforin or NHLRC1/Malin are associated with Lafora disease, a rare form of progressive myoclonus epilepsy and neurodegeneration. Therefore, autophagy impairment in Lafora disease may be a consequence of decreased PtdIns3K-C1 polyubiquitination, but also due to upregulation of MTORC1 signalling [87].

WDR45B/WIPI3 and WDR45/WIPI4 interact with ATG2 to, together with ATG9A, provide lipids for phagophore formation and elongation. Loss-of-function or missense variants of WDR45B/WIPI3 are associated with brain malformations [88, 89]. Mutations in WDR45/WIPI4 cause X-linked dominant subtype of neurodegeneration with brain iron accumulation known as beta-propeller-protein-associated neurodegeneration (BPAN). The disease affects mostly females, but germline variants in males can be viable, and result in more severe phenotypes. Clinical features include early-onset global developmental delay, and progressive parkinsonism, dementia, and dystonia during early adulthood. Reduced WDR45/WIPI4 levels in cells expressing these mutants is associated with lower autophagic activity and an accumulation of early autophagic structures. Because WDR45/WIPI4 deficiency only leads to partial defects in autophagy, it is likely that other WIPI/WDR proteins may compensate for the loss of WDR45/WIPI4 [90–94]. Indeed, both *wdr45b* knockout and *wdr45* knockout mice exhibit learning and memory defects, swollen axons, and accumulation of SQSTM1- and ubiquitin-positive aggregates in the brain, but defects in autophagy are more severe in *wdr45b wdr45* double-knockout mice [95, 96]. Accumulation of ferric iron and ferritin in BPAN is a consequence of downregulation of ferritinophagy (degradation of ferritin by autophagy). In cells from BPAN patients, compromised ferritinophagy may be a consequence of decreased levels of the cargo receptor NCOA4 due to impaired interaction with WIPI4 variants [94]. However, NCOA4 levels did not change in *WDR45* knockout SH-SY5Y cells showing defects in ferritinophagy [97].

The E1-like enzyme ATG7 activates (a) the C-terminal glycine of ATG12 for its conjugation with ATG5 to form the ATG12–ATG5-ATG16L1 complex, and (b) the members of the ATG8 family for their lipidation with phosphatidylethanolamine. Deleterious recessive ATG7 mutations were identified in five unrelated families with neurodevelopmental disorders and other clinical features. The ATG7 missense mutations identified likely interfere with its homodimerization, and a loss-of-function mutation leads to almost complete absence of lipidated LC3. Interestingly, two of the patients approached population life expectancy despite defects in autophagic flux [98]. Similarly, a homozygous missense ATG5 mutation in a conserved amino acid (E122D) causes congenital ataxia, mental retardation, and developmental delay. These patients exhibit reduced ATG12–ATG5 conjugation, and decreased autophagic flux [99]. PICALM/CALM (phosphatidylinositol binding clathrin assembly protein) is abnormally cleaved in AD. PICALM modulates autophagy through its regulation of endocytosis by affecting the formation of ATG12–ATG5-ATG16L1 vesicles to provide plasma membrane lipids to autophagosome precursor structures; and regulation of VAMP2 and VAMP3 endocytosis, required for homotypic fusion of ATG16L1 vesicles and heterotypic ATG16L1-ATG9A fusion. PICALM also regulates degradation of autophagosomes [100].

The expansion of the phagophore membrane requires small vesicles carrying the phospholipid scramblase ATG9. Therefore, ATG9 deficiency compromises autophagosome formation. For example, mutations in the central pore of ATG9A produces smaller autophagosomes [101]. Similarly, mutations that compromise ATG9A trafficking may also affect autophagy. For example, the D620N mutation in VPS35, a component of the retromer complex, impairs the ATG9 trafficking to phagophores and causes autosomal-dominant PD [102].

Loss-of-function mutations in any of the subunits of AP-4, an adaptor complex required for the incorporation of transmembrane cargo proteins into vesicles, cause a neurological disorder with early-onset spastic paraplegia and intellectual disability. At the trans-Golgi network, AP-4 packages ATG9A into vesicles for transport to the cell periphery, where they act in autophagosome biogenesis. Retention of ATG9A at the trans-Golgi network due to AP-4 deficiency leads to defects in the autophagic flux [103, 104].

Gap junction/connexin proteins (GJ) are multispan transmembrane proteins that form plasma membrane gap junctions. Mutations in GJB1/Cx32, cause an X-linked form of Charcot-Marie-Tooth disease, a disorder of the peripheral nervous system. GJ proteins inhibit autophagy through interaction with the machinery involved in the early steps of autophagosome biogenesis at the plasma membrane. Recruitment of ATG14 to the GJ-ATG complex and internalization with ATG9 allows their degradation through autophagy, which relieves the inhibition by GJ proteins [105].

## 3.2 Defects in Cargo Recruitment

Cargo recognition and incorporation into autophagosomes is a critical event during selective autophagy to ensure that damaged organelles and protein aggregates are selectively targeted for degradation. Failure in this step may lead to amplification of the defects underlying neurodegenerative diseases due to the inability of neurons to dispose of damaged biomolecules and organelles through cell division.

Selective degradation of dysfunctional mitochondria—mitophagy—is critical to avoid the generation of reactive oxygen species that damage biomolecules, prevent the depletion of ATP generation needed to sustain the high metabolic activity of neurons, or the release of proapoptotic factors. One of the pathways of stress-induced mitophagy involves the E3 ligase PRKN/Parkin and the serine-threonine kinase PINK1. Under basal conditions, PINK1 is imported into mitochondria and is constitutively cleaved by PARL (presenilin associated rhomboid like) in the inner mitochondrial membrane. The resulting fragment is subsequently degraded in the cytosol by the proteasome. Upon mitochondrial depolarization, PINK1 accumulates on the outer mitochondrial membrane due to compromised import [106–108]. PINK1-mediated phosphorylation of pre-existing ubiquitin on outer mitochondrial membrane proteins recruits PRKN, which is then also phosphorylated by PINK1 in its ubiquitin-like domain [109–111]. Active PRKN ubiquitinates several outer mitochondrial membrane proteins, triggering the recruitment of cargo receptors (such as OPTN [optineurin], CALCOCO2/NDP52, TAX1BP1 and SQSTM1) that, in turn, bring the autophagic machinery to mitochondria to initiate mitophagy [112, 113]. Mutations in PINK1 and PRKN cause autosomal-recessive early-onset PD [114–118], but Lewy bodies are not seen in most cases compared to the sporadic forms of the disease [119]. Consistent with the role of PINK1 and PRKN in mitophagy, patient-derived fibroblasts show aberrant mitochondrial morphology and function [120–123]. In contrast, PINK1-PRKN translocation onto mitochondria is less noticeable in cultured neurons and mouse brain when compared with studies in cell lines [124, 125]. These differences may result from the variability in the intracellular environment and the mitochondrial stresses in different neuronal compartments [126]. It is also possible that PINK1-PRKN mitophagy may be specifically more relevant in distal axons [127].

Mutations in VPS13D cause ataxia/spastic paraplegia and mitochondrial pathology [128, 129]. VPS13 contains a ubiquitin binding domain that is required for clearance of mitochondria through a PINK1-dependent and PRKN-independent mitophagy pathway [130], and participates in phagophore elongation in *Drosophila* [131].

SQSTM1 is a multifunctional scaffold protein containing multiple domains that regulates various processes, including acting as receptor for selective autophagy. Among its domains required for autophagy, the Phox1 and Bem1 (PB1) domain facilitates its interaction with binding partners, namely SQSTM1 for self-oligomerization and another autophagy receptor, NBR1; the LC3-interacting region (LIR) domain interacts with LC3 on the phagophore; and the C-terminal ubiquitin-associated (UBA) domain interacts with ubiquitin [132, 133]. Recognition of polyubiquitinated proteins by the UBA domains of SQSTM1 or NBR1 leads to homo- or hetero-oligomerization (between SQSTM1 and NBR1) and sequestration of polyubiquitinated proteins in condensates, which are then incorporated in phagophores through interaction with LC3 via the LIR domain [134, 135]. SQSTM1 mutations in the UBA domain or in its proximity are associated with FTD and ALS, and likely affect binding to ubiquitin, as shown for the P392L mutation [136–138]. Moreover, loss-of-function variants in SQSTM1 leading to defects in autophagosome formation cause childhood-onset neurodegeneration with ataxia, dystonia and gaze palsy [139, 140]. Furthermore, the ALS-associated mutation L341V in the LIR domain reduces the recognition of LC3 by SQSTM1 and reduces its recruitment of ubiquitinated substrates to phagophores [141].

SQSTM1 also recruits the adaptor protein WDFY3 to cytoplasmic SQSTM1 bodies, and both are required for the formation and degradation of cytoplasmic ubiquitin-positive inclusions. Mechanistically, WDFY3 exits the nucleus in response to the presence of aggregating proteins and, in the cytosol, binds both to PtdIns3P membranes through its FYVE domain and to ATG12–ATG5 through its WD-40 domain. By bringing together SQSTM1, ATG12–ATG5-ATG16L1 and LC3, WDFY3 promotes the packaging of aggregates into autophagosomes [142, 143]. Several mutations in WDFY3 cause a monogenic, autosomal dominant neurodevelopmental disorder with neurodevelopmental delay and intellectual disability. Interestingly, while missense variants in the PH domain cause microcephaly (a congenital neurodevelopmental disorder of reduced head circumference and brain volume), putative haploinsufficiency leads to large head circumference. These outcomes seem to be associated with the role of WDFY3 in the autophagic attenuation of WNT signalling—that plays important roles in brain development—by the removal of DVL3 aggregates. Indeed, whereas the missense variant R2637W in a conserved residue of the PH domain leads to upregulation of WNT signalling, *Wdfy3*$^{+/lacZ}$ mice (presenting with megalencephaly) display downregulation of the canonical WNT pathway. The variant R2823W, causing macrocephaly, destabilizes the BEACH domain. It is, therefore, possible that destabilization of the PH or BEACH domains is specifically associated with micro- or macrocephaly, respectively, through modulation of WNT signalling [144, 145]. Moreover, WDFY3 is also required for mitophagy [146]; and heterozygous depletion of WDFY3 accelerates age of onset and progression of HD [147].

The autophagic receptor OPTN contains several domains, including a LIR and a ubiquitin-like domain [148, 149]. OPTN participates in selective autophagy of pathogens, protein aggregates, and damaged mitochondria; and it is highly expressed in brain tissue [113, 148, 150–152]. Mutations in OPTN cause primary open-angle glaucoma—an eye disease that narrows the visual field due to optic neuropathy [153], ALS [150, 154], and FTD [155, 156]. The ALS homozygous nonsense mutation Q398X causes a premature stop that leads to the deletion of coiled coil 2 domain that mediates binding to ubiquitin, HTT, MYO6 (myosin VI), and the ubiquitinated receptor-interacting protein. The ALS missense mutation E478G, in a highly conserved residue, occurs within the DFxxER motif in a ubiquitin-binding domain, displays a cytoplasmic distribution that differs from the wildtype, and impairs mitophagy [150, 157]. Moreover, mutations associated with FTD significantly reduced the levels of OPTN in the cerebellum [156]. Mutations in the ubiquitin-binding domain of OPTN, including the ALS-associated mutation E478G, and in the LIR motif fail to mediate the degradation of mutant HTT inclusion bodies through K63-linked polyubiquitin-mediated autophagy. Indeed, OPTN mutants in the ubiquitin-binding domain impair autophagosome maturation by preventing the recruitment of MYO6, which is required for trafficking of autophagosomes to the lysosome [158–160]. In addition, OPTN facilitates the recruitment of ATG12–ATG5-ATG16L1 complex to WIPI2-positive phagophores, a process that is defective in the E478G mutant [161]. OPTN-induced autophagosome formation also requires interaction with RAB1A [162].

TBK1 (TANK binding kinase 1) phosphorylates the autophagy receptors required for mitophagy in the ubiquitin- and LC3-binding domains of OPTN and SQSTM1, and in the SKICH domains of CALCOCO2 and TAX1BP1. TBK1 activity is required for both recruitment of CALCOCO2 and OPTN onto mitochondria, and to increase their affinity to LC3 and ubiquitin chains [112, 163]. TBK1 may also regulate autophagy initiation through phosphorylation of STX17 (syntaxin 17) to mediate the assembly of the ULK1 complex [164]. Mutations in TBK1 cause ALS and FTD [165–167]. Patients with ALS-FTD mutations of TBK1 have TARDBP- and SQSTM1-positive inclusions in various brain regions [168, 169], suggesting an impairment of autophagy in these diseases. Several of these mutations cause protein truncation and loss of the CCD2 domain, which is required for interaction with OPTN, and a decrease in mRNA and protein levels [156, 169]. Consistent with its roles in regulating autophagy receptors, an ALS-associated TBK1 mutant that does not associate with OPTN shows impaired mitophagy [163, 170].

Mutations in RETREG1/FAM134B, a member of the reticulon-homology-domain-containing RETREG/FAM134 protein family, cause hereditary sensory and automatic neuropathy [171], a disease characterized by impaired nociception and autonomic dysfunction. RETREG1/FAM134B and other members of the family act as receptors for selective degradation of the ER (reticulophagy). The reticulon domain of RETREG1/

FAM134B mediates membrane bending to promote membrane remodelling and scission, whereas the LIR motif allows interaction with LC3. Truncated RETREG1/FAM134B mutants, associated with the disease, lack the C-terminal LIR motif and the ability to bind LC3, hindering the targeting of ER fragments for degradation by autophagy. Analysis of *retreg1*$^{-/-}$ mice suggest that defects are restricted to peripheral sensory nerves in old animals. Therefore, impaired proteostasis due to defects in RETREG1/FAM134B and aged-induced autophagy defects may contribute to progressive neurodegeneration [172, 173].

ATL (atlastin) proteins are ER-resident GTPases that have redundant roles in reticulophagy, act downstream of the RETREG1/FAM134B receptor, and in the recruitment of ULK1 complex to the ER [174, 175]. ATL1 and ATL3 specifically binds to GABARAP, and this interaction is required for its role in reticulophagy [176]. Mutations in ATL1 cause spastic paraplegia and hereditary sensory neuropathy [177–184], and the latter is also caused by mutations in ATL3 [185–187]. Notably, mutations that impair binding to GABARAP of ATL1 or ATL3, and that cause spastic paraplegia or sensory neuropathy, respectively, impair reticulophagy [176].

## 3.3 Defects in Autophagosome Maturation

Degradation of sequestered cargo requires the fusion of autophagosomes with endolysosomal vesicles to generate degradative autolysosomes. This maturation process involves tether proteins, membrane fusion driven by membrane-anchored SNARE proteins, phosphoinositides and RAB proteins. Mutations in the machinery required for autophagosome maturation are also associated with neurodegenerative diseases.

Dynein-driven transport of neuronal autophagosomes from presynaptic sites and axon terminal to the soma is required for their fusion with lysosomes. Mutations that impair dynein-mediated transport cause motor neuron disease [188], compromise autophagosome maturation and, consequently, the clearance of aggregate-prone proteins [20]. The small GTPase RAB7 participates in the bidirectional transport of autophagosomes and lysosomes along microtubules [189–191], and releases UVRAG from RUBCN/rubicon sequestration [192], thus controlling autophagosome maturation. Mutations in RAB7 cause Charcot-Marie-tooth disease, a disorder of the peripheral nervous system [193–195].

GTP-bound RAB7 together with late endosomal/lysosomal R-SNARE VAMP7/8 recruit EPG5 to late endosomes. As EPG5 also binds to LC3 and assembled STX17-SNAP29 Qabc-SNARE complexes on autophagosomes, it acts as a tether to promote the fusion of autophagosomes with late endosomes/lysosomes. Loss of EPG5 leads to abnormal fusion of autophagosomes with endocytic vesicles, compromising autophagosome maturation. Mutations in EPG5 cause the rare neurodevelopmental disorder Vici syndrome [196–198]. The HOPS complex is required for the fusion of autophagosomes with lysosomes [199]. Mutations in the HOPS complex subunit VPS11 cause genetic leukoencephalopathy, a disorder that affects the central nervous system, and leads to increased turnover of VPS11 that impairs autophagosome maturation [200].

The formation of amphisomes by autophagosome-endosome fusion is compromised by pathogenic missense ALS2 mutations, associated with ALS, due to loss of its function as guanine exchange factor for RAB5, which regulates endosome fusion and trafficking [201]. Interestingly, RAB5 also acts in a complex with BECN1 and PtdInsK3 during autophagosome biogenesis [202]. Similarly, mutations in UBQLN2 (ubiquilin 2) impair lysosome acidification (through regulation of V-ATPase function) and autophagosome maturation, MTOR signalling and expression of autophagic proteins [203, 204], and cause ALS [205].

Autophagosome maturation also requires the ESCRT machinery—a multisubunit membrane remodelling complex. Mutations in ESCRT-III subunit CHMP2B are associated with FTD and ALS and impair the formation of autolysosomes, which indicates that functional multivesicular bodies are required for autophagic degradation [206–209]. Mutations in SNX14 cause spinocerebellar ataxia and lead to enlarged lysosomes containing an accumulation of unesterified cholesterol. Effects in autophagy, however, are less clear, with data suggesting that defects in SNX14 are responsible for decreased clearance of autophagosomes or impact autophagy induction. A role in autophagosome maturation could be due to the association of SNX14 with PtdIns(3,5)-bisphosphate on the lysosome [210, 211].

Mutations in lysosomal GBA/glucocerebrosidase are the most common genetic risk factor for PD. GBA deficiency may affect autophagy, and the clearance of SNCA aggregates, by causing lysosomal dysfunction and, consequently, compromised autophagosome-lysosome fusion [212, 213]. Indeed, GBA deficiency leads to hyperactivation of MTORC1, which inhibits both autophagy and lysosome biogenesis through TFEB, due to the accumulation of glucosylsphingosine [214]. In agreement, removal of dysfunctional mitochondria is impaired by mitophagy defects triggered by heterozygous mutations in GBA [215]. Interestingly, GBA activity is also regulated by GRN/progranulin [216–218], a secreted protein that is trafficked to the lysosome upon cleavage in the extracellular space, and that plays an essential role in lysosomal function. Haploinsufficiency of progranulin causes FTD [219], abnormally enlarged lysosomes, and leads to dysfunctional autophagy [220, 221]. Two other PD-associated genes, ATP13A and SYT11, also affect autophagy through a common network. Defects in ATP13A decrease SYT11 transcriptionally through hyperactivation of MTORC1 and reduced TFEB activity, and increase SYT11 ubiquitination and degradation, leading to lysosomal dysfunction and impaired autophagy [222, 223]. ATP13A2 also recruits HDAC6 to the lysosome to deacetylate cortactin and promote fusion of autophagosomes with lysosomes [224].

## 3.4 Defects in Autophagic Lysosome Reformation

Autophagic lysosome reformation is a process that uses autolysosome membranes to generate lysosomes, especially under conditions of prolonged autophagy activation. Dysfunctions in this process are also associated with neurodegeneration. SPG15 and

SPG11 are the most prevalent autosomal recessive hereditary spastic paraplegias, a group of neurological disorders with axonopathy of corticospinal motor neurons, early-onset parkinsonism, and cognitive impairment among other features. Loss of ZFYVE26/SPG15/ spastizin or SPG11/spatacsin impairs autophagic lysosome reformation by preventing the initiation of lysosomal tubulation from autolysosomes, thus leading to an accumulation of autolysosomes and depletion of lysosomes [225, 226]. Disease-associated mutations in ZFYVE26/spastizin compromise its interaction both with RAB5A and RAB11, and with the BECN1-UVRAG-RUBCN complex, which leads to defects in fusion of autophagosomes with endosomes and lysosomes [227, 228]. SPG11 and ZFYVE26/SPG15 associate with adaptor protein complex 5 (AP-5) [229]. Mutations in AP5Z1 (adaptor related protein complex 5 subunit zeta 1) also cause hereditary spastic paraplegia, and knockout mice show defects in autophagic lysosome reformation [230].

> **Question**
> 3. Do mutations in the same protein always affect autophagy through the same mechanism?

## 3.5 Impairment of Autophagic Flux by Aggregate-prone Proteins

In AD, amyloid-β (Aβ) oligomers impair the recruitment of dynein to amphisomes by interacting with autophagic vesicles and dynein motors, which leads to the accumulation of amphisomes in distal axons due to failure in their retrograde transport to the soma [21]. Mutations in PSEN1 (presenilin 1) are a major cause of familial AD. Autophagy initiation and flux are compromised in PSEN1-deficienct cells, through inhibition of MAPK/ERK-CREB signalling and GSK3B activation, which reduces the expression of TFEB and autophagy-lysosome pathway genes [231]. PSEN1 is also required for lysosomal acidification by controlling V-ATPase targeting to the lysosome, which is compromised by disease-associated mutations [232, 233].

SNCA aggregates are resistant to degradation by autophagy, and compromise autophagosome maturation in neurons [234, 235]. Overexpression of SNCA impairs autophagy by inhibiting RAB1A, which causes mislocalization of ATG9 and decreases autophagosome biogenesis [236]. ARL6IP5 decreases in the brain during aging and this decrease is enhanced in PD. Downregulation of ARL6IP5 by SNCA inhibits the clearance of toxic aggregates by autophagy. ARL6IP5 induces autophagy through RAB1-dependent autophagosome initiation and ATG12 stabilization [237]. Accumulation of SNCA compromises autophagosome-lysosome fusion in midbrain neurons due to decreased association of YKT6 with autolysosome fusion SNARE SNAP29, and the trafficking of lysosomal hydrolases. FNT (farnesyltransferase, CAAX box) activity is elevated in neurodegenerative diseases and YKT6 is autoinhibited by farnesylation. Therefore, FNT inhibition restores the autophagic flux by promoting YKT6-SNAP29 complexes and hydrolase trafficking [238–240]. In addition, an increase in SNCA also decrease lysosome function by keeping TFEB in the cytosol [241].

Studies have suggested that HTT can act as a scaffold protein during selective autophagy, and that mutant HTT impairs cargo recognition (including mitochondria) and sequestration in HD models [242–244]. Moreover, HTT regulates dynein and kinesin motors to control autophagosome dynamics, and polyQ-HTT disrupts axonal transport of autophagosomes, thus compromising autophagosome maturation [245]. Finally, as explained above, mutant HTT can also affect autophagosome biogenesis by competing with wild-type ATXN3 in binding to BECN1, preventing BECN1 from being deubiquitinated and leading to its degradation [76]. Thus, mutant HTT may compromise autophagy at different stages of its itinerary.

Another aspect to consider is the fact that pathogenic mutations may also decrease their own lysosomal degradation, as shown for polyQ-HTT, SNCA, TARDBP and MAPT/tau [246]. This results in a positive feedback loop whereby the accumulation of these toxic aggregation-prone proteins impairs their clearance by autophagy, facilitating further accumulation, and hence further autophagic impairment. Consequently, even if autophagy impairment is not the primary cause of neurodegeneration, the accumulation of these neurodegeneration-prone proteins will impair autophagy, exacerbating the disease progression.

**Question**

4. Why does failure in the clearance of aggregate-prone proteins aggravate neurodegenerative diseases?

# 4  Autophagy as a Therapeutic Target for Neurodegenerative Diseases

As we have seen above, autophagy plays a central role in the maintenance of cellular homeostasis and the removal of aggregation-prone proteins. Therefore, upregulation of autophagy can be an efficient strategy in the clearance of protein aggregates to ameliorate symptoms of neurogenerative diseases (Table 1) [277, 278].

One of the pioneering agents in the field of autophagy inducers is rapamycin, a lipophilic macrolide antibiotic produced by the bacterium *Streptomyces hygroscopicus* found on Easter Island [253, 254, 279]. Rapamycin is a competitive inhibitor of FKBP1A/FLBP12 (FKBP prolyl isomerase 1A), resulting in inhibition of MTOR complex 1 (MTORC1) kinase activity, and thus it increases autophagosome biogenesis [280–283]. Studies have confirmed the effect of rapamycin in promoting degradation of several aggregate-prone proteins including MAPT/tau [250, 251, 284], HTT [247, 248], and SNCA [249]. Similarly, rapamycin also reduced disease-relevant pathology by upregulating autophagy in animal models of neurodegeneration [285–287]. The potential of rapamycin in alleviating neurodegenerative disorders has driven the development of rapamycin derivatives (called rapalogs) with improved pharmacological properties. These include temsirolimus (CCI-779) [252, 288–290], radaforolimus (AP23573) [291], and everolimus (RAD001) [292]. MTOR can also be inhibited by ATP-competitive com-

**Table 1** Strategies to enhance autophagy for the treatment of neurodegenerative diseases

| | | Autophagy upregulator | | Mechanism of action | Effect in Neurodegenerative diseases |
|---|---|---|---|---|---|
| MTOR-dependent | | Rapamycin | | Inhibits MTORC1 | Enhances the degradation and clearance of mutant HTT [247, 248], SNCA [249], MAPT/tau [250–252], ATXN3 [253], polyA aggregates [247], reduces neurotoxicity in corresponding disease models |
| | | Rapalogs | Everolimus Radaforolimus Sirolimus Temsirolimus | Inhibits MTORC1 | |
| | | Torin 1 | | Inhibits MTORC1 and MTORC2 (ATP-Competitive) | Promotes MAPT/tau clearance [254–256] |
| MTOR-independent | Ca²⁺ channel blockers | Amiodarone Felodipine Loperamide Nimodipine Nitrendipine Verapamil Minoxidil | | Inhibits extracellular Ca²⁺ influx | Reduces mutant HTT [257, 258], SNCA [257], MAPT/tau aggregation [257], and improves pathological phenotype |
| | Calpain inhibitor | Calpastatin Calpeptin | | Calpain inhibitor | Improves the clearance of amyloid β and hyperphosphorylated MAPT/tau [258, 259] |
| | IP₃ synthesis inhibitor | Lithium Sodium valproate (VPA) Carbamazepine (CBZ) L-690330 | | Depletes inositol 1,4,5-trisphosphate (IP₃) | Enhances the clearance of aggregate-prone proteins like mutant HTT, SNCA [260, 261] |
| | TFEB regulator | Trehalose | | Promotes TFEB nuclear translocation; activates AMPK-dependent autophagy; chemical chaperone | Accelerates the clearance of toxic protein aggregates, including mutant HTT and SNCA [262–264] |
| | | Genistein | | Promotes TFEB nuclear translocation and increases TFEB mRNA expression | Promotes the clearance of toxic HTT and MAPT/tau aggregates, protective in disease models [265–269] |

(continued)

**Table 1** (continued)

| | | Autophagy upregulator | Mechanism of action | Effect in Neurodegenerative diseases |
|---|---|---|---|---|
| | AMPK activator | Bosutinib Humanin Melatonin | AMPK stimulation | Increases the autophagy flux and promotes the clearance of toxic aggregates, alleviates disease-relevant symptoms [270–273] |
| | PPARA agonists | Gemfibrozil Wy14643 | PPARA activation | Increases autophagy and decreases amyloid β in AD mouse brains [274] |
| | | Clonidine Rilmenidine 2′5′-Dideoxyadenosine | Decreases cAMP level | Reduces mutant HTT aggregates [253, 258] |
| | | Resveratrol | Unclear | Protective effects in PD models [275] |
| | | SMER10 SMER18 SMER28 | Unclear for MER10 and SMER18. SMER28 activates VCP | Enhances mutant HTT and SNCA clearance, protective in cell and fly models of HD [276] |

pounds, such as torin 1, that target both MTORC1 and MTORC2 [255, 293]. However, given the pleiotropic actions of MTOR in regulating many autophagy-independent processes [255, 294], further application of rapamycin and rapalogs is limited with concerns regarding their adverse effects including immunosuppression, oral ulcers, and impaired wound healing [295–297], since these are large molecules that do not penetrate the brain well, necessitating high peripheral exposures. MTORC2 inhibition by ATP-competitive compounds may also have some undesired effects with chronic use.

This has driven great effort being invested to identify drugs capable of upregulating autophagy independent of the MTOR pathway, highlighting the significance of understanding MTOR-independent mechanisms regulating autophagy [253]. Repurposing of existing drugs by screening their effects on autophagy has provided valuable insights into MTOR-independent signaling mechanisms regulating autophagy [298, 299]. For example, several FDA-approved $Ca^{2+}$ channel antagonists, including felodipine [257, 300], verapamil [301, 302], and amiodarone [303], enhance autophagy and promote the clearance of toxic aggregate-prone proteins [257, 258]. These $Ca^{2+}$ channel blockers exert their effect by inhibiting the influx of extracellular $Ca^{2+}$ into the cell, as elevated intracellular $Ca^{2+}$ would impair autophagy. Intracellular $Ca^{2+}$ levels play a crucial role in regulating the activity of various $Ca^{2+}$-binding proteins, including calpain. Calpain is a calcium-dependent cysteine protease that negatively regulates autophagy through the $Ca^{2+}$-calpain-GNAS/ Gsα (a subunit of heterotrimeric G proteins) cAMP pathway [258, 304] independent of

MTOR inhibition [258, 305]. Specific calpain inhibitors, such as calpeptin and calpastatin, have also been shown to promote autophagic clearance [258, 259].

In addition to strategies that alter $Ca^{2+}$ influx to the cell, drugs that affect the release of $Ca^{2+}$ from the ER also show promise for altering intracellular $Ca^{2+}$ levels and autophagy. Inositol 1,4,5-trisphosphate ($IP_3$) regulates ER $Ca^{2+}$ release through binding to ITPR/IP3R (inositol 1,4,5-triphosphate receptor) on the ER membrane. The resulting $Ca^{2+}$ flux from the ER to mitochondria plays a crucial role in ATP production [260, 306]. Deficiency in this $Ca^{2+}$ flux leads to cellular energy depletion, which activates autophagy through AMP-activated protein kinase (AMPK). Mood-stabilizing drugs, such as lithium, sodium valproate (VPA), and carbamazepine (CBZ), have been shown to decrease intracellular levels of $IP_3$ by interfering with the phosphoinositide cycle, disrupt the spatiotemporal availability of $IP_3$. Repurposing these mood stabilizers has been associated with improved autophagic clearance of substrates, such as mutant HTT and SNCA fragments, providing a therapeutic avenue for upregulating autophagy beyond direct MTOR inhibition [260, 261].

An emerging class of engineered molecules, referred to as molecular glues, operate by inducing proximity between tethered parts with the aim of increasing relevant interactions (Fig. 3) [307]. These novel therapeutics were inspired by proteasome-targeting chimeras (PROTACs), which bring E3 ubiquitin ligases and target substrates together to promote K48 ubiquitination-mediated proteasomal clearance [308]. Chimeras targeting substrates to the autophagic machinery have also been designed as an approach to increase clearance [309–311]. These chimeras are generally composed of three moieties: (a) a ligand for the substrate of interest; (b) a warhead targeting components of autophagy; and (c) a linker connecting these two components. Autophagosome tethering compounds (ATTEC)

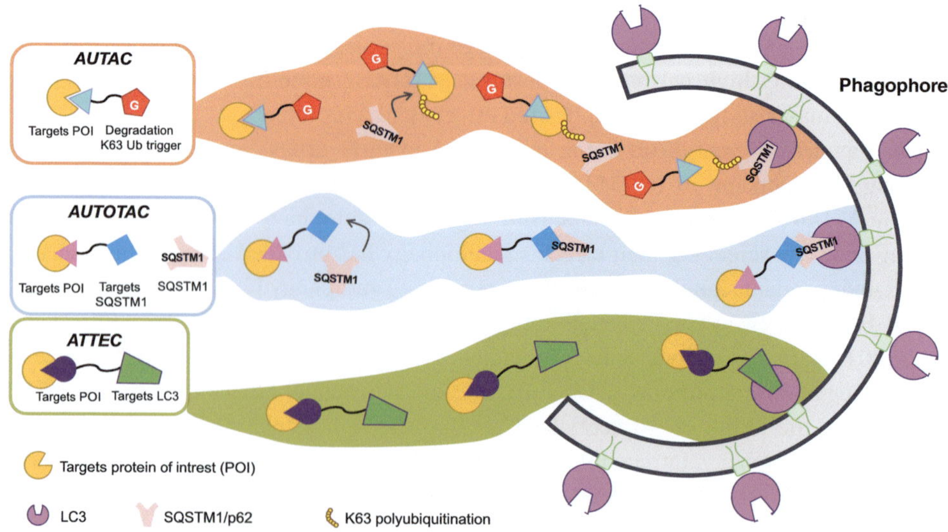


**Fig. 3** Schematic of different autophagy-targeting chimeras

directly exploit the advantage of lipidated LC3, and directly enhance substrate recruitment to the autophagosome through tethering with LC3 [312]. An ATTEC that binds to both LC3 and the expanded polyQ stretch of mutant HTT has demonstrated the ability to specifically lower mutant HTT through autophagic degradation in mouse and fly models of HD. This led to the rescue of neuronal health and the improvement of HD-relevant deficits [312]. The autophagy cargo receptor SQSTM1 has also been utilized to drive substrate recruitment to autophagosomes through AUTOphagy-TArgeting Chimera (AUTOTAC), which includes a module that directly interacts with SQSTM1. Such interaction with SQSTM1 enables efficient targeting to the autophagic membrane through interaction between SQSTM1 and LC3. By bridging protein substrates with SQSTM1, AUTOTACs can efficiently mediate the clearance of not only monomeric but also bulky and aggregation-prone substrates, such as MAPT/tau fibrils [313]. Furthermore, as substrate K63 polyubiquitination is readily recognized by SQSTM1 [314], autophagy targeting chimeras (AUTAC) employing a cGMP-based degradation tag which promotes the K63 polyubiquitination of substrates was developed [310, 315]. AUTACs have also been shown to effectively promote the degradation of damaged organelles like mitochondria. These chimeras hold great promise as powerful drug candidates due to their high target selectivity and minimal side effects.

However, it should be noted that upregulating autophagosome biogenesis does not necessarily represent a valid strategy for all neurodegenerative diseases, and increasing autophagosome formation could even be deleterious when impaired aggregate clearance is due to lysosomal dysfunction [7, 316, 317]. Thus, studies have also been conducted to investigate the possibility of improving autophagic-lysosomal proteolysis with the aim of ameliorating disease-relevant phenotype. A promising target is TFEB, whose nuclear localization tightly regulates the expression of genes governing lysosome biogenesis and autophagic-lysosomal proteolytic activity [318, 319]. Trehalose, a natural disaccharide found in many non-mammalian species, has gained increasing attention as an MTOR-independent autophagy enhancer in recent decades [262, 320]. This natural product acts in a complex manner. In addition to its chemical chaperone properties [321, 322], trehalose was also shown to promote TFEB nuclear localization, leading to upregulation of autophagy and accelerated clearance of toxic aggregate-prone proteins, including mutant HTT and SNCA [262–264]. Beyond its TFEB-dependent functions, trehalose was recently found to induce autophagy in an AMPK-dependent manner, which is achieved through inhibiting glucose uptake to create functionally "starved" status [323]. Another natural compound promoting TFEB activation and nuclear localization is genistein, which was also shown to promote toxic aggregation clearance in an autophagy-dependent manner in fibroblasts derived from HD patients [265], and in AD animal models [266–269]. Moreover, the ability of genistein to penetrate the blood-brain barrier enables it to directly access neurons and exert its potential therapeutic effects [324]. This highlights the potential of genistein as a valuable therapeutic agent for neurodegenerative disorders and warrants further investigations into its mechanisms of action and clinical applications.

In conclusion, the modulation of autophagy represents a promising therapeutic strategy for neurodegenerative disorders. The wide range of autophagy-inducing agents, including those targeting both MTOR-dependent and MTOR-independent pathways, as well as emerging modalities like AUTACs, offer exciting possibilities for the development of effective treatments. Furthermore, beyond conventional pharmacological interventions, genetic manipulation of core autophagic genes, such as *BECN1* [325–328], *ATG5* [329], and *ATG7* [330], has shown beneficial effects in various animal models of neurodegenerative diseases [331]. However, challenges remain, including the need for more selective and targeted autophagy-inducing agents, a deeper understanding towards the complex mechanisms regulating autophagy, and the translation of autophagy-based therapies into clinical practice. Addressing these challenges will pave the way for the development of more innovative and efficacious interventions for neurodegenerative disease, offering hope for improved patient outcomes and enhanced quality of life.

**Question**

5. Give two strategies that can enhance autophagy to ameliorate neurodegenerative diseases.
6. Are autophagy enhancing drugs always a suitable approach for treating neurodegenerative diseases? Why?

## 5 Introduction to Cardinal Articles

Cardinal article 1: Ravikumar B, Duden R, Rubinsztein DC. Aggregate-prone proteins with polyglutamine and polyalanine expansions are degraded by autophagy. Hum Mol Genet. 2002;11(9):1107–17.

Cardinal article 2: Ravikumar B, Vacher C, Berger Z, Davies JE, Luo S, Oroz LG, et al. Inhibition of mTOR induces autophagy and reduces toxicity of polyglutamine expansions in fly and mouse models of Huntington disease. Nat Genet. 2004;36(6):585–95.

Cardinal article 3: Komatsu M, Waguri S, Chiba T, Murata S, Iwata J, Tanida I, et al. Loss of autophagy in the central nervous system causes neurodegeneration in mice. Nature. 2006;441(7095):880–4.

Cardinal article 4: Hara T, Nakamura K, Matsui M, Yamamoto A, Nakahara Y, Suzuki-Migishima R, et al. Suppression of basal autophagy in neural cells causes neurodegenerative disease in mice. Nature. 2006;441(7095):885–9.

In an early study in the field, we showed that aggregate-prone proteins with polyQ and polyA expansions are degraded by autophagy [6]. Previous studies had revealed the presence of HTT within autophagosomes [332], and that nigral neurons of patients with PD contain elevated levels of autophagic structures [333]. However, it was generally assumed that aggregate-prone proteins were degraded by the ubiquitin-proteasome system, and that the accumulation of these autophagic structures could be an indication of autophagic cell death. By inhibiting the pathway at different stages, Ravikumar et al. [6] demonstrated that increased aggregate formation and cell death could result from failure in the clearance of mutant proteins by autophagy. Indeed, induction of autophagy produced the opposite result, leading to a decrease in aggregation and cell death. The role of autophagy in the clearance of other aggregate-prone proteins, including SNCA [334], was demonstrated in subsequent studies. These experiments relied on manipulation of the autophagic pathway to study the outcomes in protein aggregation instead of simply drawing conclusions from presence of the proteins within autophagosomes, a read-out that can be misinterpreted if changes in the autophagic flux are not considered.

Ravikumar et al. [6] helped to move the field forward because it directly implicated autophagy in the clearance of aggregate-prone proteins and suggested that stimulation of autophagy could be exploited for the treatment of neurodegenerative diseases. This prompted further research to test this potential role. In a following report [247], we found that MTOR is sequestered in HTT aggregates in brains of HD mice models and individuals with HD. Notably, experiments in fly and mice models of HD disease have shown that MTOR inhibition has a protective effect against neurodegeneration. Indeed, treatment with the rapamycin analog CCI-779 improved behaviour and motor performance in a mouse model of HD. Similar protection was subsequently shown for additional aggregate-prone proteins, such as other polyQ and polyA proteins, and MAPT/tau [335].

In two key articles released in the same edition of Nature, Keiji Tanaka's lab [3] and Noboru Mizushima's lab [4] show that conditional loss of autophagy in the central nervous system results in neurodegeneration. Both groups exploit a conditional Cre recombinase system under a *Nes* (nestin) promoter to cause knockout of *Atg7* [3] or *Atg5* [4] exclusively in the cells of the central nervous system in mice. By restricting the autophagic loss to the central nervous system only, this model system enabled the groups to overcome the technical challenge of the metabolic insufficiency-mediated perinatal death associated with ubiquitous autophagic loss. Both the $atg5^{-/-}$ and $atg7^{-/-}$ mice develop impaired motor function, including an abnormal limb-clasping reflex and reduced coordinated movement. This was associated with the accumulation of polyubiquitinated proteins in inclusion bodies, and loss of neurons throughout the cerebral and cerebellar cortices.

These two papers [3, 4] made a critical contribution to the autophagy and neurodegeneration fields by showing that loss of autophagic function is sufficient to induce neurodegeneration in mice with the absence of any known neurodegeneration-causing mutations. Importantly, this was the first demonstration that autophagic impairment alone can cause neurodegeneration. It raised many exciting questions for the neurodegeneration and autophagy fields, including the mechanism by which autophagic impairment drives neurodegeneration and whether the accumulation of autophagic substrates such as aggregation-prone proteins may drive neurodegeneration. These papers also created many new avenues for scientific exploration, including whether autophagy is impaired in human neurodegeneration in the absence of mutations in the core autophagy genes, and whether upregulation of autophagy can mitigate neurodegeneration. Overall, these trailblazing studies highlighted the critical role of autophagy in clearing aggregation-prone proteins and preventing neurodegeneration and pioneered the autophagy and neurodegeneration subfield.

## 5.1 Take Note Of

Neurodegenerative diseases are often complex and involve a multitude of cellular changes. You must consider that while some of these changes may be causing the disease, other changes may be a consequence of the disease process itself. For example, the accumulation of aggregation-prone proteins results in the impairment of autophagy, which further exacerbates the disease progression, even if autophagic impairment was not the primary cause for the disease.

You must take care when interpreting data that suggests that a certain mutation is involved in disease through defects in autophagy. Certain mutations may lead to a wide range of outcomes, with autophagy defects only being a minor part of a much bigger picture. Many autophagy genes also function in other cellular pathways. It can be challenging to deconvolute which effects are responsible for the disease and assessing this requires appropriate controls.

Remember that there are non-cell-autonomous regulatory mechanisms between neurons and glia. For example, cytokines secreted from microglia and astrocytes bind to receptors on neurons to alter neuronal signalling pathways and autophagy. Consequently, impairment in the health and/or function of one cell type is likely to also affect the other cell types in the brain.

Special attention should be paid to understanding how autophagy is affected at different stages of a disease, before recommending the use of autophagy-enhancing drugs. For example, in cases where the fusion of autophagosomes with lysosomes is impaired, or the lysosomes themselves are dysfunctional, upregulation of autophagy will lead only to an increase in the number of autophagosomes, without enhancing the clearance of autophagic substrates. Furthermore, it is important that autophagy-enhancing drugs are given early in the disease progression, preceding widespread neuronal death. Also, sporadic forms of neurodegenerative diseases are difficult to mimic using animal models, hindering our ability to design suitable therapeutic strategies using autophagy.

> **Take Home Message**
> - Autophagy is required for normal neurodevelopment and for the maintenance of neuronal health and function.
> - The functions of autophagy in the nervous system include the clearance of aggregation-prone proteins, maintenance of synaptic function, regulation of neurodevelopmental pathways, and critical functions in glia.
> - Neurons are particularly vulnerable to impaired autophagy due to their long, post-mitotic lifetimes, their elongated morphology that exacerbates cellular trafficking defects, and their high energetic demand.
> - Due to the critical sensitivity of neurons to autophagic impairment, mutations in many autophagy proteins are associated with neurodevelopmental and/or neurodegenerative diseases.

(continued)

- Mutations associated with disease may occur in core autophagy genes or in the machinery that controls their stability, recruitment, or trafficking.
- Aggregation-prone proteins may exacerbate autophagy defects and lead to worsening of the progression of neurodegenerative diseases.
- Upregulation of autophagy is a promising strategy for treating neurodegeneration.
- Common strategies leveraging autophagy for the clearance of toxic substrates underlying neurodegenerative diseases include: (a) MTOR inhibition; (b) decrease cytosolic $Ca^{2+}$ level and calpain activity; (c) decrease inositol 1,4,5-trisphosphate level; (d) molecular chimeras targeting substrates to autophagy; (e) lysosomal proteolytic activity upregulator.

**Answer Guide to Questions**

1. Neurons are selectively vulnerable to impaired autophagy for the following main reasons:
   (a) Neurons are post-mitotic so cannot dilute accumulated autophagy substrates by cell division. This means that autophagy substrates can accumulate over the long lifetime of a neuron.
   (b) Autophagosomes must traffic long distances from the axons and dendrites where they are made, back to the cell body where they fuse with lysosomes to degrade the cargo. Even a minor impairment in this trafficking can significantly reduce the clearance of autophagy substrates.
   (c) Neurons have a high energetic demand that requires an efficient turnover of mitochondria by mitophagy.
2. Autophagy is important in neurons and glia for:
   (a) The maintenance and modulation of synaptic function, learning and memory.
   (b) Critical neurodevelopmental pathways that affect neurogenesis and neural progenitor cell differentiation.
   (c) In astrocytes, for metabolic support to neurons and preventing a pro-inflammatory activation of astrocytes.
   (d) In microglia, for efficient synaptic pruning and the clearance of extracellular debris, and for preventing a pro-inflammatory activation of microglia.
   (e) In oligodendrocytes and Schwann cells, for removing excess organelles to form the myelin sheath, and for maintaining their function and survival.
3. Mutations may occur in different domains of the protein and affect more than one of its functions. For example, mutations in WIPI2 may affect its ability to bind PtdIns3P or to ATG16L1, with both affecting autophagosome biogenesis.

(continued)

(continued)

4. Failure in the clearance of aggregate-prone proteins may exacerbate defects in autophagy by interfering with the autophagic machinery. Consequently, autophagy will be impaired and the ability of the cells to clear these aggregate-prone proteins will be further compromised. Because aggregate-prone proteins are directly involved in neurodegenerative diseases, their accumulation will contribute to a faster progression of the disease.

5. Strategies that can enhance autophagy to ameliorate neurodegenerative diseases include:

    (a) Inhibition of MTOR activity using rapamycin, rapalogs or torin 1.

    (b) Decreasing cytosolic $Ca^{2+}$ levels and calpain activity using $Ca^{2+}$ channel blockers, such as felodipine.

    (c) Reduction of intracellular inositol 1,4,5-trisphosphate ($IP_3$) availability using mood-stabilizing drugs, such as lithium, sodium valproate, and carbamazepine.

    (d) Using molecular chimeras to directly target substrates to autophagy.

    (e) Increasing autophagic-lysosomal proteolysis activity. Drugs such as trehalose and genistein upregulate lysosomal biogenesis and autophagic-lysosomal proteolytic activity.

6. Upregulation of autophagosome biogenesis is not suitable for all neurodegenerative diseases. Lysosomal function is required for autophagy. Therefore, if lysosomal function is impaired, an increase in autophagosome biogenesis will only result in accumulation of autophagosomes.

# References

1. Kuma A, Komatsu M, Mizushima N. Autophagy-monitoring and autophagy-deficient mice. Autophagy. 2017;13(10):1619–28.

2. Yoshii SR, Kuma A, Mizushima N. Transgenic rescue of Atg5-null mice from neonatal lethality with neuron-specific expression of ATG5: systemic analysis of adult Atg5-deficient mice. Autophagy. 2017;13(4):763–4.

3. Komatsu M, Waguri S, Chiba T, Murata S, Iwata J, Tanida I, et al. Loss of autophagy in the central nervous system causes neurodegeneration in mice. Nature. 2006;441(7095):880–4.

4. Hara T, Nakamura K, Matsui M, Yamamoto A, Nakahara Y, Suzuki-Migishima R, et al. Suppression of basal autophagy in neural cells causes neurodegenerative disease in mice. Nature. 2006;441(7095):885–9.

5. Karsli-Uzunbas G, Guo JY, Price S, Teng X, Laddha SV, Khor S, et al. Autophagy is required for glucose homeostasis and lung tumor maintenance. Cancer Discov. 2014;4(8):914–27.

6. Ravikumar B, Duden R, Rubinsztein DC. Aggregate-prone proteins with polyglutamine and polyalanine expansions are degraded by autophagy. Hum Mol Genet. 2002;11(9):1107–17.

7. Menzies FM, Fleming A, Rubinsztein DC. Compromised autophagy and neurodegenerative diseases. Nat Rev Neurosci. 2015;16(6):345–57.

8. Nixon RA. The role of autophagy in neurodegenerative disease. Nat Med. 2013;19(8):983–97.

9. Wong E, Cuervo AM. Autophagy gone awry in neurodegenerative diseases. Nat Neurosci. 2010;13(7):805–11.

10. Fleming A, Bourdenx M, Fujimaki M, Karabiyik C, Krause GJ, Lopez A, et al. The different autophagy degradation pathways and neurodegeneration. Neuron. 2022;110(6):935–66.

11. Wilson DM 3rd, Cookson MR, Van Den Bosch L, Zetterberg H, Holtzman DM, Dewachter I. Hallmarks of neurodegenerative diseases. Cell. 2023;186(4):693–714.

12. Cai Q, Ganesan D. Regulation of neuronal autophagy and the implications in neurodegenerative diseases. Neurobiol Dis. 2022;162:105582.

13. Cason SE, Mogre SS, Holzbaur ELF, Koslover EF. Spatiotemporal analysis of axonal autophagosome-lysosome dynamics reveals limited fusion events and slow maturation. Mol Biol Cell. 2022;33(13):ar123.

14. Katsumata K, Nishiyama J, Inoue T, Mizushima N, Takeda J, Yuzaki M. Dynein- and activity-dependent retrograde transport of autophagosomes in neuronal axons. Autophagy. 2010;6(3):378–85.

15. Maday S, Wallace KE, Holzbaur EL. Autophagosomes initiate distally and mature during transport toward the cell soma in primary neurons. J Cell Biol. 2012;196(4):407–17.

16. Cheng XT, Zhou B, Lin MY, Cai Q, Sheng ZH. Axonal autophagosomes recruit dynein for retrograde transport through fusion with late endosomes. J Cell Biol. 2015;209(3):377–86.

17. Cai Q, Lu L, Tian JH, Zhu YB, Qiao H, Sheng ZH. Snapin-regulated late endosomal transport is critical for efficient autophagy-lysosomal function in neurons. Neuron. 2010;68(1):73–86.

18. Ikenaka K, Kawai K, Katsuno M, Huang Z, Jiang YM, Iguchi Y, et al. dnc-1/dynactin 1 knockdown disrupts transport of autophagosomes and induces motor neuron degeneration. PLoS One. 2013;8(2):e54511.

19. Khobrekar NV, Quintremil S, Dantas TJ, Vallee RB. The dynein adaptor RILP controls neuronal autophagosome biogenesis, transport, and clearance. Dev Cell. 2020;53(2):141–53.e4.

20. Ravikumar B, Acevedo-Arozena A, Imarisio S, Berger Z, Vacher C, O'Kane CJ, et al. Dynein mutations impair autophagic clearance of aggregate-prone proteins. Nat Genet. 2005;37(7):771–6.

21. Tammineni P, Ye X, Feng T, Aikal D, Cai Q. Impaired retrograde transport of axonal autophagosomes contributes to autophagic stress in Alzheimer's disease neurons. elife. 2017;6:e21776.

22. Martinez-Vicente M. Neuronal mitophagy in neurodegenerative diseases. Front Mol Neurosci. 2017;10:64.

23. Misgeld T, Schwarz TL. Mitostasis in neurons: maintaining mitochondria in an extended cellular architecture. Neuron. 2017;96(3):651–66.

24. Goldsmith J, Ordureau A, Harper JW, Holzbaur ELF. Brain-derived autophagosome profiling reveals the engulfment of nucleoid-enriched mitochondrial fragments by basal autophagy in neurons. Neuron. 2022;110(6):967–76.e8.

25. Kallergi E, Siva Sankar D, Matera A, Kolaxi A, Paolicelli RC, Dengjel J, et al. Profiling of purified autophagic vesicle degradome in the maturing and aging brain. Neuron. 2023;111:2329–2347.e7.

26. Fritsch LE, Moore ME, Sarraf SA, Pickrell AM. Ubiquitin and receptor-dependent mitophagy pathways and their implication in neurodegeneration. J Mol Biol. 2020;432(8):2510–24.

27. Katayama H, Hama H, Nagasawa K, Kurokawa H, Sugiyama M, Ando R, et al. Visualizing and modulating mitophagy for therapeutic studies of neurodegeneration. Cell. 2020;181(5):1176–87.e16.

28. Birdsall V, Waites CL. Autophagy at the synapse. Neurosci Lett. 2019;697:24–8.

29. Compans B, Camus C, Kallergi E, Sposini S, Martineau M, Butler C, et al. NMDAR-dependent long-term depression is associated with increased short term plasticity through autophagy mediated loss of PSD-95. Nat Commun. 2021;12(1):2849.

30. Domise M, Sauve F, Didier S, Caillerez R, Begard S, Carrier S, et al. Neuronal AMP-activated protein kinase hyper-activation induces synaptic loss by an autophagy-mediated process. Cell Death Dis. 2019;10(3):221.
31. Gundelfinger ED, Karpova A, Pielot R, Garner CC, Kreutz MR. Organization of presynaptic autophagy-related processes. Front Synaptic Neurosci. 2022;14:829354.
32. Nikoletopoulou V, Tavernarakis N. Regulation and roles of autophagy at synapses. Trends Cell Biol. 2018;28(8):646–61.
33. Kuijpers M, Kochlamazashvili G, Stumpf A, Puchkov D, Swaminathan A, Lucht MT, et al. Neuronal autophagy regulates presynaptic neurotransmission by controlling the axonal endoplasmic reticulum. Neuron. 2021;109(2):299–313.e9.
34. Kallergi E, Daskalaki AD, Kolaxi A, Camus C, Ioannou E, Mercaldo V, et al. Dendritic autophagy degrades postsynaptic proteins and is required for long-term synaptic depression in mice. Nat Commun. 2022;13(1):680.
35. Glatigny M, Moriceau S, Rivagorda M, Ramos-Brossier M, Nascimbeni AC, Lante F, et al. Autophagy is required for memory formation and reverses age-related memory decline. Curr Biol. 2019;29(3):435–48.e8.
36. Kulkarni VV, Maday S. Compartment-specific dynamics and functions of autophagy in neurons. Dev Neurobiol. 2018;78(3):298–310.
37. Deng Z, Zhou X, Lu JH, Yue Z. Autophagy deficiency in neurodevelopmental disorders. Cell Biosci. 2021;11(1):214.
38. Fleming A, Rubinsztein DC. Autophagy in neuronal development and plasticity. Trends Neurosci. 2020;43(10):767–79.
39. Wang S, Li B, Qiao H, Lv X, Liang Q, Shi Z, et al. Autophagy-related gene Atg5 is essential for astrocyte differentiation in the developing mouse cortex. EMBO Rep. 2014;15(10):1053–61.
40. Paolicelli RC, Bolasco G, Pagani F, Maggi L, Scianni M, Panzanelli P, et al. Synaptic pruning by microglia is necessary for normal brain development. Science. 2011;333(6048):1456–8.
41. Kim HJ, Cho MH, Shim WH, Kim JK, Jeon EY, Kim DH, et al. Deficient autophagy in microglia impairs synaptic pruning and causes social behavioral defects. Mol Psychiatry. 2017;22(11):1576–84.
42. Tang G, Gudsnuk K, Kuo SH, Cotrina ML, Rosoklija G, Sosunov A, et al. Loss of mTOR-dependent macroautophagy causes autistic-like synaptic pruning deficits. Neuron. 2014;83(5):1131–43.
43. Allen NJ, Barres BA. Neuroscience: glia – more than just brain glue. Nature. 2009;457(7230):675–7.
44. Madill M, McDonagh K, Ma J, Vajda A, McLoughlin P, O'Brien T, et al. Amyotrophic lateral sclerosis patient iPSC-derived astrocytes impair autophagy via non-cell autonomous mechanisms. Mol Brain. 2017;10(1):22.
45. Sung K, Jimenez-Sanchez M. Autophagy in astrocytes and its implications in neurodegeneration. J Mol Biol. 2020;432(8):2605–21.
46. Takano T, Tian GF, Peng W, Lou N, Libionka W, Han X, et al. Astrocyte-mediated control of cerebral blood flow. Nat Neurosci. 2006;9(2):260–7.
47. Khakh BS, Sofroniew MV. Diversity of astrocyte functions and phenotypes in neural circuits. Nat Neurosci. 2015;18(7):942–52.
48. Di Malta C, Fryer JD, Settembre C, Ballabio A. Autophagy in astrocytes: a novel culprit in lysosomal storage disorders. Autophagy. 2012;8(12):1871–2.
49. Chandrasekaran A, Dittlau KS, Corsi GI, Haukedal H, Doncheva NT, Ramakrishna S, et al. Astrocytic reactivity triggered by defective autophagy and metabolic failure causes neurotoxicity in frontotemporal dementia type 3. Stem Cell Reports. 2021;16(11):2736–51.

50. Motori E, Puyal J, Toni N, Ghanem A, Angeloni C, Malaguti M, et al. Inflammation-induced alteration of astrocyte mitochondrial dynamics requires autophagy for mitochondrial network maintenance. Cell Metab. 2013;18(6):844–59.
51. Morales I, Sanchez A, Puertas-Avendano R, Rodriguez-Sabate C, Perez-Barreto A, Rodriguez M. Neuroglial transmitophagy and Parkinson's disease. Glia. 2020;68(11):2277–99.
52. Davis CH, Kim KY, Bushong EA, Mills EA, Boassa D, Shih T, et al. Transcellular degradation of axonal mitochondria. Proc Natl Acad Sci USA. 2014;111(26):9633–8.
53. Perry VH, Nicoll JA, Holmes C. Microglia in neurodegenerative disease. Nat Rev Neurol. 2010;6(4):193–201.
54. Hickman S, Izzy S, Sen P, Morsett L, El Khoury J. Microglia in neurodegeneration. Nat Neurosci. 2018;21(10):1359–69.
55. Cho MH, Cho K, Kang HJ, Jeon EY, Kim HS, Kwon HJ, et al. Autophagy in microglia degrades extracellular beta-amyloid fibrils and regulates the NLRP3 inflammasome. Autophagy. 2014;10(10):1761–75.
56. Choi I, Zhang Y, Seegobin SP, Pruvost M, Wang Q, Purtell K, et al. Microglia clear neuron-released alpha-synuclein via selective autophagy and prevent neurodegeneration. Nat Commun. 2020;11(1):1386.
57. Choi I, Wang M, Yoo S, Xu P, Seegobin SP, Li X, et al. Autophagy enables microglia to engage amyloid plaques and prevents microglial senescence. Nat Cell Biol. 2023;25(7):963–74.
58. Cheng J, Liao Y, Dong Y, Hu H, Yang N, Kong X, et al. Microglial autophagy defect causes parkinson disease-like symptoms by accelerating inflammasome activation in mice. Autophagy. 2020;16(12):2193–205.
59. Xu Y, Propson NE, Du S, Xiong W, Zheng H. Autophagy deficiency modulates microglial lipid homeostasis and aggravates tau pathology and spreading. Proc Natl Acad Sci USA. 2021;118(27):e2023418118.
60. Festa BP, Siddiqi FH, Jimenez-Sanchez M, Won H, Rob M, Djajadikerta A, et al. Microglial-to-neuronal CCR5 signaling regulates autophagy in neurodegeneration. Neuron. 2023;111(13):2021–37.e12.
61. Bradl M, Lassmann H. Oligodendrocytes: biology and pathology. Acta Neuropathol. 2010;119(1):37–53.
62. Jang SY, Shin YK, Park SY, Park JY, Rha SH, Kim JK, et al. Autophagy is involved in the reduction of myelinating Schwann cell cytoplasm during myelin maturation of the peripheral nerve. PLoS One. 2015;10(1):e0116624.
63. Bankston AN, Forston MD, Howard RM, Andres KR, Smith AE, Ohri SS, et al. Autophagy is essential for oligodendrocyte differentiation, survival, and proper myelination. Glia. 2019;67(9):1745–59.
64. Mizushima N, Levine B. Autophagy in mammalian development and differentiation. Nat Cell Biol. 2010;12(9):823–30.
65. Wang G, Liu X, Gaertig MA, Li S, Li XJ. Ablation of huntingtin in adult neurons is non-deleterious but its depletion in young mice causes acute pancreatitis. Proc Natl Acad Sci USA. 2016;113(12):3359–64.
66. Webster CP, Smith EF, Bauer CS, Moller A, Hautbergue GM, Ferraiuolo L, et al. The C9orf72 protein interacts with Rab1a and the ULK1 complex to regulate initiation of autophagy. EMBO J. 2016;35(15):1656–76.
67. Yang M, Liang C, Swaminathan K, Herrlinger S, Lai F, Shiekhattar R, et al. A C9ORF72/SMCR8-containing complex regulates ULK1 and plays a dual role in autophagy. Sci Adv. 2016;2(9):e1601167.

68. Farg MA, Sundaramoorthy V, Sultana JM, Yang S, Atkinson RA, Levina V, et al. C9ORF72, implicated in amytrophic lateral sclerosis and frontotemporal dementia, regulates endosomal trafficking. Hum Mol Genet. 2014;23(13):3579–95.

69. Sullivan PM, Zhou X, Robins AM, Paushter DH, Kim D, Smolka MB, et al. The ALS/FTLD associated protein C9orf72 associates with SMCR8 and WDR41 to regulate the autophagy-lysosome pathway. Acta Neuropathol Commun. 2016;4(1):51.

70. Sellier C, Campanari ML, Julie Corbier C, Gaucherot A, Kolb-Cheynel I, Oulad-Abdelghani M, et al. Loss of C9ORF72 impairs autophagy and synergizes with polyQ Ataxin-2 to induce motor neuron dysfunction and cell death. EMBO J. 2016;35(12):1276–97.

71. Liang C, Shao Q, Zhang W, Yang M, Chang Q, Chen R, et al. Smcr8 deficiency disrupts axonal transport-dependent lysosomal function and promotes axonal swellings and gain of toxicity in C9ALS/FTD mouse models. Hum Mol Genet. 2019;28(23):3940–53.

72. Wilson GR, Sim JC, McLean C, Giannandrea M, Galea CA, Riseley JR, et al. Mutations in RAB39B cause X-linked intellectual disability and early-onset Parkinson disease with alpha-synuclein pathology. Am J Hum Genet. 2014;95(6):729–35.

73. Amick J, Roczniak-Ferguson A, Ferguson SM. C9orf72 binds SMCR8, localizes to lysosomes, and regulates mTORC1 signaling. Mol Biol Cell. 2016;27(20):3040–51.

74. Shao Q, Yang M, Liang C, Ma L, Zhang W, Jiang Z, et al. C9orf72 and smcr8 mutant mice reveal MTORC1 activation due to impaired lysosomal degradation and exocytosis. Autophagy. 2020;16(9):1635–50.

75. Gstrein T, Edwards A, Pristoupilova A, Leca I, Breuss M, Pilat-Carotta S, et al. Mutations in Vps15 perturb neuronal migration in mice and are associated with neurodevelopmental disease in humans. Nat Neurosci. 2018;21(2):207–17.

76. Ashkenazi A, Bento CF, Ricketts T, Vicinanza M, Siddiqi F, Pavel M, et al. Polyglutamine tracts regulate beclin 1-dependent autophagy. Nature. 2017;545(7652):108–11.

77. Hill SM, Wrobel L, Ashkenazi A, Fernandez-Estevez M, Tan K, Burli RW, et al. VCP/p97 regulates Beclin-1-dependent autophagy initiation. Nat Chem Biol. 2021;17(4):448–55.

78. Tresse E, Salomons FA, Vesa J, Bott LC, Kimonis V, Yao TP, et al. VCP/p97 is essential for maturation of ubiquitin-containing autophagosomes and this function is impaired by mutations that cause IBMPFD. Autophagy. 2010;6(2):217–27.

79. Darwich NF, Phan JM, Kim B, Suh E, Papatriantafyllou JD, Changolkar L, et al. Autosomal dominant VCP hypomorph mutation impairs disaggregation of PHF-tau. Science. 2020;370(6519):eaay8826.

80. Wang L, Ren A, Tian T, Li N, Cao X, Zhang P, et al. Whole-exome sequencing identifies damaging de novo variants in anencephalic cases. Front Neurosci. 2019;13:1285.

81. Maroofian R, Gubas A, Kaiyrzhanov R, Scala M, Hundallah K, Severino M, et al. Homozygous missense WIPI2 variants cause a congenital disorder of autophagy with neurodevelopmental impairments of variable clinical severity and disease course. Brain Commun. 2021;3(3):fcab183.

82. Jelani M, Dooley HC, Gubas A, Mohamoud HSA, Khan MTM, Ali Z, et al. A mutation in the major autophagy gene, WIPI2, associated with global developmental abnormalities. Brain. 2019;142(5):1242–54.

83. Palamiuc L, Ravi A, Emerling BM. Phosphoinositides in autophagy: current roles and future insights. FEBS J. 2020;287(2):222–38.

84. Stavoe AK, Gopal PP, Gubas A, Tooze SA, Holzbaur EL. Expression of WIPI2B counteracts age-related decline in autophagosome biogenesis in neurons. elife. 2019;8:e44219.

85. Park SJ, Frake RA, Karabiyik C, Son SM, Siddiqi FH, Bento CF, et al. Vinexin contributes to autophagic decline in brain ageing across species. Cell Death Differ. 2022;29(5):1055–70.

86. Sanchez-Martin P, Lahuerta M, Viana R, Knecht E, Sanz P. Regulation of the autophagic PI3KC3 complex by laforin/malin E3-ubiquitin ligase, two proteins involved in Lafora disease. Biochim Biophys Acta, Mol Cell Res. 2020;1867(2):118613.
87. Aguado C, Sarkar S, Korolchuk VI, Criado O, Vernia S, Boya P, et al. Laforin, the most common protein mutated in Lafora disease, regulates autophagy. Hum Mol Genet. 2010;19(14):2867–76.
88. Suleiman J, Allingham-Hawkins D, Hashem M, Shamseldin HE, Alkuraya FS, El-Hattab AW. WDR45B-related intellectual disability, spastic quadriplegia, epilepsy, and cerebral hypoplasia: a consistent neurodevelopmental syndrome. Clin Genet. 2018;93(2):360–4.
89. Almannai M, Marafi D, Abdel-Salam GMH, Zaki MS, Duan R, Calame D, et al. El-Hattab-Alkuraya syndrome caused by biallelic WDR45B pathogenic variants: further delineation of the phenotype and genotype. Clin Genet. 2022;101(5–6):530–40.
90. Haack TB, Hogarth P, Kruer MC, Gregory A, Wieland T, Schwarzmayr T, et al. Exome sequencing reveals de novo WDR45 mutations causing a phenotypically distinct, X-linked dominant form of NBIA. Am J Hum Genet. 2012;91(6):1144–9.
91. Saitsu H, Nishimura T, Muramatsu K, Kodera H, Kumada S, Sugai K, et al. De novo mutations in the autophagy gene WDR45 cause static encephalopathy of childhood with neurodegeneration in adulthood. Nat Genet. 2013;45(4):445–9, 449e1.
92. Zarate YA, Jones JR, Jones MA, Millan F, Juusola J, Vertino-Bell A, et al. Lessons from a pair of siblings with BPAN. Eur J Hum Genet. 2016;24(7):1080–3.
93. Carvill GL, Liu A, Mandelstam S, Schneider A, Lacroix A, Zemel M, et al. Severe infantile onset developmental and epileptic encephalopathy caused by mutations in autophagy gene WDR45. Epilepsia. 2018;59(1):e5–e13.
94. Tsukida K, Muramatsu SI, Osaka H, Yamagata T, Muramatsu K. WDR45 variants cause ferrous iron loss due to impaired ferritinophagy associated with nuclear receptor coactivator 4 and WD repeat domain phosphoinositide interacting protein 4 reduction. Brain Commun. 2022;4(6):fcac304.
95. Zhao YG, Sun L, Miao G, Ji C, Zhao H, Sun H, et al. The autophagy gene Wdr45/Wipi4 regulates learning and memory function and axonal homeostasis. Autophagy. 2015;11(6):881–90.
96. Ji C, Zhao H, Li D, Sun H, Hao J, Chen R, et al. Role of Wdr45b in maintaining neural autophagy and cognitive function. Autophagy. 2020;16(4):615–25.
97. Aring L, Choi EK, Kopera H, Lanigan T, Iwase S, Klionsky DJ, et al. A neurodegeneration gene, WDR45, links impaired ferritinophagy to iron accumulation. J Neurochem. 2022;160(3):356–75.
98. Collier JJ, Guissart C, Olahova M, Sasorith S, Piron-Prunier F, Suomi F, et al. Developmental consequences of defective ATG7-mediated autophagy in humans. N Engl J Med. 2021;384(25):2406–17.
99. Kim M, Sandford E, Gatica D, Qiu Y, Liu X, Zheng Y, et al. Mutation in ATG5 reduces autophagy and leads to ataxia with developmental delay. elife. 2016;5:e12245.
100. Moreau K, Fleming A, Imarisio S, Lopez Ramirez A, Mercer JL, Jimenez-Sanchez M, et al. PICALM modulates autophagy activity and tau accumulation. Nat Commun. 2014;5:4998.
101. Maeda S, Yamamoto H, Kinch LN, Garza CM, Takahashi S, Otomo C, et al. Structure, lipid scrambling activity and role in autophagosome formation of ATG9A. Nat Struct Mol Biol. 2020;27(12):1194–201.
102. Zavodszky E, Seaman MN, Moreau K, Jimenez-Sanchez M, Breusegem SY, Harbour ME, et al. Mutation in VPS35 associated with Parkinson's disease impairs WASH complex association and inhibits autophagy. Nat Commun. 2014;5:3828.
103. Davies AK, Itzhak DN, Edgar JR, Archuleta TL, Hirst J, Jackson LP, et al. AP-4 vesicles contribute to spatial control of autophagy via RUSC-dependent peripheral delivery of ATG9A. Nat Commun. 2018;9(1):3958.

104. Ivankovic D, Drew J, Lesept F, White IJ, Lopez Domenech G, Tooze SA, et al. Axonal autophagosome maturation defect through failure of ATG9A sorting underpins pathology in AP-4 deficiency syndrome. Autophagy. 2020;16(3):391–407.

105. Bejarano E, Yuste A, Patel B, Stout RF Jr, Spray DC, Cuervo AM. Connexins modulate autophagosome biogenesis. Nat Cell Biol. 2014;16(5):401–14.

106. Jin SM, Lazarou M, Wang C, Kane LA, Narendra DP, Youle RJ. Mitochondrial membrane potential regulates PINK1 import and proteolytic destabilization by PARL. J Cell Biol. 2010;191(5):933–42.

107. Meissner C, Lorenz H, Weihofen A, Selkoe DJ, Lemberg MK. The mitochondrial intramembrane protease PARL cleaves human Pink1 to regulate Pink1 trafficking. J Neurochem. 2011;117(5):856–67.

108. Yamano K, Youle RJ. PINK1 is degraded through the N-end rule pathway. Autophagy. 2013;9(11):1758–69.

109. Kane LA, Lazarou M, Fogel AI, Li Y, Yamano K, Sarraf SA, et al. PINK1 phosphorylates ubiquitin to activate Parkin E3 ubiquitin ligase activity. J Cell Biol. 2014;205(2):143–53.

110. Kazlauskaite A, Kondapalli C, Gourlay R, Campbell DG, Ritorto MS, Hofmann K, et al. Parkin is activated by PINK1-dependent phosphorylation of ubiquitin at Ser65. Biochem J. 2014;460(1):127–39.

111. Koyano F, Okatsu K, Kosako H, Tamura Y, Go E, Kimura M, et al. Ubiquitin is phosphorylated by PINK1 to activate parkin. Nature. 2014;510(7503):162–6.

112. Heo JM, Ordureau A, Paulo JA, Rinehart J, Harper JW. The PINK1-PARKIN mitochondrial ubiquitylation pathway drives a program of OPTN/NDP52 recruitment and TBK1 activation to promote mitophagy. Mol Cell. 2015;60(1):7–20.

113. Lazarou M, Sliter DA, Kane LA, Sarraf SA, Wang C, Burman JL, et al. The ubiquitin kinase PINK1 recruits autophagy receptors to induce mitophagy. Nature. 2015;524(7565):309–14.

114. Valente EM, Abou-Sleiman PM, Caputo V, Muqit MM, Harvey K, Gispert S, et al. Hereditary early-onset Parkinson's disease caused by mutations in PINK1. Science. 2004;304(5674):1158–60.

115. Matsumine H, Saito M, Shimoda-Matsubayashi S, Tanaka H, Ishikawa A, Nakagawa-Hattori Y, et al. Localization of a gene for an autosomal recessive form of juvenile Parkinsonism to chromosome 6q25.2-27. Am J Hum Genet. 1997;60(3):588–96.

116. Kitada T, Asakawa S, Hattori N, Matsumine H, Yamamura Y, Minoshima S, et al. Mutations in the parkin gene cause autosomal recessive juvenile parkinsonism. Nature. 1998;392(6676):605–8.

117. Valente EM, Bentivoglio AR, Dixon PH, Ferraris A, Ialongo T, Frontali M, et al. Localization of a novel locus for autosomal recessive early-onset parkinsonism, PARK6, on human chromosome 1p35-p36. Am J Hum Genet. 2001;68(4):895–900.

118. Shimura H, Hattori N, Kubo S, Mizuno Y, Asakawa S, Minoshima S, et al. Familial Parkinson disease gene product, parkin, is a ubiquitin-protein ligase. Nat Genet. 2000;25(3):302–5.

119. Schneider SA, Alcalay RN. Neuropathology of genetic synucleinopathies with parkinsonism: review of the literature. Mov Disord. 2017;32(11):1504–23.

120. Grunewald A, Gegg ME, Taanman JW, King RH, Kock N, Klein C, et al. Differential effects of PINK1 nonsense and missense mutations on mitochondrial function and morphology. Exp Neurol. 2009;219(1):266–73.

121. Hoepken HH, Gispert S, Morales B, Wingerter O, Del Turco D, Mulsch A, et al. Mitochondrial dysfunction, peroxidation damage and changes in glutathione metabolism in PARK6. Neurobiol Dis. 2007;25(2):401–11.

122. Exner N, Treske B, Paquet D, Holmstrom K, Schiesling C, Gispert S, et al. Loss-of-function of human PINK1 results in mitochondrial pathology and can be rescued by parkin. J Neurosci. 2007;27(45):12413–8.

123. Zanellati MC, Monti V, Barzaghi C, Reale C, Nardocci N, Albanese A, et al. Mitochondrial dysfunction in Parkinson disease: evidence in mutant PARK2 fibroblasts. Front Genet. 2015;6:78.
124. Van Laar VS, Arnold B, Cassady SJ, Chu CT, Burton EA, Berman SB. Bioenergetics of neurons inhibit the translocation response of Parkin following rapid mitochondrial depolarization. Hum Mol Genet. 2011;20(5):927–40.
125. Rakovic A, Shurkewitsch K, Seibler P, Grunewald A, Zanon A, Hagenah J, et al. Phosphatase and tensin homolog (PTEN)-induced putative kinase 1 (PINK1)-dependent ubiquitination of endogenous Parkin attenuates mitophagy: study in human primary fibroblasts and induced pluripotent stem cell-derived neurons. J Biol Chem. 2013;288(4):2223–37.
126. Ge P, Dawson VL, Dawson TM. PINK1 and Parkin mitochondrial quality control: a source of regional vulnerability in Parkinson's disease. Mol Neurodegener. 2020;15(1):20.
127. Ashrafi G, Schlehe JS, LaVoie MJ, Schwarz TL. Mitophagy of damaged mitochondria occurs locally in distal neuronal axons and requires PINK1 and Parkin. J Cell Biol. 2014;206(5):655–70.
128. Seong E, Insolera R, Dulovic M, Kamsteeg EJ, Trinh J, Bruggemann N, et al. Mutations in VPS13D lead to a new recessive ataxia with spasticity and mitochondrial defects. Ann Neurol. 2018;83(6):1075–88.
129. Koh K, Ishiura H, Shimazaki H, Tsutsumiuchi M, Ichinose Y, Nan H, et al. VPS13D-related disorders presenting as a pure and complicated form of hereditary spastic paraplegia. Mol Genet Genomic Med. 2020;8(3):e1108.
130. Anding AL, Wang C, Chang TK, Sliter DA, Powers CM, Hofmann K, et al. Vps13D encodes a ubiquitin-binding protein that is required for the regulation of mitochondrial size and clearance. Curr Biol. 2018;28(2):287–95.e6.
131. Insolera R, Lorincz P, Wishnie AJ, Juhasz G, Collins CA. Mitochondrial fission, integrity and completion of mitophagy require separable functions of Vps13D in Drosophila neurons. PLoS Genet. 2021;17(8):e1009731.
132. Pankiv S, Clausen TH, Lamark T, Brech A, Bruun JA, Outzen H, et al. p62/SQSTM1 binds directly to Atg8/LC3 to facilitate degradation of ubiquitinated protein aggregates by autophagy. J Biol Chem. 2007;282(33):24131–45.
133. Ciuffa R, Lamark T, Tarafder AK, Guesdon A, Rybina S, Hagen WJ, et al. The selective autophagy receptor p62 forms a flexible filamentous helical scaffold. Cell Rep. 2015;11(5):748–58.
134. Lamark T, Perander M, Outzen H, Kristiansen K, Overvatn A, Michaelsen E, et al. Interaction codes within the family of mammalian Phox and Bem1p domain-containing proteins. J Biol Chem. 2003;278(36):34568–81.
135. Birgisdottir AB, Lamark T, Johansen T. The LIR motif – crucial for selective autophagy. J Cell Sci. 2013;126(Pt 15):3237–47.
136. Fecto F, Yan J, Vemula SP, Liu E, Yang Y, Chen W, et al. SQSTM1 mutations in familial and sporadic amyotrophic lateral sclerosis. Arch Neurol. 2011;68(11):1440–6.
137. Cavey JR, Ralston SH, Hocking LJ, Sheppard PW, Ciani B, Searle MS, et al. Loss of ubiquitin-binding associated with Paget's disease of bone p62 (SQSTM1) mutations. J Bone Miner Res. 2005;20(4):619–24.
138. Le Ber I, Camuzat A, Guerreiro R, Bouya-Ahmed K, Bras J, Nicolas G, et al. SQSTM1 mutations in French patients with frontotemporal dementia or frontotemporal dementia with amyotrophic lateral sclerosis. JAMA Neurol. 2013;70(11):1403–10.
139. Haack TB, Ignatius E, Calvo-Garrido J, Iuso A, Isohanni P, Maffezzini C, et al. Absence of the autophagy adaptor SQSTM1/p62 causes childhood-onset neurodegeneration with ataxia, dystonia, and gaze palsy. Am J Hum Genet. 2016;99(3):735–43.
140. Muto V, Flex E, Kupchinsky Z, Primiano G, Galehdari H, Dehghani M, et al. Biallelic SQSTM1 mutations in early-onset, variably progressive neurodegeneration. Neurology. 2018;91(4):e319–e30.

141. Goode A, Butler K, Long J, Cavey J, Scott D, Shaw B, et al. Defective recognition of LC3B by mutant SQSTM1/p62 implicates impairment of autophagy as a pathogenic mechanism in ALS-FTLD. Autophagy. 2016;12(7):1094–104.

142. Clausen TH, Lamark T, Isakson P, Finley K, Larsen KB, Brech A, et al. p62/SQSTM1 and ALFY interact to facilitate the formation of p62 bodies/ALIS and their degradation by autophagy. Autophagy. 2010;6(3):330–44.

143. Filimonenko M, Isakson P, Finley KD, Anderson M, Jeong H, Melia TJ, et al. The selective macroautophagic degradation of aggregated proteins requires the PI3P-binding protein Alfy. Mol Cell. 2010;38(2):265–79.

144. Kadir R, Harel T, Markus B, Perez Y, Bakhrat A, Cohen I, et al. ALFY-controlled DVL3 autophagy regulates Wnt signaling, determining human brain size. PLoS Genet. 2016;12(3):e1005919.

145. Le Duc D, Giulivi C, Hiatt SM, Napoli E, Panoutsopoulos A, Harlan De Crescenzo A, et al. Pathogenic WDFY3 variants cause neurodevelopmental disorders and opposing effects on brain size. Brain. 2019;142(9):2617–30.

146. Napoli E, Song G, Panoutsopoulos A, Riyadh MA, Kaushik G, Halmai J, et al. Beyond autophagy: a novel role for autism-linked Wdfy3 in brain mitophagy. Sci Rep. 2018;8(1):11348.

147. Fox LM, Kim K, Johnson CW, Chen S, Croce KR, Victor MB, et al. Huntington's disease pathogenesis is modified in vivo by Alfy/Wdfy3 and selective macroautophagy. Neuron. 2020;105(5):813–21.e6.

148. Wild P, Farhan H, McEwan DG, Wagner S, Rogov VV, Brady NR, et al. Phosphorylation of the autophagy receptor optineurin restricts Salmonella growth. Science. 2011;333(6039):228–33.

149. Wagner S, Carpentier I, Rogov V, Kreike M, Ikeda F, Lohr F, et al. Ubiquitin binding mediates the NF-kappaB inhibitory potential of ABIN proteins. Oncogene. 2008;27(26):3739–45.

150. Maruyama H, Morino H, Ito H, Izumi Y, Kato H, Watanabe Y, et al. Mutations of optineurin in amyotrophic lateral sclerosis. Nature. 2010;465(7295):223–6.

151. Hortobagyi T, Troakes C, Nishimura AL, Vance C, van Swieten JC, Seelaar H, et al. Optineurin inclusions occur in a minority of TDP-43 positive ALS and FTLD-TDP cases and are rarely observed in other neurodegenerative disorders. Acta Neuropathol. 2011;121(4):519–27.

152. Osawa T, Mizuno Y, Fujita Y, Takatama M, Nakazato Y, Okamoto K. Optineurin in neurodegenerative diseases. Neuropathology. 2011;31(6):569–74.

153. Rezaie T, Child A, Hitchings R, Brice G, Miller L, Coca-Prados M, et al. Adult-onset primary open-angle glaucoma caused by mutations in optineurin. Science. 2002;295(5557):1077–9.

154. Deng HX, Bigio EH, Zhai H, Fecto F, Ajroud K, Shi Y, et al. Differential involvement of optineurin in amyotrophic lateral sclerosis with or without SOD1 mutations. Arch Neurol. 2011;68(8):1057–61.

155. Feng SM, Che CH, Feng SY, Liu CY, Li LY, Li YX, et al. Novel mutation in optineurin causing aggressive ALS+/-frontotemporal dementia. Ann Clin Transl Neurol. 2019;6(12):2377–83.

156. Pottier C, Bieniek KF, Finch N, van de Vorst M, Baker M, Perkersen R, et al. Whole-genome sequencing reveals important role for TBK1 and OPTN mutations in frontotemporal lobar degeneration without motor neuron disease. Acta Neuropathol. 2015;130(1):77–92.

157. Wong YC, Holzbaur EL. Optineurin is an autophagy receptor for damaged mitochondria in parkin-mediated mitophagy that is disrupted by an ALS-linked mutation. Proc Natl Acad Sci USA. 2014;111(42):E4439–48.

158. Shen WC, Li HY, Chen GC, Chern Y, Tu PH. Mutations in the ubiquitin-binding domain of OPTN/optineurin interfere with autophagy-mediated degradation of misfolded proteins by a dominant-negative mechanism. Autophagy. 2015;11(4):685–700.

159. Tumbarello DA, Waxse BJ, Arden SD, Bright NA, Kendrick-Jones J, Buss F. Autophagy receptors link myosin VI to autophagosomes to mediate Tom1-dependent autophagosome maturation and fusion with the lysosome. Nat Cell Biol. 2012;14(10):1024–35.

160. Sundaramoorthy V, Walker AK, Tan V, Fifita JA, McCann EP, Williams KL, et al. Defects in optineurin- and myosin VI-mediated cellular trafficking in amyotrophic lateral sclerosis. Hum Mol Genet. 2017;26(17):3452.

161. Bansal M, Moharir SC, Sailasree SP, Sirohi K, Sudhakar C, Sarathi DP, et al. Optineurin promotes autophagosome formation by recruiting the autophagy-related Atg12-5-16L1 complex to phagophores containing the Wipi2 protein. J Biol Chem. 2018;293(1):132–47.

162. Song GJ, Jeon H, Seo M, Jo M, Suk K. Interaction between optineurin and Rab1a regulates autophagosome formation in neuroblastoma cells. J Neurosci Res. 2018;96(3):407–15.

163. Richter B, Sliter DA, Herhaus L, Stolz A, Wang C, Beli P, et al. Phosphorylation of OPTN by TBK1 enhances its binding to Ub chains and promotes selective autophagy of damaged mitochondria. Proc Natl Acad Sci USA. 2016;113(15):4039–44.

164. Kumar S, Gu Y, Abudu YP, Bruun JA, Jain A, Farzam F, et al. Phosphorylation of syntaxin 17 by TBK1 controls autophagy initiation. Dev Cell. 2019;49(1):130–44.e6.

165. Iida A, Hosono N, Sano M, Kamei T, Oshima S, Tokuda T, et al. Novel deletion mutations of OPTN in amyotrophic lateral sclerosis in Japanese. Neurobiol Aging. 2012;33(8):1843.e19–24.

166. van Blitterswijk M, van Vught PW, van Es MA, Schelhaas HJ, van der Kooi AJ, de Visser M, et al. Novel optineurin mutations in sporadic amyotrophic lateral sclerosis patients. Neurobiol Aging. 2012;33(5):1016.e1–7.

167. Cirulli ET, Lasseigne BN, Petrovski S, Sapp PC, Dion PA, Leblond CS, et al. Exome sequencing in amyotrophic lateral sclerosis identifies risk genes and pathways. Science. 2015;347(6229):1436–41.

168. Van Mossevelde S, van der Zee J, Gijselinck I, Engelborghs S, Sieben A, Van Langenhove T, et al. Clinical features of TBK1 carriers compared with C9orf72, GRN and non-mutation carriers in a Belgian cohort. Brain. 2016;139(Pt 2):452–67.

169. Freischmidt A, Wieland T, Richter B, Ruf W, Schaeffer V, Muller K, et al. Haploinsufficiency of TBK1 causes familial ALS and fronto-temporal dementia. Nat Neurosci. 2015;18(5):631–6.

170. Moore AS, Holzbaur EL. Dynamic recruitment and activation of ALS-associated TBK1 with its target optineurin are required for efficient mitophagy. Proc Natl Acad Sci USA. 2016;113(24):E3349–58.

171. Kurth I, Pamminger T, Hennings JC, Soehendra D, Huebner AK, Rotthier A, et al. Mutations in FAM134B, encoding a newly identified Golgi protein, cause severe sensory and autonomic neuropathy. Nat Genet. 2009;41(11):1179–81.

172. Khaminets A, Heinrich T, Mari M, Grumati P, Huebner AK, Akutsu M, et al. Regulation of endoplasmic reticulum turnover by selective autophagy. Nature. 2015;522(7556):354–8.

173. Bhaskara RM, Grumati P, Garcia-Pardo J, Kalayil S, Covarrubias-Pinto A, Chen W, et al. Curvature induction and membrane remodeling by FAM134B reticulon homology domain assist selective ER-phagy. Nat Commun. 2019;10(1):2370.

174. Liang JR, Lingeman E, Ahmed S, Corn JE. Atlastins remodel the endoplasmic reticulum for selective autophagy. J Cell Biol. 2018;217(10):3354–67.

175. Liu N, Zhao H, Zhao YG, Hu J, Zhang H. Atlastin 2/3 regulate ER targeting of the ULK1 complex to initiate autophagy. J Cell Biol. 2021;220(7):e202012091.

176. Chen Q, Xiao Y, Chai P, Zheng P, Teng J, Chen J. ATL3 is a tubular ER-phagy receptor for GABARAP-mediated selective autophagy. Curr Biol. 2019;29(5):846–55.e6.

177. Zhao X, Alvarado D, Rainier S, Lemons R, Hedera P, Weber CH, et al. Mutations in a newly identified GTPase gene cause autosomal dominant hereditary spastic paraplegia. Nat Genet. 2001;29(3):326–31.

178. Abel A, Fonknechten N, Hofer A, Durr A, Cruaud C, Voit T, et al. Early onset autosomal dominant spastic paraplegia caused by novel mutations in SPG3A. Neurogenetics. 2004;5(4):239–43.

179. Muglia M, Magariello A, Nicoletti G, Patitucci A, Gabriele AL, Conforti FL, et al. Further evidence that SPG3A gene mutations cause autosomal dominant hereditary spastic paraplegia. Ann Neurol. 2002;51(6):794–5.
180. Tessa A, Casali C, Damiano M, Bruno C, Fortini D, Patrono C, et al. SPG3A: an additional family carrying a new atlastin mutation. Neurology. 2002;59(12):2002–5.
181. D'Amico A, Tessa A, Sabino A, Bertini E, Santorelli FM, Servidei S. Incomplete penetrance in an SPG3A-linked family with a new mutation in the atlastin gene. Neurology. 2004;62(11):2138–9.
182. Guelly C, Zhu PP, Leonardis L, Papic L, Zidar J, Schabhuttl M, et al. Targeted high-throughput sequencing identifies mutations in atlastin-1 as a cause of hereditary sensory neuropathy type I. Am J Hum Genet. 2011;88(1):99–105.
183. Orlacchio A, Montieri P, Babalini C, Gaudiello F, Bernardi G, Kawarai T. Late-onset hereditary spastic paraplegia with thin corpus callosum caused by a new SPG3A mutation. J Neurol. 2011;258(7):1361–3.
184. Khan TN, Klar J, Tariq M, Anjum Baig S, Malik NA, Yousaf R, et al. Evidence for autosomal recessive inheritance in SPG3A caused by homozygosity for a novel ATL1 missense mutation. Eur J Hum Genet. 2014;22(10):1180–4.
185. Kornak U, Mademan I, Schinke M, Voigt M, Krawitz P, Hecht J, et al. Sensory neuropathy with bone destruction due to a mutation in the membrane-shaping atlastin GTPase 3. Brain. 2014;137(Pt 3):683–92.
186. Fischer D, Schabhuttl M, Wieland T, Windhager R, Strom TM, Auer-Grumbach M. A novel missense mutation confirms ATL3 as a gene for hereditary sensory neuropathy type 1. Brain. 2014;137(Pt 7):e286.
187. Xu H, Zhang C, Cao L, Song J, Xu X, Zhang B, et al. ATL3 gene mutation in a Chinese family with hereditary sensory neuropathy type 1F. J Peripher Nerv Syst. 2019;24(1):150–5.
188. Puls I, Jonnakuty C, LaMonte BH, Holzbaur EL, Tokito M, Mann E, et al. Mutant dynactin in motor neuron disease. Nat Genet. 2003;33(4):455–6.
189. Pankiv S, Alemu EA, Brech A, Bruun JA, Lamark T, Overvatn A, et al. FYCO1 is a Rab7 effector that binds to LC3 and PI3P to mediate microtubule plus end-directed vesicle transport. J Cell Biol. 2010;188(2):253–69.
190. Jordens I, Fernandez-Borja M, Marsman M, Dusseljee S, Janssen L, Calafat J, et al. The Rab7 effector protein RILP controls lysosomal transport by inducing the recruitment of dynein-dynactin motors. Curr Biol. 2001;11(21):1680–5.
191. Wijdeven RH, Janssen H, Nahidiazar L, Janssen L, Jalink K, Berlin I, et al. Cholesterol and ORP1L-mediated ER contact sites control autophagosome transport and fusion with the endocytic pathway. Nat Commun. 2016;7:11808.
192. Sun Q, Westphal W, Wong KN, Tan I, Zhong Q. Rubicon controls endosome maturation as a Rab7 effector. Proc Natl Acad Sci USA. 2010;107(45):19338–43.
193. Verhoeven K, De Jonghe P, Coen K, Verpoorten N, Auer-Grumbach M, Kwon JM, et al. Mutations in the small GTP-ase late endosomal protein RAB7 cause Charcot-Marie-Tooth type 2B neuropathy. Am J Hum Genet. 2003;72(3):722–7.
194. Houlden H, King RH, Muddle JR, Warner TT, Reilly MM, Orrell RW, et al. A novel RAB7 mutation associated with ulcero-mutilating neuropathy. Ann Neurol. 2004;56(4):586–90.
195. Meggouh F, Bienfait HM, Weterman MA, de Visser M, Baas F. Charcot-Marie-Tooth disease due to a de novo mutation of the RAB7 gene. Neurology. 2006;67(8):1476–8.
196. Wang Z, Miao G, Xue X, Guo X, Yuan C, Wang Z, et al. The Vici syndrome protein EPG5 is a Rab7 effector that determines the fusion specificity of autophagosomes with late endosomes/lysosomes. Mol Cell. 2016;63(5):781–95.

197. Cullup T, Kho AL, Dionisi-Vici C, Brandmeier B, Smith F, Urry Z, et al. Recessive mutations in EPG5 cause Vici syndrome, a multisystem disorder with defective autophagy. Nat Genet. 2013;45(1):83–7.

198. Hori I, Otomo T, Nakashima M, Miya F, Negishi Y, Shiraishi H, et al. Defects in autophagosome-lysosome fusion underlie Vici syndrome, a neurodevelopmental disorder with multisystem involvement. Sci Rep. 2017;7(1):3552.

199. Wartosch L, Gunesdogan U, Graham SC, Luzio JP. Recruitment of VPS33A to HOPS by VPS16 is required for lysosome fusion with endosomes and autophagosomes. Traffic. 2015;16(7):727–42.

200. Zhang J, Lachance V, Schaffner A, Li X, Fedick A, Kaye LE, et al. A founder mutation in VPS11 causes an autosomal recessive leukoencephalopathy linked to autophagic defects. PLoS Genet. 2016;12(4):e1005848.

201. Otomo A, Kunita R, Suzuki-Utsunomiya K, Ikeda JE, Hadano S. Defective relocalization of ALS2/alsin missense mutants to Rac1-induced macropinosomes accounts for loss of their cellular function and leads to disturbed amphisome formation. FEBS Lett. 2011;585(5):730–6.

202. Ravikumar B, Imarisio S, Sarkar S, O'Kane CJ, Rubinsztein DC. Rab5 modulates aggregation and toxicity of mutant huntingtin through macroautophagy in cell and fly models of Huntington disease. J Cell Sci. 2008;121(Pt 10):1649–60.

203. Wu JJ, Cai A, Greenslade JE, Higgins NR, Fan C, Le NTT, et al. ALS/FTD mutations in UBQLN2 impede autophagy by reducing autophagosome acidification through loss of function. Proc Natl Acad Sci USA. 2020;117(26):15230–41.

204. Senturk M, Lin G, Zuo Z, Mao D, Watson E, Mikos AG, et al. Ubiquilins regulate autophagic flux through mTOR signalling and lysosomal acidification. Nat Cell Biol. 2019;21(3):384–96.

205. Deng HX, Chen W, Hong ST, Boycott KM, Gorrie GH, Siddique N, et al. Mutations in UBQLN2 cause dominant X-linked juvenile and adult-onset ALS and ALS/dementia. Nature. 2011;477(7363):211–5.

206. Skibinski G, Parkinson NJ, Brown JM, Chakrabarti L, Lloyd SL, Hummerich H, et al. Mutations in the endosomal ESCRTIII-complex subunit CHMP2B in frontotemporal dementia. Nat Genet. 2005;37(8):806–8.

207. Parkinson N, Ince PG, Smith MO, Highley R, Skibinski G, Andersen PM, et al. ALS phenotypes with mutations in CHMP2B (charged multivesicular body protein 2B). Neurology. 2006;67(6):1074–7.

208. Filimonenko M, Stuffers S, Raiborg C, Yamamoto A, Malerod L, Fisher EM, et al. Functional multivesicular bodies are required for autophagic clearance of protein aggregates associated with neurodegenerative disease. J Cell Biol. 2007;179(3):485–500.

209. Lee JA, Beigneux A, Ahmad ST, Young SG, Gao FB. ESCRT-III dysfunction causes autophagosome accumulation and neurodegeneration. Curr Biol. 2007;17(18):1561–7.

210. Akizu N, Cantagrel V, Zaki MS, Al-Gazali L, Wang X, Rosti RO, et al. Biallelic mutations in SNX14 cause a syndromic form of cerebellar atrophy and lysosome-autophagosome dysfunction. Nat Genet. 2015;47(5):528–34.

211. Bryant D, Liu Y, Datta S, Hariri H, Seda M, Anderson G, et al. SNX14 mutations affect endoplasmic reticulum-associated neutral lipid metabolism in autosomal recessive spinocerebellar ataxia 20. Hum Mol Genet. 2018;27(11):1927–40.

212. Schondorf DC, Aureli M, McAllister FE, Hindley CJ, Mayer F, Schmid B, et al. iPSC-derived neurons from GBA1-associated Parkinson's disease patients show autophagic defects and impaired calcium homeostasis. Nat Commun. 2014;5:4028.

213. Bae EJ, Yang NY, Lee C, Lee HJ, Kim S, Sardi SP, et al. Loss of glucocerebrosidase 1 activity causes lysosomal dysfunction and alpha-synuclein aggregation. Exp Mol Med. 2015;47(3):e153.

214. Kumar M, Srikanth MP, Deleidi M, Hallett PJ, Isacson O, Feldman RA. Acid ceramidase involved in pathogenic cascade leading to accumulation of alpha-synuclein in iPSC model of GBA1-associated Parkinson's disease. Hum Mol Genet. 2023;32(11):1888–900.

215. Li H, Ham A, Ma TC, Kuo SH, Kanter E, Kim D, et al. Mitochondrial dysfunction and mitophagy defect triggered by heterozygous GBA mutations. Autophagy. 2019;15(1):113–30.

216. Arrant AE, Roth JR, Boyle NR, Kashyap SN, Hoffmann MQ, Murchison CF, et al. Impaired beta-glucocerebrosidase activity and processing in frontotemporal dementia due to progranulin mutations. Acta Neuropathol Commun. 2019;7(1):218.

217. Valdez C, Ysselstein D, Young TJ, Zheng J, Krainc D. Progranulin mutations result in impaired processing of prosaposin and reduced glucocerebrosidase activity. Hum Mol Genet. 2020;29(5):716–26.

218. Zhou X, Sun L, Bracko O, Choi JW, Jia Y, Nana AL, et al. Impaired prosaposin lysosomal trafficking in frontotemporal lobar degeneration due to progranulin mutations. Nat Commun. 2017;8:15277.

219. Kao AW, McKay A, Singh PP, Brunet A, Huang EJ. Progranulin, lysosomal regulation and neurodegenerative disease. Nat Rev Neurosci. 2017;18(6):325–33.

220. Chang MC, Srinivasan K, Friedman BA, Suto E, Modrusan Z, Lee WP, et al. Progranulin deficiency causes impairment of autophagy and TDP-43 accumulation. J Exp Med. 2017;214(9):2611–28.

221. Elia LP, Mason AR, Alijagic A, Finkbeiner S. Genetic regulation of neuronal progranulin reveals a critical role for the autophagy-lysosome pathway. J Neurosci. 2019;39(17):3332–44.

222. Bento CF, Ashkenazi A, Jimenez-Sanchez M, Rubinsztein DC. The Parkinson's disease-associated genes ATP13A2 and SYT11 regulate autophagy via a common pathway. Nat Commun. 2016;7:11803.

223. Dehay B, Ramirez A, Martinez-Vicente M, Perier C, Canron MH, Doudnikoff E, et al. Loss of P-type ATPase ATP13A2/PARK9 function induces general lysosomal deficiency and leads to Parkinson disease neurodegeneration. Proc Natl Acad Sci USA. 2012;109(24):9611–6.

224. Wang R, Tan J, Chen T, Han H, Tian R, Tan Y, et al. ATP13A2 facilitates HDAC6 recruitment to lysosome to promote autophagosome-lysosome fusion. J Cell Biol. 2019;218(1):267–84.

225. Chang J, Lee S, Blackstone C. Spastic paraplegia proteins spastizin and spatacsin mediate autophagic lysosome reformation. J Clin Invest. 2014;124(12):5249–62.

226. Varga RE, Khundadze M, Damme M, Nietzsche S, Hoffmann B, Stauber T, et al. In vivo evidence for lysosome depletion and impaired autophagic clearance in hereditary spastic paraplegia type SPG11. PLoS Genet. 2015;11(8):e1005454.

227. Vantaggiato C, Panzeri E, Castelli M, Citterio A, Arnoldi A, Santorelli FM, et al. ZFYVE26/SPASTIZIN and SPG11/SPATACSIN mutations in hereditary spastic paraplegia types AR-SPG15 and AR-SPG11 have different effects on autophagy and endocytosis. Autophagy. 2019;15(1):34–57.

228. Vantaggiato C, Crimella C, Airoldi G, Polishchuk R, Bonato S, Brighina E, et al. Defective autophagy in spastizin mutated patients with hereditary spastic paraparesis type 15. Brain. 2013;136(Pt 10):3119–39.

229. Hirst J, Barlow LD, Francisco GC, Sahlender DA, Seaman MN, Dacks JB, et al. The fifth adaptor protein complex. PLoS Biol. 2011;9(10):e1001170.

230. Khundadze M, Ribaudo F, Hussain A, Rosentreter J, Nietzsche S, Thelen M, et al. A mouse model for SPG48 reveals a block of autophagic flux upon disruption of adaptor protein complex five. Neurobiol Dis. 2019;127:419–31.

231. Chong CM, Ke M, Tan Y, Huang Z, Zhang K, Ai N, et al. Presenilin 1 deficiency suppresses autophagy in human neural stem cells through reducing gamma-secretase-independent ERK/CREB signaling. Cell Death Dis. 2018;9(9):879.

232. Lee JH, Yu WH, Kumar A, Lee S, Mohan PS, Peterhoff CM, et al. Lysosomal proteolysis and autophagy require presenilin 1 and are disrupted by Alzheimer-related PS1 mutations. Cell. 2010;141(7):1146–58.
233. Lee JH, McBrayer MK, Wolfe DM, Haslett LJ, Kumar A, Sato Y, et al. Presenilin 1 maintains lysosomal Ca(2+) homeostasis via TRPML1 by regulating vATPase-mediated lysosome acidification. Cell Rep. 2015;12(9):1430–44.
234. Tanik SA, Schultheiss CE, Volpicelli-Daley LA, Brunden KR, Lee VM. Lewy body-like alpha-synuclein aggregates resist degradation and impair macroautophagy. J Biol Chem. 2013;288(21):15194–210.
235. Volpicelli-Daley LA, Gamble KL, Schultheiss CE, Riddle DM, West AB, Lee VM. Formation of alpha-synuclein Lewy neurite-like aggregates in axons impedes the transport of distinct endosomes. Mol Biol Cell. 2014;25(25):4010–23.
236. Winslow AR, Chen CW, Corrochano S, Acevedo-Arozena A, Gordon DE, Peden AA, et al. Alpha-Synuclein impairs macroautophagy: implications for Parkinson's disease. J Cell Biol. 2010;190(6):1023–37.
237. Siddique I, Kamble K, Gupta S, Solanki K, Bhola S, Ahsan N, et al. ARL6IP5 ameliorates alpha-synuclein burden by inducing autophagy via preventing ubiquitination and degradation of ATG12. Int J Mol Sci. 2023;24(13):10499.
238. Cuddy LK, Wani WY, Morella ML, Pitcairn C, Tsutsumi K, Fredriksen K, et al. Stress-induced cellular clearance is mediated by the SNARE protein ykt6 and disrupted by alpha-synuclein. Neuron. 2019;104(5):869–84.e11.
239. Pitcairn C, Murata N, Zalon AJ, Stojkovska I, Mazzulli JR. Impaired autophagic-lysosomal fusion in Parkinson's patient midbrain neurons occurs through loss of ykt6 and is rescued by farnesyltransferase inhibition. J Neurosci. 2023;43(14):2615–29.
240. Jeong A, Cheng S, Zhong R, Bennett DA, Bergo MO, Li L. Protein farnesylation is upregulated in Alzheimer's human brains and neuron-specific suppression of farnesyltransferase mitigates pathogenic processes in Alzheimer's model mice. Acta Neuropathol Commun. 2021;9(1):129.
241. Decressac M, Mattsson B, Weikop P, Lundblad M, Jakobsson J, Bjorklund A. TFEB-mediated autophagy rescues midbrain dopamine neurons from alpha-synuclein toxicity. Proc Natl Acad Sci USA. 2013;110(19):E1817–26.
242. Martinez-Vicente M, Talloczy Z, Wong E, Tang G, Koga H, Kaushik S, et al. Cargo recognition failure is responsible for inefficient autophagy in Huntington's disease. Nat Neurosci. 2010;13(5):567–76.
243. Ochaba J, Lukacsovich T, Csikos G, Zheng S, Margulis J, Salazar L, et al. Potential function for the Huntingtin protein as a scaffold for selective autophagy. Proc Natl Acad Sci USA. 2014;111(47):16889–94.
244. Rui YN, Xu Z, Patel B, Chen Z, Chen D, Tito A, et al. Huntingtin functions as a scaffold for selective macroautophagy. Nat Cell Biol. 2015;17(3):262–75.
245. Wong YC, Holzbaur EL. The regulation of autophagosome dynamics by huntingtin and HAP1 is disrupted by expression of mutant huntingtin, leading to defective cargo degradation. J Neurosci. 2014;34(4):1293–305.
246. Sampognaro PJ, Arya S, Knudsen GM, Gunderson EL, Sandoval-Perez A, Hodul M, et al. Correction: mutations in alpha-synuclein, TDP-43 and tau prolong protein half-life through diminished degradation by lysosomal proteases. Mol Neurodegener. 2023;18(1):34.
247. Ravikumar B, Vacher C, Berger Z, Davies JE, Luo S, Oroz LG, et al. Inhibition of mTOR induces autophagy and reduces toxicity of polyglutamine expansions in fly and mouse models of Huntington disease. Nat Genet. 2004;36(6):585–95.

248. Sarkar S, Krishna G, Imarisio S, Saiki S, O'Kane CJ, Rubinsztein DC. A rational mechanism for combination treatment of Huntington's disease using lithium and rapamycin. Hum Mol Genet. 2008;17(2):170–8.

249. Malagelada C, Jin ZH, Jackson-Lewis V, Przedborski S, Greene LA. Rapamycin protects against neuron death in in vitro and in vivo models of Parkinson's disease. J Neurosci. 2010;30(3):1166–75.

250. Spilman P, Podlutskaya N, Hart MJ, Debnath J, Gorostiza O, Bredesen D, et al. Inhibition of mTOR by rapamycin abolishes cognitive deficits and reduces amyloid-beta levels in a mouse model of Alzheimer's disease. PLoS One. 2010;5(4):e9979.

251. Ozcelik S, Fraser G, Castets P, Schaeffer V, Skachokova Z, Breu K, et al. Rapamycin attenuates the progression of tau pathology in P301S tau transgenic mice. PLoS One. 2013;8(5):e62459.

252. Jiang T, Yu JT, Zhu XC, Zhang QQ, Cao L, Wang HF, et al. Temsirolimus attenuates tauopathy in vitro and in vivo by targeting tau hyperphosphorylation and autophagic clearance. Neuropharmacology. 2014;85:121–30.

253. Sarkar S, Ravikumar B, Floto RA, Rubinsztein DC. Rapamycin and mTOR-independent autophagy inducers ameliorate toxicity of polyglutamine-expanded huntingtin and related proteinopathies. Cell Death Differ. 2009;16(1):46–56.

254. Kim YC, Guan KL. mTOR: a pharmacologic target for autophagy regulation. J Clin Invest. 2015;125(1):25–32.

255. Thoreen CC, Kang SA, Chang JW, Liu Q, Zhang J, Gao Y, et al. An ATP-competitive mammalian target of rapamycin inhibitor reveals rapamycin-resistant functions of mTORC1. J Biol Chem. 2009;284(12):8023–32.

256. Chong CM, Tan Y, Tong J, Ke M, Zhang K, Yan L, et al. Presenilin-1 F105C mutation leads to tau accumulation in human neurons via the Akt/mTORC1 signaling pathway. Cell Biosci. 2022;12(1):131.

257. Siddiqi FH, Menzies FM, Lopez A, Stamatakou E, Karabiyik C, Ureshino R, et al. Felodipine induces autophagy in mouse brains with pharmacokinetics amenable to repurposing. Nat Commun. 2019;10(1):1817.

258. Williams A, Sarkar S, Cuddon P, Ttofi EK, Saiki S, Siddiqi FH, et al. Novel targets for Huntington's disease in an mTOR-independent autophagy pathway. Nat Chem Biol. 2008;4(5):295–305.

259. Medeiros R, Kitazawa M, Chabrier MA, Cheng D, Baglietto-Vargas D, Kling A, et al. Calpain inhibitor A-705253 mitigates Alzheimer's disease-like pathology and cognitive decline in aged 3xTgAD mice. Am J Pathol. 2012;181(2):616–25.

260. Sarkar S, Rubinsztein DC. Inositol and IP3 levels regulate autophagy: biology and therapeutic speculations. Autophagy. 2006;2(2):132–4.

261. Sarkar S, Floto RA, Berger Z, Imarisio S, Cordenier A, Pasco M, et al. Lithium induces autophagy by inhibiting inositol monophosphatase. J Cell Biol. 2005;170(7):1101–11.

262. Sarkar S, Davies JE, Huang Z, Tunnacliffe A, Rubinsztein DC. Trehalose, a novel mTOR-independent autophagy enhancer, accelerates the clearance of mutant huntingtin and alpha-synuclein. J Biol Chem. 2007;282(8):5641–52.

263. Rusmini P, Cortese K, Crippa V, Cristofani R, Cicardi ME, Ferrari V, et al. Trehalose induces autophagy via lysosomal-mediated TFEB activation in models of motoneuron degeneration. Autophagy. 2019;15(4):631–51.

264. Palmieri M, Pal R, Nelvagal HR, Lotfi P, Stinnett GR, Seymour ML, et al. mTORC1-independent TFEB activation via Akt inhibition promotes cellular clearance in neurodegenerative storage diseases. Nat Commun. 2017;8:14338.

265. Pierzynowska K, Gaffke L, Cyske Z, Wegrzyn G. Genistein induces degradation of mutant huntingtin in fibroblasts from Huntington's disease patients. Metab Brain Dis. 2019;34(3):715–20.

266. Arguello G, Balboa E, Tapia PJ, Castro J, Yanez MJ, Mattar P, et al. Genistein activates transcription factor EB and corrects Niemann-Pick C phenotype. Int J Mol Sci. 2021;22(8):4220.
267. Moskot M, Montefusco S, Jakobkiewicz-Banecka J, Mozolewski P, Wegrzyn A, Di Bernardo D, et al. The phytoestrogen genistein modulates lysosomal metabolism and transcription factor EB (TFEB) activation. J Biol Chem. 2014;289(24):17054–69.
268. Jiao F, Zhou B, Meng L. The regulatory mechanism and therapeutic potential of transcription factor EB in neurodegenerative diseases. CNS Neurosci Ther. 2023;29(1):37–59.
269. Pierzynowska K, Podlacha M, Gaffke L, Majkutewicz I, Mantej J, Wegrzyn A, et al. Autophagy-dependent mechanism of genistein-mediated elimination of behavioral and biochemical defects in the rat model of sporadic Alzheimer's disease. Neuropharmacology. 2019;148:332–46.
270. Lee JK, Jin HK, Park MH, Kim BR, Lee PH, Nakauchi H, et al. Acid sphingomyelinase modulates the autophagic process by controlling lysosomal biogenesis in Alzheimer's disease. J Exp Med. 2014;211(8):1551–70.
271. Luengo E, Buendia I, Fernandez-Mendivil C, Trigo-Alonso P, Negredo P, Michalska P, et al. Pharmacological doses of melatonin impede cognitive decline in tau-related Alzheimer models, once tauopathy is initiated, by restoring the autophagic flux. J Pineal Res. 2019;67(1):e12578.
272. Lonskaya I, Hebron ML, Selby ST, Turner RS, Moussa CE. Nilotinib and bosutinib modulate pre-plaque alterations of blood immune markers and neuro-inflammation in Alzheimer's disease models. Neuroscience. 2015;304:316–27.
273. Lonskaya I, Hebron ML, Desforges NM, Schachter JB, Moussa CE. Nilotinib-induced autophagic changes increase endogenous parkin level and ubiquitination, leading to amyloid clearance. J Mol Med (Berl). 2014;92(4):373–86.
274. Luo R, Su LY, Li G, Yang J, Liu Q, Yang LX, et al. Activation of PPARA-mediated autophagy reduces Alzheimer disease-like pathology and cognitive decline in a murine model. Autophagy. 2020;16(1):52–69.
275. Ferretta A, Gaballo A, Tanzarella P, Piccoli C, Capitanio N, Nico B, et al. Effect of resveratrol on mitochondrial function: implications in parkin-associated familiar Parkinson's disease. Biochim Biophys Acta. 2014;1842(7):902–15.
276. Sarkar S, Perlstein EO, Imarisio S, Pineau S, Cordenier A, Maglathlin RL, et al. Small molecules enhance autophagy and reduce toxicity in Huntington's disease models. Nat Chem Biol. 2007;3(6):331–8.
277. Rubinsztein DC. The roles of intracellular protein-degradation pathways in neurodegeneration. Nature. 2006;443(7113):780–6.
278. Ciccocioppo F, Bologna G, Ercolino E, Pierdomenico L, Simeone P, Lanuti P, et al. Neurodegenerative diseases as proteinopathies-driven immune disorders. Neural Regen Res. 2020;15(5):850–6.
279. Sehgal SN, Baker H, Vezina C. Rapamycin (AY-22,989), a new antifungal antibiotic. II. Fermentation, isolation and characterization. J Antibiot (Tokyo). 1975;28(10):727–32.
280. Yuan HX, Guan KL. Structural insights of mTOR complex 1. Cell Res. 2016;26(3):267–8.
281. Aylett CH, Sauer E, Imseng S, Boehringer D, Hall MN, Ban N, et al. Architecture of human mTOR complex 1. Science. 2016;351(6268):48–52.
282. Yip CK, Murata K, Walz T, Sabatini DM, Kang SA. Structure of the human mTOR complex I and its implications for rapamycin inhibition. Mol Cell. 2010;38(5):768–74.
283. Yang H, Rudge DG, Koos JD, Vaidialingam B, Yang HJ, Pavletich NP. mTOR kinase structure, mechanism and regulation. Nature. 2013;497(7448):217–23.
284. Caccamo A, Majumder S, Richardson A, Strong R, Oddo S. Molecular interplay between mammalian target of rapamycin (mTOR), amyloid-beta, and Tau: effects on cognitive impairments. J Biol Chem. 2010;285(17):13107–20.

285. Wander SA, Hennessy BT, Slingerland JM. Next-generation mTOR inhibitors in clinical oncology: how pathway complexity informs therapeutic strategy. J Clin Invest. 2011;121(4):1231–41.
286. Cortes CJ, Qin K, Cook J, Solanki A, Mastrianni JA. Rapamycin delays disease onset and prevents PrP plaque deposition in a mouse model of Gerstmann-Straussler-Scheinker disease. J Neurosci. 2012;32(36):12396–405.
287. Menzies FM, Huebener J, Renna M, Bonin M, Riess O, Rubinsztein DC. Autophagy induction reduces mutant ataxin-3 levels and toxicity in a mouse model of spinocerebellar ataxia type 3. Brain. 2010;133(Pt 1):93–104.
288. Jiang T, Yu JT, Zhu XC, Tan MS, Wang HF, Cao L, et al. Temsirolimus promotes autophagic clearance of amyloid-beta and provides protective effects in cellular and animal models of Alzheimer's disease. Pharmacol Res. 2014;81:54–63.
289. Siracusa R, Paterniti I, Cordaro M, Crupi R, Bruschetta G, Campolo M, et al. Neuroprotective effects of temsirolimus in animal models of Parkinson's disease. Mol Neurobiol. 2018;55(3):2403–19.
290. Frederick C, Ando K, Leroy K, Heraud C, Suain V, Buee L, et al. Rapamycin ester analog CCI-779/Temsirolimus alleviates tau pathology and improves motor deficit in mutant tau transgenic mice. J Alzheimers Dis. 2015;44(4):1145–56.
291. Rivera VM, Squillace RM, Miller D, Berk L, Wardwell SD, Ning Y, et al. Ridaforolimus (AP23573; MK-8669), a potent mTOR inhibitor, has broad antitumor activity and can be optimally administered using intermittent dosing regimens. Mol Cancer Ther. 2011;10(6):1059–71.
292. Lebwohl D, Anak O, Sahmoud T, Klimovsky J, Elmroth I, Haas T, et al. Development of everolimus, a novel oral mTOR inhibitor, across a spectrum of diseases. Ann N Y Acad Sci. 2013;1291:14–32.
293. Zhao J, Zhai B, Gygi SP, Goldberg AL. mTOR inhibition activates overall protein degradation by the ubiquitin proteasome system as well as by autophagy. Proc Natl Acad Sci USA. 2015;112(52):15790–7.
294. Kim DH, Sarbassov DD, Ali SM, King JE, Latek RR, Erdjument-Bromage H, et al. mTOR interacts with raptor to form a nutrient-sensitive complex that signals to the cell growth machinery. Cell. 2002;110(2):163–75.
295. Frake RA, Ricketts T, Menzies FM, Rubinsztein DC. Autophagy and neurodegeneration. J Clin Invest. 2015;125(1):65–74.
296. Huang H, Cui W, Qiu W, Zhu M, Zhao R, Zeng D, et al. Impaired wound healing results from the dysfunction of the Akt/mTOR pathway in diabetic rats. J Dermatol Sci. 2015;79(3):241–51.
297. Soefje SA, Karnad A, Brenner AJ. Common toxicities of mammalian target of rapamycin inhibitors. Target Oncol. 2011;6(2):125–9.
298. Panda PK, Fahrner A, Vats S, Seranova E, Sharma V, Chipara M, et al. Chemical screening approaches enabling drug discovery of autophagy modulators for biomedical applications in human diseases. Front Cell Dev Biol. 2019;7:38.
299. Zhang L, Yu J, Pan H, Hu P, Hao Y, Cai W, et al. Small molecule regulators of autophagy identified by an image-based high-throughput screen. Proc Natl Acad Sci USA. 2007;104(48):19023–8.
300. Saltiel E, Ellrodt AG, Monk JP, Langley MS. Felodipine. A review of its pharmacodynamic and pharmacokinetic properties, and therapeutic use in hypertension. Drugs. 1988;36(4):387–428.
301. Park HW, Park H, Semple IA, Jang I, Ro SH, Kim M, et al. Pharmacological correction of obesity-induced autophagy arrest using calcium channel blockers. Nat Commun. 2014;5:4834.
302. Zhang X, Chen S, Lu K, Wang F, Deng J, Xu Z, et al. Verapamil ameliorates motor neuron degeneration and improves lifespan in the SOD1(G93A) mouse model of ALS by enhancing autophagic flux. Aging Dis. 2019;10(6):1159–73.
303. Lin CW, Chen YS, Lin CC, Chen YJ, Lo GH, Lee PH, et al. Amiodarone as an autophagy promoter reduces liver injury and enhances liver regeneration and survival in mice after partial hepatectomy. Sci Rep. 2015;5:15807.

304. Russo R, Berliocchi L, Adornetto A, Varano GP, Cavaliere F, Nucci C, et al. Calpain-mediated cleavage of Beclin-1 and autophagy deregulation following retinal ischemic injury in vivo. Cell Death Dis. 2011;2(4):e144.

305. Ugland H, Naderi S, Brech A, Collas P, Blomhoff HK. cAMP induces autophagy via a novel pathway involving ERK, cyclin E and Beclin 1. Autophagy. 2011;7(10):1199–211.

306. Taylor CW. Regulation of IP(3) receptors by cyclic AMP. Cell Calcium. 2017;63:48–52.

307. Schreiber SL. The rise of molecular glues. Cell. 2021;184(1):3–9.

308. Paiva SL, Crews CM. Targeted protein degradation: elements of PROTAC design. Curr Opin Chem Biol. 2019;50:111–9.

309. Ji CH, Lee MJ, Kim HY, Heo AJ, Park DY, Kim YK, et al. Targeted protein degradation via the autophagy-lysosome system: AUTOTAC (AUTOphagy-TArgeting Chimera). Autophagy. 2022;18(9):2259–62.

310. Takahashi D, Moriyama J, Nakamura T, Miki E, Takahashi E, Sato A, et al. AUTACs: cargo-specific degraders using selective autophagy. Mol Cell. 2019;76(5):797–810.e10.

311. Zhao L, Zhao J, Zhong K, Tong A, Jia D. Targeted protein degradation: mechanisms, strategies and application. Signal Transduct Target Ther. 2022;7(1):113.

312. Li Z, Wang C, Wang Z, Zhu C, Li J, Sha T, et al. Allele-selective lowering of mutant HTT protein by HTT-LC3 linker compounds. Nature. 2019;575(7781):203–9.

313. Ji CH, Kim HY, Lee MJ, Heo AJ, Park DY, Lim S, et al. The AUTOTAC chemical biology platform for targeted protein degradation via the autophagy-lysosome system. Nat Commun. 2022;13(1):904.

314. Dosa A, Csizmadia T. The role of K63-linked polyubiquitin in several types of autophagy. Biol Futur. 2022;73(2):137–48.

315. Sawa T, Zaki MH, Okamoto T, Akuta T, Tokutomi Y, Kim-Mitsuyama S, et al. Protein S-guanylation by the biological signal 8-nitroguanosine 3′,5′-cyclic monophosphate. Nat Chem Biol. 2007;3(11):727–35.

316. Dehay B, Bove J, Rodriguez-Muela N, Perier C, Recasens A, Boya P, et al. Pathogenic lysosomal depletion in Parkinson's disease. J Neurosci. 2010;30(37):12535–44.

317. Koh JY, Kim HN, Hwang JJ, Kim YH, Park SE. Lysosomal dysfunction in proteinopathic neurodegenerative disorders: possible therapeutic roles of cAMP and zinc. Mol Brain. 2019;12(1):18.

318. Medina DL, Fraldi A, Bouche V, Annunziata F, Mansueto G, Spampanato C, et al. Transcriptional activation of lysosomal exocytosis promotes cellular clearance. Dev Cell. 2011;21(3):421–30.

319. Settembre C, Di Malta C, Polito VA, Garcia Arencibia M, Vetrini F, Erdin S, et al. TFEB links autophagy to lysosomal biogenesis. Science. 2011;332(6036):1429–33.

320. Lee HJ, Yoon YS, Lee SJ. Mechanism of neuroprotection by trehalose: controversy surrounding autophagy induction. Cell Death Dis. 2018;9(7):712.

321. Crowe JH. Trehalose as a "chemical chaperone": fact and fantasy. Adv Exp Med Biol. 2007;594:143–58.

322. Tapia H, Koshland DE. Trehalose is a versatile and long-lived chaperone for desiccation tolerance. Curr Biol. 2014;24(23):2758–66.

323. DeBosch BJ, Heitmeier MR, Mayer AL, Higgins CB, Crowley JR, Kraft TE, et al. Trehalose inhibits solute carrier 2A (SLC2A) proteins to induce autophagy and prevent hepatic steatosis. Sci Signal. 2016;9(416):ra21.

324. Tsai TH. Concurrent measurement of unbound genistein in the blood, brain and bile of anesthetized rats using microdialysis and its pharmacokinetic application. J Chromatogr A. 2005;1073(1–2):317–22.

325. Pickford F, Masliah E, Britschgi M, Lucin K, Narasimhan R, Jaeger PA, et al. The autophagy-related protein beclin 1 shows reduced expression in early Alzheimer disease and regulates amyloid beta accumulation in mice. J Clin Invest. 2008;118(6):2190–9.

326. Rocchi A, Yamamoto S, Ting T, Fan Y, Sadleir K, Wang Y, et al. A Becn1 mutation mediates hyperactive autophagic sequestration of amyloid oligomers and improved cognition in Alzheimer's disease. PLoS Genet. 2017;13(8):e1006962.
327. Spencer B, Potkar R, Trejo M, Rockenstein E, Patrick C, Gindi R, et al. Beclin 1 gene transfer activates autophagy and ameliorates the neurodegenerative pathology in alpha-synuclein models of Parkinson's and Lewy body diseases. J Neurosci. 2009;29(43):13578–88.
328. Nascimento-Ferreira I, Santos-Ferreira T, Sousa-Ferreira L, Auregan G, Onofre I, Alves S, et al. Overexpression of the autophagic beclin-1 protein clears mutant ataxin-3 and alleviates Machado-Joseph disease. Brain. 2011;134(Pt 5):1400–15.
329. Lopez A, Lee SE, Wojta K, Ramos EM, Klein E, Chen J, et al. A152T tau allele causes neurodegeneration that can be ameliorated in a zebrafish model by autophagy induction. Brain. 2017;140(4):1128–46.
330. Crews L, Spencer B, Desplats P, Patrick C, Paulino A, Rockenstein E, et al. Selective molecular alterations in the autophagy pathway in patients with Lewy body disease and in models of alpha-synucleinopathy. PLoS One. 2010;5(2):e9313.
331. Ejlerskov P, Ashkenazi A, Rubinsztein DC. Genetic enhancement of macroautophagy in vertebrate models of neurodegenerative diseases. Neurobiol Dis. 2019;122:3–8.
332. Kegel KB, Kim M, Sapp E, McIntyre C, Castano JG, Aronin N, et al. Huntingtin expression stimulates endosomal-lysosomal activity, endosome tubulation, and autophagy. J Neurosci. 2000;20(19):7268–78.
333. Anglade P, Vyas S, Javoy-Agid F, Herrero MT, Michel PP, Marquez J, et al. Apoptosis and autophagy in nigral neurons of patients with Parkinson's disease. Histol Histopathol. 1997;12(1):25–31.
334. Webb JL, Ravikumar B, Atkins J, Skepper JN, Rubinsztein DC. Alpha-Synuclein is degraded by both autophagy and the proteasome. J Biol Chem. 2003;278(27):25009–13.
335. Berger Z, Ravikumar B, Menzies FM, Oroz LG, Underwood BR, Pangalos MN, et al. Rapamycin alleviates toxicity of different aggregate-prone proteins. Hum Mol Genet. 2006;15(3):433–42.

## Further Reading

Fleming A, Bourdenx M, Fujimaki M, Karabiyik C, Krause GJ, Lopez A, et al. The different autophagy degradation pathways and neurodegeneration. Neuron. 2022;110(6):935–66.
Griffey CJ, Yamamoto A. Macroautophagy in CNS health and disease. Nat Rev. Neurosci. 2022;23(7):411–27.
Festa BP, Barbosa AD, Rob M, Rubinsztein DC. The pleiotropic roles of autophagy in Alzheimer's disease: from pathophysiology to therapy. Curr Opin Pharmacol. 2021;60:149–57.
Stamatakou E, Wrobel L, Hill SM, Puri C, Son SM, Fujimaki M, et al. Mendelian neurodegenerative disease genes involved in autophagy. Cell Discov. 2020;6:24.
Menzies FM, et al. Autophagy and neurodegeneration: pathogenic mechanisms and therapeutic opportunities. Neuron. 2017;93:1015–34.
Menzies FM, Fleming A, Caricasole A, Bento CF, Andrews SP, Ashkenazi A, et al. Autophagy and neurodegeneration: pathogenic mechanisms and therapeutic opportunities. Neuron. 2017;93(5):1015–34.
Nixon RA. The role of autophagy in neurodegenerative disease. Nat Med. 2013;19(8):983–97.
Harris H, Rubinsztein DC. Control of autophagy as a therapy for neurodegenerative disease. Nat Rev. Neurol. 2011;8(2):108–17.
Wong E, Cuervo AM. Autophagy gone awry in neurodegenerative diseases. Nat Neurosci. 2010;13(7):805–11.

# Autophagy and Intracellular Membrane Dynamics in Aging and Age-Related Diseases

Satoshi Minami, Shuhei Nakamura, and Tamotsu Yoshimori

**What You Will Learn in This Chapter**

Aging is a biologically programmed process in which the functions of cells and organs deteriorate over time and is closely associated with the development of various age-related diseases; therefore, overcoming aging is important for achieving a healthy longevity. Autophagy is a mechanism that maintains intracellular homeostasis by degrading and recycling intracellular macromolecules and organelles in lysosomes, and recent studies have revealed that "disabled autophagy" is one of the hallmarks of aging. This chapter first outlines the mechanisms by which disabled autophagy occurs with aging, and then describes the impact of autophagy regulation on lifespan and health span based on the latest findings. Second, the molecular mechanisms by which autophagy extends lifespan and healthy life expectancy will be reviewed, broadly classifying them into (1) lifespan extension mechanism by selective autophagy and (2) mechanism to suppress inflammaging by autophagy. Third, the recently revealed organ-specific autophagy and non-canonical autophagy in the suppression of aging will be outlined. Finally, we would like to discuss the possibility of anti-aging therapy through autophagy regulation and future challenges.

S. Minami
Department of Genetics, Graduate School of Medicine, Osaka University, Suita, Osaka, Japan
e-mail: minami@kid.med.osaka-u.ac.jp

S. Nakamura (✉)
Department of Biochemistry, Nara Medical University, Kashihara, Nara, Japan
e-mail: shuhei.nakamura@naramed-u.ac.jp

T. Yoshimori (✉)
Health Promotion System Science, Graduate School of Medicine, Osaka University, Suita, Osaka, Japan

Integrated Frontier Research for Medical Science Division, Institute for Open and Transdisciplinary Research Initiatives (OTRI), , Osaka University, Suita, Osaka, Japan



B. Loos, D. J. Klionsky (eds.), *Autophagy - From Molecular Mechanisms to Flux Control in Health and Disease*, Learning Materials in Biosciences, https://doi.org/10.1007/978-3-031-88121-3_7

# 1    Introduction

Aging is a biologically complicated programmed process characterized by the decline of cellular and organ functions over time, ultimately leading to the death of the organism [1]. Aging is a major risk factor for the development of many diseases, including neurodegenerative diseases, cancer, and cardiovascular disease. The development of these age-related diseases not only leads directly to a decline in an individual's quality of life and shortened healthy life expectancy, but also poses a major medical challenge in the recent global aging society [2]. Therefore, it is important to elucidate the mechanisms that lead to the onset of aging and age-related diseases in living organisms and to develop therapies to combat aging in order to realize "healthy longevity" that enables people to enjoy independent daily living even in old age, as well as from a socioeconomic perspective [2].

Macroautophagy (hereafter referred to as autophagy) is an evolutionarily conserved mechanism that degrades intracellular components and maintains intracellular homeostasis. Recently, disabled autophagy has been highlighted as one of the hallmarks of aging common to various organisms [3]. Previous studies have revealed that (1) autophagy declines with aging, (2) autophagy is activated in various longevity models, and activation of autophagy is necessary for life extension in these models, and (3) disabled autophagy exacerbates aging and age-related diseases, while its activation attenuates aging and age-related diseases, indicating that autophagy is a central regulatory system in aging. Recent research has revealed the mechanism that autophagy decreases with aging, and the mechanisms that autophagy regulates aging through the degradation of selective substrates and the suppression of inflammaging (age-related inflammation), as well as the mechanism that organ-specific autophagy regulates aging, providing insight into part of the complex, multifactorial mechanism that autophagy regulates individual aging. Furthermore, noncanonical autophagy has recently been indicated to play an important role in the progression of aging.

Here, we summarize the relationship between aging and autophagy and outline the currently known mechanisms by which autophagy regulates aging. Finally, we would like to discuss the path toward the development of anti-aging therapies through the regulation of autophagy.

# 2    Autophagy in Aging

## 2.1    Autophagy Declines with Age

Autophagy is a mechanism by which cytoplasmic material is enclosed within a double-membrane vesicle, the autophagosome, which fuses with lysosomes to degrade its contents [4]. Autophagy was previously thought to be a bulk degradation system responsible for nonspecific and massive cellular degradation; however, subsequent studies have revealed that autophagy is a highly selective intracellular degradation mechanism and an important mechanism for maintaining a cellular and organ homeostasis through removal of protein aggregates and injured organelles [5]. The molecular mechanism underlying the progression of autophagy is described in detail in other reviews [4, 5]. Importantly, it has been demonstrated that autophagic activity develops disabled with aging in various organisms [6], and age-related disabled autophagy are considered to be one of the hallmarks of aging [3]. Specifically, the expression of several autophagy-related genes involved in autophagosome formation is decreased with age in Drosophila [7–9], brain and muscle tissues of mice [10, 11], as well as in brain tissue of rats and naked mole-rats [12, 13]. Consistently, findings in humans also indicate that the expression of autophagy-related genes involved in autophagosome formation declines with age in brain and muscle tissues [11, 14]. These studies imply that the gradual decrease in autophagy-related genes with age and impaired autophagosome formation are key features of aging in organisms. However, it has been reported that disabled autophagy with aging is not limited to a decrease in autophagosome formation. For example, in a study with *Caenorhabditis elegans*, autophagic flux analysis using fluorescent-tagged LGG-1 (a homolog of yeast Atg8 [autophagy related 8]) in combination with lysosomal inhibitors reveals a decrease in autophagy activity accompanied by autophagosome and/or autolysosome accumulation with aging [15, 16]. Analyses of liver and hypothalamic neurons in aging mice indicate that the process of autophagosome removal as well as the formation of autophagosomes is impaired with aging [10, 17]. Then, what are the mechanisms by which the process of autophagosome removal is impaired with aging? Increased expression of RUBCN (rubicon autophagy regulator) could play a part [18]. Homologs of RUBCN (referred to herein as RUBCN for simplicity), a negative regulator of autophagy by suppressing autophagosome maturation thereby resulting in autophagosome accumulation [19, 20], is increased in *C. elegans*, Drosophila, and mouse liver and kidney with aging. Further analysis revealed that the deficiency of RUBCN ameliorates the age-related decline in autophagy activity and age-related diseases, indicating that increased expression of RUBCN is part of the pathogenesis of the age-related decline in autophagy activity and subsequent exacerbation of age-related diseases [18]. Age-related impairment of autophagosome removal is also caused by impaired lysosome function, as reported from studies using yeast, *C. elegans* and rodent cells [21–23]. Age-related lysosomal dysfunction has been also identified to contribute to the decline in cellular function and the subsequent development of age-related diseases [6, 15, 16]. Furthermore, it has been found that in humans, disabled autophagy due to impaired autophagosome removal as well as reduced autophagosome formation with aging is significantly associated with the onset and progression of several age-related diseases [24, 25]. Based on the above findings from multiple model organisms, disabled autophagy is a hallmark of aging that is conserved among species. Conversely, age-related disabled autophagy are caused by a combination of several pathological conditions, including abnormal autophagosome formation and impairment of autophagosome removal which is due to increased RUBCN or lysosomal dysfunction.

## 3    Longevity Paradigm and Autophagy

Extensive studies using model organisms ranging from yeast to mice have revealed a number of genetic pathways and environmental interventions that delay aging and consequently extend the lifespan of organisms is conserved across species, and these interventions are also referred to as the longevity paradigm. These longevity paradigms consist of dietary restriction, activation of the energy deficiency-sensing signal AMP- activated protein kinase (AMPK), decreased activity of the nutrient surplus signals the MTOR (mechanistic target of rapamycin kinase) and insulin/IGF1 (insulin like growth factor 1) signaling pathways, decreased mitochondrial respiratory levels, hormesis heat shock, and germ cell removal, which prolong lifespan in various species [26]. Studies of the longevity paradigm have long been conducted to identify the underlying molecular mechanisms mediating lifespan extension [26], and have revealed that long-lived mutants generally exhibit activation of autophagy and increased expression of autophagy-related and lysosomal genes compared to wild-type organisms [15, 27]. Importantly, all of the longevity paradigms that extend lifespan in various species were identified to require autophagy and lysosome-associated genes for their lifespan extension [28, 29]. These findings indicate a model of enhanced autophagy activity as a possible mediator in lifespan extension in the longevity paradigm. Furthermore, RNAseq analysis of peripheral blood leukocytes reveals increased expression of autophagy-lysosome-related genes in centenarians and their children [30], and CD4+ T lymphocytes isolated from offspring of parents with exceptional longevity show enhanced autophagy activity compared to age-matched controls [31], suggesting that the association between increased autophagic activity and longevity is also conserved in humans.

## 4    Impaired Autophagy Exacerbates Aging

Experiments using various model organisms have shown that genetically defective autophagy leads to accelerated aging, organ dysfunction, and exacerbation of age-related diseases. Specifically, mutations or knockdown of genes encoding autophagy machinery in *C. elegans* and *Drosophila* shorten lifespan and health span [6, 7, 29]. Knockdown of transcription factors that regulate autophagy, such as TFEB (transcription factor EB (HLH-30 in *C. elegans*) and FOXO (forkhead box O; DAF-16 in *C. elegans*), also shorten the lifespan of both wild-type and long-lived DAF-2/INSR/IGF1R mutant *C. elegans* [32]. In mice, systemic knockout of autophagy-related genes leads to death during development or neonatal period [33, 34]; thus, the impact of systemic autophagy deficiency on mouse aging had long been unknown. However, in a recent doxycycline-inducible shRNA-based mouse model to suppress ATG5 expression in the whole body except the brain, revealed autophagy deficiency from 2-months of age results in age-related changes in multiple organs and shortened lifespan (death at a median age of 6 months) [35]. Furthermore, studies on autophagy and age-related diseases in mammals have been conducted using tissue-specific autophagy-related gene-deficient mice and show that autophagy deficiency accelerates age-related organ dysfunction and age-related disease models [36, 37]. These tissue-specific autophagy-deficient mouse models also show accumulation of dysfunctional organelles and protein aggregates due to dysfunctional autophagy, and importantly, these changes are also observed in tissues of aged wild-type animals, indicating that autophagy deficiency is compatible as a model for accelerated natural aging.

## 5 Genetic Activation of Autophagy Prolongs Lifespan and Healthy Life Expectancy

The role of autophagy genes in determining lifespan is also supported by the evidence that genetic activation of autophagy prolongs lifespan in various model organisms. In *C. elegans*, overexpression of HLH-30/TFEB prolongs lifespan with activation of autophagy [38, 39], and knockdown of the nuclear export protein XPO-1/exportin-1 results in nuclear enrichment of HLH-30 and promotes autophagy, maintaining protein homeostasis and extending lifespan [40]. In *Drosophila*, systemic overexpression of Atg1 activates autophagy and extends lifespan [39]. Furthermore, overexpression of Atg1 and Atg8a in the nervous system [7, 41], Atg8a in muscle [9], and AMPK in the nervous or intestine [41] in *Drosophila* promotes autophagy, resulting in an extended lifespan. Similarly, systemic overexpression of Atg5 in mice activates autophagy and extends healthy life expectancy, including increased exercise capacity and improved metabolism, and also extends lifespan [42]. A knockin mutation of BECN1 that inhibits BECN1-BCL2 complex formation (BECN1$^{F121A/F121A}$) activates autophagic flux at the basal level, resulting in improved lifespan and health span and fewer spontaneous tumors [43], increased neurogenesis [44] and improved cognitive function in mice [45]. Deletion of *RUBCN*, a negative regulator of autophagy, also improves motility and extends lifespan of *C. elegans* and Drosophila in an autophagy-dependent manner [18]. Furthermore, RUBCN-deficient mice ameliorates kidney fibrosis, SNCA/α-synuclein accumulation that causes Parkinson disease, osteoporosis, non-alcoholic fatty liver disease, and age-related macular degeneration [18, 46–48]. Taken together, regulation of autophagy is an effective approach to slow aging and promote healthy life expectancy in diverse species, including mammals.

## 6 Overactivation of Autophagy Exacerbates Aging

Disabled autophagy is closely associated with the aging process and the development of age-related diseases. Conversely, excessive autophagy can also be detrimental to the individual under certain circumstances, thus it is important that autophagic activity should be regulated within an appropriate level. For example, in *Drosophila*, mildly enhancing Atg1 expression activates autophagy and prolongs lifespan, while highly enhancing Atg1 expression reduces mitochondrial metabolism, enhances the immune response system, and shortens lifespan [39]. In *C. elegans*, mutants lacking SGK-1 (Serum and Glucocorticoid- Inducible Kinase homolog-1) show activation of autophagy and shortening of lifespan due to increased mitochondrial permeability, whereas suppression of autophagy as well as inhibition of mitochondrial permeability restores lifespan of the *C. elegans* [49]. Then why is excessive induction of autophagy associated with a shortened lifespan? One possible explanation is that autophagy excessively eliminates important intracellular components. In the above SGK-1-deficient mutant, autophagy induced by increased mitochondrial permeability might have resulted in the disruption of mitochondrial homeostasis by excessive removal of permeabilized mitochondria, leading to the shortening of lifespan by autophagy. Actually, it is known that excessive induction of autophagy in mammalian cells can lead to cell death [50–52]. Furthermore, there are reports that suppressing autophagy in the systemic or nervous [16] or intestine [53] in *C. elegans* after reproductive age prolongs lifespan and health span, suggesting that activation of autophagy at an older age exacerbates aging. Then what are the possible mechanisms how the activation of autophagy after reproductive age shortens lifespan? One is that autophagy after reproductive age might promote aging by selectively degrading substrates that inhibit aging. It has been reported that autophagy degrade SIRT1 (sirtuin1) [54] as well as nuclear lamina protein LMNB1 (lamin B1) [55], thereby promote cellular senescence, which is closely correlate with organismal aging. Furthermore, in aging individuals, it is necessary to pay close attention to the status of autophagy. It has been reported that in aging individuals, autophagosome biogenesis is impaired, as well as autophagosome removal capacity is compromised due to impaired autophagosome maturation caused by increased RUBCN [18] as well as lysosomal dysfunction [56]. Under these conditions, enhancing autophagosome biogenesis alone leads to significant autophagosome accumulation because autophagosomes cannot be degraded after their formation. Recently, it has been revealed that excessive enhancement of autophagosome biogenesis leads to autosis, a $Na^+$, $K^+$-ATPase-dependent programmed cell death [57, 58]. Autosis is a cell death mechanism in which the intracellular membrane, which constitutes an intracellular organelle, is consumed by excessive autophagosome formation, resulting in organelle dysfunction. For prevention, it is essential to improve the accumulation of autophagosomes by inhibiting RUBCN and/or suppressing autophagosome biogenesis [50, 58]. It is also known that when autophagosome biogenesis is enhanced during lysosomal dysfunction, an increase in the amount of substrate transported to lysosomes leads to lysosomal overburden, which exacerbates lysosomal dysfunction and leads to cellular damage and lysosomal cell death due to increased lysosomal membrane permeability [59–61]. In conclusion, simply enhancing autophagosome biogenesis could be counterproductive in extending lifespan in individuals after reproductive age, and approaches to suppress RUBCN or improve lysosome function might be important.

# 7 Mechanisms of Autophagy to Extend Lifespan and Healthy Life Expectancy

The above genetic associations indicate that autophagy is a central regulator of aging. Then how does autophagy prevent aging of organs and individuals? In recent years, there has been a growing body of evidence that specific cargo turnover via selective autophagy plays an important role in mitigating the progression of aging and age-related diseases. There is also accumulating evidence that autophagy regulates a chronic, low-grade inflammatory state associated with aging called "inflammaging", thereby preventing the progression of aging and age-related diseases. In the following, the molecular mechanisms of autophagy in preventing the progression of aging and age-related diseases will be reviewed with a focus on selective autophagy and the regulation of inflammaging by autophagy (Fig. 1).

# 8 Selective Autophagy in Aging

Aging and age-related diseases are closely related to the accumulation of damaged organelles, and it is necessary to properly remove dysfunctional damaged organelles in order to maintain cellular and organ homeostasis. The following outlines each type of selective autophagy and its relationship to aging and age-related diseases.

# 9 Mitophagy and Aging

Accumulation of dysfunctional mitochondria is a common feature of aging and various age-related diseases [3, 62]. Various molecular mechanisms have been reported to be involved in mitochondrial dysfunction during aging, however, there is no doubt that a decrease in mitophagy plays a central role [62]. Mitophagy is a mechanism that selectively removes damaged and excess mitochondria by autophagy and plays an important role in maintaining cellular and organ homeostasis and is protective against age-related diseases [63]. Studies in *C. elegans* [64], *Drosophila* [65], mice [66], and humans [67] have reported that mitophagy declines with aging as well as typical age-related diseases including Alzheimer disease [68] and Parkinson disease [69]. In a study examining the causal relationship between mitophagy and aging, deletion of *Pink1* and *Prkn*, encoding molecules involved in representative mitophagy, causes early behavioral decline and shortened lifespan [70, 71], whereas induction of mitophagy prolongs lifespan [65, 72] as well as improves the Alzheimer disease model [68, 73]. In addition, studies in *C. elegans* have examined the significance of mitophagy in long-lived mutants and found that inhibition of PINK-1 or DCT-1 (orthologs of the mammalian mitophagy receptors BNIP3 and BNIP3L) shortens the lifespan of long-lived *daf-2* mutants, *eat-2* mutants (replicating experimental dietary restriction) and mutants with mitochondrial dysfunction [64, 74], indicating that

## Selective autophagy in aging

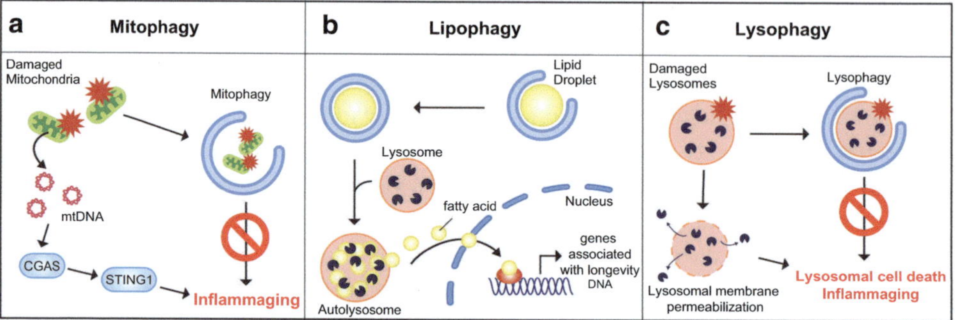


## Inflammaging and autophagy

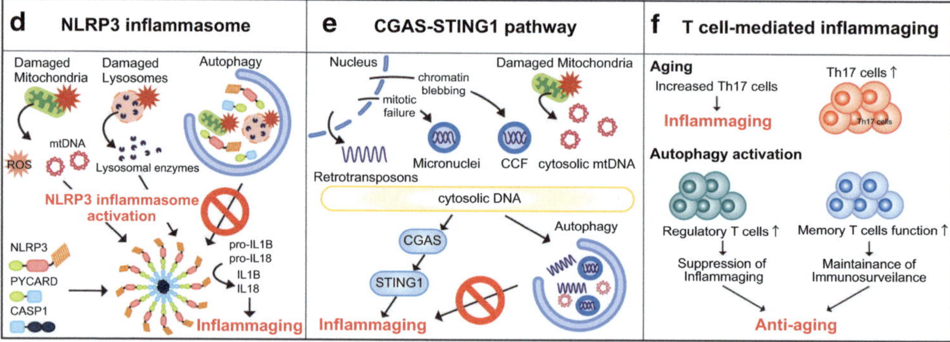


**Fig. 1** Mechanisms by which autophagy extends lifespan and health span. (**a**–**c**) Mechanisms of lifespan and healthy life extension by selective autophagy. (**a**) Mitophagy prevents the release of mitochondrial DNA into the cytoplasm by removing damaged mitochondria, thereby inhibiting the inflammaging associated with aging. (**b**) Lipophagy and lipolysis in lysosomes are involved in lifespan extension by regulating transcription through lipid signaling. (**c**) Lysophagy inhibits lysosomal cell death and inflammaging by preventing the leakage of lysosomal enzymes into the cytoplasm following lysosomal damage. (**d**–**f**) Mechanisms of lifespan and health span extension by autophagic regulation of inflammaging. (**d**) Mitochondrial and lysosomal damage activates the NLRP3 inflammasome during the aging process via mitochondrial reactive oxygen species as well as leakage of mtDNA and lysosomal enzymes into the cytosol. Autophagy inhibits NLRP3 inflammasome activation through removal of damaged mitochondria and lysosomes as well as NLRP3 inflammasome components. (**e**) The increase in cytoplasmic DNA such as micronuclei, CCFs, mtDNA, and retrotransposons activates the CGAS-STING1 pathway, resulting in inflammaging during the aging process. Autophagy inhibits inflammaging by removing these cytoplasmic DNAs. (**f**) Impaired autophagy during aging exacerbates inflammaging by promoting differentiation into Th17 cells. Activation of autophagy suppresses inflamming by promoting differentiation into regulatory T cells, and also enhances memory T cell function to maintain immune surveillance, thereby promoting removal of senescent cells and suppressing exacerbation of inflammaging. mtDNA, mitochondrial DNA; CCF, cytoplasmic chromatin fragment

mitophagy plays a causal role in lifespan extension in the long-lived mutant model of *C. elegans*. Recently, it has been recently reported that mitochondrial DNA release from damaged mitochondria into the cytoplasm is a trigger for the exacerbation of inflammaging during aging [75], thus preventing mitochondrial DNA release into the cytoplasm by removing damaged mitochondria through mitophagy is essential for suppressing inflammaging associated with aging [76]. Indeed, when validated using knockout mice for *pink1* and *prkn*, inflammaging is enhanced in mice deficient in mitophagy [76]. Collectively, it is thought that a decrease in mitophagy with aging causes age-related mitochondrial dysfunction, leading to inflammaging, shortened lifespan and the development of age-related diseases (Fig. 1a).

## 10    Lipophagy and Aging

Studies in diverse organisms suggest that dietary lipids are associated with longevity [77–79] and that specific changes in lipid metabolism are associated with various longevity interventions [80]. Autophagy regulates lipid metabolism by degrading intracellular lipid droplets through a mechanism called lipophagy [81]. Lipids transported to lysosomes by lipophagy are degraded by lysosomal lipases. The lysosomal lipase LIPL-4 is highly expressed in long-lived mutants [82, 83], suggesting that the long-lived mutants could have high lipophagy activity. Aged *C. elegans* show lipid deposition in non-adipose tissues, including the nervous system [84], and dietary restriction, an intervention that promotes longevity, reduces this ectopic lipid accumulation, whereas inhibition of LGG-1 or HLH-30/TFEB exacerbates the ectopic lipid accumulation [84]. These results suggest that *C. elegans* develops abnormalities in lipophagy with aging and that dietary restriction activates lipophagy. It is also known that autophagy is closely related to lipid droplet formation [85, 86] as well as lipid droplet degradation, and it is necessary to further examine the significance of autophagy on lifespan extension through lipid droplet regulation in the future. Recently, lysosome-derived lipid signaling resulting from lipolysis by the lysosomal lipase LIPL-4 has received attention as being directly related to lifespan extension in *C. elegans* [83, 87]. As a mechanism, lysosomal lipids degraded by LIPL-4 bind to the chaperone LBP-8 (Lipid Binding Protein 8) and are translocated into the nucleus, activating NHR-49 (Nuclear Hormone Receptor family 49) and NHR-80, thereby enhancing mitochondrial lipid metabolism and leading to longevity [82, 88]. Lysosome-derived lipid signaling regulates lifespan via interactions not only between intracellular organelles but also between tissues of the organism [89]. Lysosomal lipolysis in adipose tissue produces dihomo-γ-linolenic acid, a polyunsaturated fatty acid, which binds to the lipid chaperone protein LBP-3 and is transported to the nervous system, where dihomo-γ-linolenic acid activates NHR-49 and promotes production of the neuropeptide NLP-11, thereby extending lifespan. Thus, lipophagy and lysosomal lipolysis could regulate lifespan in a cell-autonomous or non-autonomous manner via lipid signaling, and further studies in mammals are desirable in the future (Fig. 1b).

## 11 Lysophagy and Aging

The findings that lysosomal degradation capacity is higher in long-lived mutants of *C. elegans* [21] and that loss of a specific lysosomal membrane protein (SCAV-3 [mammalian SCARB2/LIMP-2]) in *C. elegans* is associated with shorter lifespan [90], suggest that maintenance of lysosomal function and its integrity are important for longevity. In mammals, lysosomal damage was previously thought to be a rare phenomenon observed only in limited pathological conditions such as crystalline nephropathy, gout, and silicosis; however, recent reports have revealed that accumulation of damaged lysosomes is a characteristic feature of senescent cells [56], and subsequent analysis suggested that insoluble polymers within lysosomes that accumulate in senescent cells cause lysosomal damage. Leakage of intralysosomal proteolytic enzymes (such as CTSB [cathepsin B]) into the cytoplasm following lysosomal damage is closely associated with aging and the progression of various neurodegenerative diseases by causing lysosomal cell death, activation of NLRP3 (NLR family pyrin domain containing 3) inflammasome and activation of macrophages through mitochondrial metabolic alterations [91–96]. Thus, while the maintenance of lysosomal function and its integrity is an important factor for longevity, lysosomal damage is widely observed in senescent cells and age-related diseases, and lysosomal damage induces lysosomal cell death and inflammaging, which are strongly associated with the development of aging and age-related diseases. Therefore, appropriate removal of damaged lysosomes by lysophagy [97] could be an important mechanism for longevity and healthy longevity, although further studies are needed (Fig. 1c).

## 12 Nucleophagy and Aging

Alterations in nuclear microstructure are a universal feature of aging, premature aging, and other age-related diseases. Recently, it has become clear that regulation of nuclear microstructure by nucleophagy can play an anti-aging role. During nulceophagy, the nuclear membrane anchor protein ANC-1 in *C. elegans* (SYNE2/NESPRIN-2 in mammals) works cooperatively with autophagy components to degrade nuclear components by autophagy. This nucleophagy is involved in lifespan extension and germ cell immortalization by regulating the size of the nucleolus, a well-known biomarker of aging [98].

## 13 Inflammaging and Autophagy

Inflammation is an evolutionarily conserved defense mechanism, designed to maintain homeostasis of the organism in the face of acute and localized injury, and serves as an adaptive response to infection and damage [99]. Conversely, aging is associated with a chronic low-grade inflammatory state (referred to as inflammaging), which is associated with biological aging and the development of age-related diseases [100]. Inflammaging has been thought to be essentially the activation of innate immune responses (responses to microbial motifs, endogenous danger signals, and environmental stimuli) sensed by the NLRP3 inflammasome, CGAS (cyclic GMP-AMP synthase)-STING1 (stimulator of interferon response cGAMP interactor 1) pathway, and TLR (toll like receptor) pathway; however, T cells also regulate inflammaging [101, 102]. Recently, autophagy has been shown to regulate inflammaging in close association with innate immune responses and the maintenance of T cell homeostasis.

## 14    NLRP3 Inflammasome and Autophagy

The NLRP3 inflammasome has been identified as an important component of the innate immune response and integrates host immune homeostasis [103]. This intracellular protein complex activates CASP1 (caspase 1), resulting in the secretion of inflammatory cytokines such as IL1B/IL-1β (interleukin 1 beta), which is associated with exacerbation of inflammaging and the development of age-related diseases, including neurodegenerative disorders [100, 103, 104]. Recently, autophagy has been shown to regulate the NLRP3 inflammasome by various mechanisms. First, it was reported that mitochondrial damage activates the NLRP3 inflammasome via mitochondrial reactive oxygen species/ROS and mitochondrial DNA released into the cytoplasm, while autophagy suppresses the inflammasome by eliminating damaged mitochondria [105–107]. It was also reported that autophagy degrades inflammasome components to prevent inflammasome overactivation. PYCARD/ASC, a component of the NLRP3 inflammasome, was ubiquitinated and degraded by SQSTM1/p62-dependent autophagy [108]. It was also found that MEFV/TRIM20 (MEFV innate immunity regulator, pyrin) interacts with NLRP3 inflammasome components and degrades them by serving as a platform for the assembly of ULK1 (unc-51 like autophagy activating kinase 1)-BECN1-ATG16L1 [109]. In addition, autophagy inactivates NLRP3 via cytoplasmic sequestration [110]. Furthermore, autophagy can also suppress NLRP3 inflammasomes via autophagic removal of damaged lysosomes by lysophagy [97], since NLRP3 inflammasome is known to be activated by lysosomal membrane destabilization and subsequent leakage of CTSB into the cytoplasm [92–94] (Fig. 1d).

## 15    CGAS-STING1 Pathway and Autophagy

The CGAS-STING1 pathway senses cytoplasmic DNA and leads to the production of type 1 interferons and activation of the transcription factor NFKB/NFκB (nuclear factor kappa B) for the production of various inflammatory cytokines. The CGAS-STING1 pathway has been studied as an essential pathway for the innate immune response to exogenous cytoplasmic DNA during viral infection. Recently, however, it has become clear that the CGAS-STING1 pathway is also activated by endogenous cytoplasmic DNA in aging and age-related diseases, resulting in inflammaging even in the absence of pathogens. Endogenous cytoplasmic DNA is divided into four major categories: micronuclei, cytoplasmic chromatin fragments (CCFs), mitochondrial DNA, and retrotransposons [111], all of which activate the CGAS-STING1 pathway [75, 112, 113] and are closely related to autophagy. Micronuclei are chromosomes (fragments) enclosed in the nuclear membrane in the cytoplasm and are produced by mitotic failure. Once micronuclei are formed, CGAS, a cytosolic DNA sensor, accumulates in the micronuclei and recruits Atg8-family proteins via its LC3-interacting region (LIR), which is responsible for autophagic removal of the micronuclei [114]. CCF results from chromatin blebbing into the cytoplasm, independent of cellular division. CCF formation is caused by a decrease in nuclear membrane integrity and is formed by autophagic degradation of LMNB1 [55], and CCFs appear to be degraded by autophagy [115, 116]. Mitochondrial DNA is released into the cytoplasm by increased mitochondrial outer membrane permeability, mitochondrial permeability transition pore/mPTP opening, and mitochondrial outer membrane disruption to activate the CGAS-STING1 pathway [111], whereas autophagy suppresses cytosplasmic mitochondrial DNA release by eliminating damaged mitochondria [76]. Retrotransposons are activated by DNA damage and translocate to the cytoplasm to activate the CGAS-STING1 pathway, while autophagy maintains genomic stability by degrading retrotransposon RNA [117]. Collectively, autophagy represses activation of the CGAS-STING1 pathway by degrading cytoplasmic DNA (Fig. 1e). Although it was reported that CGAS has a LIR and is an adaptor for autophagic degradation of micronuclei as described above [114], whether the CGAS-autophagy pathway is also involved in the removal of CCF, mtDNA, and retrotransposons is unknown, although it is tempting to speculate that the CGAS-autophagy pathway is the common pathway to degrade cytoplasmic DNA.

## 16    T Cell-Mediated Inflammaging and Autophagy

It has recently been revealed that senescent T cells exacerbate inflammaging and are involved in the progression of age-related diseases as part of the following mechanisms: First, senescent T cells secrete inflammatory cytokines such as IFNG (interferon gamma) and TNF (tumor necrosis factor), which induce cellular senescence in the whole body by inducing inflammaging [101]. Second, senescent T cells have a disrupted immune surveillance system, and their ability to eliminate senescent cells in the whole body is impaired [118], allowing them to accumulate and exacerbate inflammaging.

Recently, the close relationship between autophagy and T cell senescence has become clear. T cells isolated from the offspring of parents with exceptional longevity show enhanced autophagic activity compared with age-matched controls [31]. It was also found that age-related impairment of autophagy is most likely to affect regulatory T cells and memory T cells, which are the most mitochondria-rich of all immune cells [119]. Regulatory T cells suppress excessive immune responses and inflammaging. Activation of T cell autophagy leads to increased differentiation into regulatory T cells and ameliorates inflammaging, while impairment of T cell autophagy results in increased differentiation into Th17 rather than regulatory T cells, which exacerbates inflammaging [120–122]. Memory T cells play an important role in the immune surveillance mechanism, and the elderly are more susceptible to infection and less responsive to vaccines as a result of impaired function of these cells. Autophagic activation induces memory T cells and improves infection and vaccine responses in the elderly [123, 124]. Based on the above, although further investigation is needed, autophagy could suppress individual aging by maintaining the homeostasis of regulatory T cells and suppressing inflammaging, as well as by maintaining the homeostasis of memory T cells and their immunosurveillance ability (Fig. 1f).

Collectively, autophagy appears to suppress aging and the progression of age-related diseases through the regulation of various selective autophagies and the precise multistep regulation of inflammaging.

## 17    Organ- Specific Autophagy and Aging

During the aging process, a gradual decline in function occurs at both the organism and organ levels. Recent studies have revealed that the aging of individual organs affects the aging of other organs and the aging of the entire organism. Moreover, recent studies with *C. elegans*, *Drosophila*, and mice have revealed that the organ-specific function of autophagy could regulate aging in the whole individual [29].

Consistent with recent findings that the nervous system regulates lifespan [125, 126], several reports have implicated that neuronal autophagy is involved in the regulation of lifespan. In *C. elegans*, restoration of neuronal-specific ATG-18 expression in the *atg-18;daf-2* double mutant completely restores their short lifespan, indicating the importance of neuronal autophagy in individual aging [127]. Lifespan extension was also examined in *C. elegans* by inhibiting the autophagy suppressor RUBCN specifically in each tissue (nerve, intestine, muscle, and subcutis), and the most efficient lifespan extension was achieved by neuron-specific RUBCN inhibition [18]. Furthermore, overexpression of autophagy-related factors (Atg1 and Atg8a) and mitophagy-related factor BNIP3 throughout the neurons extends the lifespan of Drosophila [7, 41, 128]. Mechanistically, activation of neuronal autophagy would regulate individual aging by acting in a cell non-autonomous manner, based on the following reports: the effects of ATG-18 on lifespan recovery in neurons are mediated by the release of neurotransmitters and neuropeptides [127]; overexpression of Atg1 in neurons activates intestinal autophagy and prevents age-related intestinal barrier dysfunction [41]; and overexpression of BNIP3 in neurons suppresses intestinal stem cell senescence and maintains intestinal barrier function, muscle mitochondrial and protein homeostasis [128]. However, the detailed cell non-autonomous mechanism that neuronal autophagy regulate individual aging remains unclear and awaits further verification. It is also unclear in which regions of the brain autophagy regulates individual aging, although recent reports have been revealed that the hypothalamus is an important brain region that regulates individual aging [125, 126], thus further studies with mammals are needed to verify the details.

Autophagy in the intestine also regulates lifespan. It has been reported that in *C. elegans*, activation of autophagy in the intestine is indispensable for the lifespan-extending effects of dietary restriction [127] and long-lived mutants *(daf-2* mutant [127], *eat-2* mutant [129], and germline-deficient *glp-1* mutant [15]). In *Drosophila*, specific activation of intestinal autophagy also extends lifespan [41, 130]. Then, how does intestinal autophagy extend lifespan? One possible explanation is the function of intestinal autophagy in maintaining the intestinal barrier function. Previous studies have shown that the intestinal barrier function is maintained by dietary restriction (*C. elegans* [129], *Drosophila* [131, 132], protein restriction (pig [133]) and long-lived mutants (*daf-2* [129]), whereas intestinal barrier function is impaired with aging, and that aging-associated intestinal barrier dysfunction induces systemic inflammaging that exacerbates individual aging and various age-related diseases [134]. Furthermore, previous studies have revealed that autophagy is an important mechanism for maintaining intestinal barrier function [129], and it was reported that deficiency of ATG protein is related to the pathogenesis of Crohn disease, which is characterized by intestinal barrier dysfunction, in humans [135]. Collectively, the maintenance of intestinal barrier function by intestinal autophagy would be an important lifespan extension mechanism that is conserved across species. Furthermore, a recent report demonstrated that specific activation of intestinal autophagy activates neuronal autophagy as well [41], suggesting that intestinal autophagy can extend lifespan through non-autonomous cellular functions. Consistently, the intestine has been found to interact with various organs including the brain and has recently been referred to as "the second brain" [136–138]. At present, however, it is still unclear what kind of cellular non-autonomous function of intestinal autophagy regulates systemic aging, and this remains to be verified in the future.

Consistent with reports that muscle counteracts systemic aging via muscle-derived cytokines called myokines [139], studies in *C. elegans*, *Drosophila* and mice have found that muscle-specific autophagy is associated with longevity. In *C. elegans*, inhibition of LGG-1 and ATG-18 in muscle shortens the lifespan of long-lived mutants (*eat-2* and *daf-2* mutants) [15, 129]. Overexpression of Atg8a in *Drosophila* muscle also extends lifespan [9]. In Drosophila, muscle specific overexpression of the transcription factor foxo maintains muscle proteostasis and prolongs lifespan at least in part by promoting autophagy [8, 9]. In mice, muscle-specific ATG7 deficiency impairs muscle function and shortens lifespan [11]. Reduced muscle autophagy leads to muscle weakness called sarcopenia [11], and muscle weakness significantly impairs healthy life expectancy, thus maintenance of muscle function through activation of muscle autophagy is important for healthy longevity.

Aging of the immune system is also involved in the regulation of lifespan. Based on the recent analysis of a mouse model in which DNA damage and subsequent cellular senescence were induced specifically in the immune system (immune system-specific deficiency of ERCC1, a DNA repair protein [140]), and a mouse model in which mitochondrial dysfunction [141], a signature of T-cell senescence, was induced (T-cell-specific TFAM deficiency model [101]), it was found that the induction of immune senescence is associated with the development of cellular senescence and inflammaging in the whole body organs, shortening of lifespan, development of age-related diseases and decreased physical activity. These studies revealed a close relationship between the immune system and aging. As described in the section on "inflammaging and autophagy", autophagy in the immune system is thought to regulate NLRP3 inflammasome and CGAS-STING1 pathway, as well as T-cell homeostasis, thereby regulating inflammaging and consequently individual aging.

As described above, autophagy in the nervous system, intestines, muscles, and immune system regulates aging in the whole organism, and organ-specific autophagy regulate aging of the entire organism through cell non-autonomous functions, however, the cell non-autonomous functions of autophagy that regulate individual aging are largely unknown at present and await further elucidation. In mammals, it has been reported that not only the above-mentioned organs but also organs such as heart, liver, adipose tissue, and kidney is closely related to longevity, thus the relationship between autophagy of these organs and individual aging is also awaiting further verification.

## 18    Non-canonical Autophagy and Aging

Non-canonical autophagy is an emerging mechanism that plays an important role in the maintenance of cellular homeostasis through the remodeling of intracellular membranes induced by stresses to the intracellular membrane, and is characterized by the binding of lipidated Atg8-family proteins to single-layer intracellular membranes, such as phagosomes and endolysosomes [142, 143]. The binding of lipidated Atg8-family proteins to the intracellular membrane in response to stress is called "Atg8ylation," because the binding of lipidated Atg8-family proteins to the intracellular membrane in response to stress is similar to ubiquitination, a protein important for cellular stress responses, and because ubiquitin and Atg8-family proteins are similar in sequence and structure [142, 143]. Non-canonical autophagy is initiated by the binding of V-ATPase on the intracellular membrane to the WD40-domain of ATG16L1, a domain that is dispensable for canonical autophagy, resulting in Atg8ylation. The process of non-canonical autophagy requires the ATG-conjugation systems (ATG7, ATG3, ATG10, ATG12, ATG5, ATG16L1), which is essential for lipidation of Atg8-family proteins as in canonical autophagy, whereas the upstream regulators of autophagy (RB1CC1/FIP200, ULK1, ATG13) are not required. Furthermore, in canonical autophagy, phosphatidylethanolamine (PE) is the predominant lipid that binds to Atg8-family proteins, whereas in non-canonical autophagy, phosphatidylserine (PS) as well as PE is found to bind to Atg8-family proteins [144].

Non-canonical autophagy associated with phagocytosis and endocytosis processes are called LC3-associated phagocytosis (LAP) [145] and LC3-associated endocytosis (LANDO) [146], respectively. LAP and LANDO processes involve Atg8ylation of phagosomal and endosomal membranes in recognition of membrane stress signals, including oxidative stress [147]. Atg8ylated phagosomal and endosomal membranes can efficiently fuse with lysosomal membranes, which facilitate the processes of phagocytosis and endocytosis [145]. Indeed, LAP attenuates age-related vision loss via efficient phagocytosis of photoreceptor outer segments by retinal pigment epithelial cells [148], and prevents exacerbation of inflammaging via efficient phagocytosis of dying cells by macrophages [149, 150]. Furthermore, LAP counteracts bacterial infection via efficient phagocytosis of bacteria by macrophages; however, their activity declines with age, and the decreased activity of LAP is a cause of vulnerability to bacterial infection in the elderly [151]. LANDO inhibits the progression of Alzheimer disease through its involvement in amyloid-β degradation and amyloid-β receptor recycling in microglia [146, 152].

Non-canonical autophagy is also activated by STING1: STING1 induces non-canonical autophagy on single membrane vesicles near the endoplasmic reticulum (ER) and Golgi apparatus [153]. This type of non-canonical autophagy suppresses inflammation by degrading STING1 as well as removing cytoplasmic DNA [153–155]. Considering that during individual aging, cytoplasmic DNA increases due to nuclear and mitochondrial DNA damage, and subsequent activation of the CGAS-STING1 pathway exacerbates inflammaging [111], non-canonical autophagy activated by STING1 can suppress individual aging through suppressing inflammaging.

Non-canonical autophagy is also activated during lysosomal stress and plays a central role in the lysosomal stress response. Upon lysosomal stress, Atg8ylation on lysosomal membranes is closely associated with lysosomal homeostasis by inducing microlysophagy to maintain lysosomal function as well as by activating TFEB to promote lysosomal biogenesis. Microlysophagy is induced depending on the Atg8-family protein lipidation machinery during lysosomal stress and regulates lysosome size and enzyme activity by selectively turning over the lysosomal membrane [156]. TFEB activation by Atg8ylation is mediated by two mechanisms: MCOLN1/TRPML1, a lysosomal calcium channel, is opened by Atg8ylation, which enhances calcium release from the lysosome and activates TFEB [157], and Atg8ylation on lysosomal membranes activates TFEB by sequestering FLCN (folliculin), which acts as a GTPase-activating protein/GAP for the RRAGC-RRAGD GTPase and plays a central role in TFEB suppression by MTORC1 [158]. Considering the marked increase in damaged and dysfunctional lysosomes with aging [56], it is conceivable that noncanonical autophagy, which is activated during lysosomal stress, can counteract aging stress via microlysophagy and TFEB activation, although detailed future verification is needed.

Non-canonical autophagy also plays a role in secretion as well as degradation. In particular, non-canonical autophagy is involved in the biosynthesis of extracellular vesicles [159–161] and regulation of their contents [162]. As it is well known that extracellular vesicles are closely related to individual aging [125, 163], it is necessary to examine whether the regulation of extracellular vesicle production and contents by Atg8ylation regulates individual aging in the future. Furthermore, Atg8ylation may be associated with the severity of infections in the elderly through the production of extracellular vesicles, because the extracellular vesicles produced by Atg8ylation serve as decoys for bacterial toxins [161].

Collectively, non-canonical autophagy is involved in various mechanisms such as phagocytosis, endocytosis, inflammaging activation by STING1, lysosomal stress response (microlysophagy activation, TFEB activation), and extracellular vesicle production, and is predicted to be closely associated with aging. However, many questions currently exist: How does the activity of non-canonical autophagy change with aging? How much does non-canonical autophagy play a role in aging? Which type of noncanonical autophagy is involved in aging? Finally, is it possible to control aging by regulating non-canonical autophagy? These questions need to be clarified in detail by future research.

## 19     Interventions to Combat Aging Via Autophagic Activation

### 19.1     Lifestyle Interventions

Lifestyle habits such as diet and exercise have been shown to be closely related to longevity and healthy longevity by regulating autophagy activity. This section outlines the close relationship between lifestyle and longevity in the context of autophagy.

## 20     Dietary Interventions

Calorie restriction is considered the gold standard for extending life expectancy and healthy life expectancy [164]. Caloric restriction extends lifespan in a wide range of species from yeast to primates as as well as prevents a variety of age-related diseases [165, 166]. Caloric restriction also regulates the aging process in humans [167, 168]. However, it is difficult to incorporate caloric restriction as an anti-aging therapy for humans. This is because maintaining strict caloric restriction for many years in daily clinical practice is difficult in terms of patient compliance, and prolonged caloric restriction can lead to loss of muscle strength [169], impaired reproductive function [170], and compromised immune function [171]. Therefore, it is important to elucidate the molecular mechanisms underlying the regulation of the aging process by calorie restriction and to develop therapies that are as effective as caloric restriction.

The molecular mechanisms by which caloric restriction regulates the aging process are now known to be mediated through multiple pathways, including nutrient-sensing mechanisms such as decreased MTORC1 signalling, increased AMPK, and increased sirtuin pathway, and at least portion of the life-extending effect of caloric restriction is mediated by the activation of autophagy [164, 172]. Autophagy is activated via nutrient sensing mechanisms [166], TFEB [38] and FOXA (forkhead box A) [173] under calorie restriction, and the protective effects of calorie restriction on lifespan extension and age-related diseases are abolished by deficiency of autophagy [38, 129, 173–175]. Meanwhile, the question has recently been raised as to whether the effects of caloric restriction on aging are truly due to caloric restriction. Unlike most humans practicing caloric restriction, who take in fewer calories at each meal, calorie-restricted mice eat one restricted meal per day within two hours, followed by a long period of fasting each day [176], suggesting that time-restricted feeding (TRF), which restricts food intake to specific times of the day rather than calorie restriction, is what improves metabolism and extends lifespan. It is now clear that many of the effects of caloric restriction on healthy life expectancy and life extension are due to TRF, especially TRF along the circadian rhythm [177–181]. It has also been reported that TRF activates autophagy in organs throughout the body and that the protective effect of TRF on age-related metabolic abnormalities is dependent on activated autophagy [182]. Furthermore, recent studies with Drosophila have reported that the health and longevity-enhancing effects of TRF are due to the activation of autophagy in accordance with circadian rhythms [183]. The report elucidated that lifespan-extending effect of TRF is abolished by knockdown of circadian rhythm genes and nocturnal-specific autophagy genes, and that overexpression of autophagy genes specifically during the nighttime prolongs lifespan even without TRF. Based on the above, TRF-mediated lifespan extension effects are concluded to be mediated by nocturnal autophagy. This study indicates that behavioral or pharmacological interventions that promote nighttime-specific autophagy may help to achieve healthy longevity. Clinically, TRF is likely to be more likely to achieve patient compliance than calorie restriction and it is predicted to have fewer side effects than sustained caloric restriction, thus its potential for clinical application in daily practice is promising. However, this study was conducted using Drosophila, and verification in mammals and humans awaits. Especially in humans, it has been reported that skipping breakfast is associated with shortened life expectancy [184, 185], indicating that the clinical application of TRF, which requires fasting for breakfast, needs to be thoroughly verified. At any rate, further investigation of the mechanism by which "circadian autophagy" promotes longevity and healthy life expectancy would pave the way for autophagy-mediated anti-aging therapies.

## 21  Exercise Interventions

Exercise is also a typical intervention to prevent aging. Previous clinical studies have shown that exercise tolerance is directly related to longevity [186, 187], and that exercise is beneficial for various age-related diseases, including frailty, cardiovascular disease, metabolic diseases, and neurodegenerative diseases [188, 189]. Although exercise can be expected to attenuate aging, for many patients, especially older patients with severe symptoms, poor physical condition and physical fitness often preclude specific exercise regimens. Therefore, it is important to identify mechanisms associated with exercise in order to develop alternative treatment strategies that can be as effective as exercise therapy.

It has been reported that the health benefits of exercise, such as improved exercise tolerance [190] and improved glucose metabolism [191], are mediated by autophagy. As for the effect of exercise on endurance improvement, it has been indicated that the effect is due to autophagy induction via activation of AMPK and PPARGC1A/PGC1α (PPARG coactivator 1 alpha) in muscles [192, 193]. The effect of exercise on glucose metabolism is reported to be due to the activation of autophagy in the metabolic organs including liver, which is induced by FN1 (fibronection 1) secreted from muscles during exercise [194]. Furthermore, exercise exerts neuroprotective effects such as preventing neurodegenerative diseases and improving cognition [189, 195], and at least part of these neuroprotective effects of exercise is mediated by autophagy activation. Exercise has been reported to be neuroprotective by activating the autophagy-lysosome pathway in cerebral nerves [196, 197] via the PPARA/PPARα (peroxisome proliferator activated receptor alpha)-TFEB pathway [196] and the AMPK-SIRT1-TFEB pathway [198].

Collectively, it is evident that lifestyle habits such as diet and exercise are closely related to longevity and healthy longevity through the regulation of autophagy. It is desirable to develop a treatment that is as effective as diet and exercise therapy by clarifying the molecular mechanisms underlying lifestyle habits activate autophagy and contribute to longevity through future research.

## 22    Autophagy Modulators

The recent evidence that disabled autophagy is a signature of aging [3] has prompted a strong research effort on the development of compounds that facilitate autophagy [6]. This section outlines the compounds that have been identified to regulate autophagy.

## 23    MTOR Inhibitor

Rapamycin, an inhibitor of MTOR, has demonstrated lifespan-extending effects in model organisms ranging from yeast to mice, at least in part through autophagy [199]. However, MTOR inhibition has side effects such as immunosuppression and impaired glucose tolerance because MTOR plays an important role in immune and metabolic maintenance. Furthermore, rapamycin is unable to cross the blood-brain barrier properly due to its large molecular size. Therefore, it is necessary in the future to develop drugs that are better transferred into the brain and have fewer side effects.

## 24    TFEB Modulator

Lysosomal dysfunction is one possible cause of disabled autophagy during aging, and therefore an approach to improve lysosomal function is promising for restoring autophagy. Indeed, studies in *C. elegans* have shown that the lysosomal master regulator HLH-30/TFEB positively regulates lifespan, at least in part through autophagy [38], thus targeting TFEB is a promising therapeutic approach to promote longevity. Indeed, modulators of

TFEB homologs, ouabain and fisetin, have been demonstrated to regulate autophagy and prevent pathophysiological aging. Ouabain is a cardiac glycoside that enhances TFEB activation and induces downstream autophagy-lysosomal gene expression by inhibiting the MTOR pathway, and inhibits the accumulation of abnormal toxic MPT/tau in the brain [200]. Fisetin, a flavonol, promotes clearance of endogenous MAPT/tau via activation of TFEB and NFE2L2/NRF2 (NFE2 like bZIP transcription factor 2) through inhibition of the MTOR pathway [201]. Recently, it has been found that itaconic acid alkylates the cysteine 212 residue of TFEB, allowing TFEB to escape phosphorylation by MTORC1 and activating TFEB in an MTORC1-independent manner [202]. Thus, itaconic acid and its derivatives can activate TFEB avoiding the side effects of MTORC1 inhibition. Future studies are expected to verify the therapeutic effects of these compounds on aging and age-related diseases.

## 25	NAD$^+$

Nicotinamide adenine dinucleotide (NAD$^+$) plays an important role in the maintenance of biological homeostasis by acting as a substrate for NAD$^+$-consuming enzymes (NADase). In particular, it acts directly as a substrate of sirtuins for activation of autophagy and mitophagy [203]. Among surtuins, SIRT1 is an NAD$^+$-dependent deacetylase present in the nucleus and activates autophagy via deacetylation of key autophagy proteins (ATG5, ATG7, Atg8-family proteins) [204, 205] and autophagy-related transcription factors TFEB and FOXO [206, 207]. Furthermore, SIRT1 activates mitophagy via deacetylation of PPARGC1A and FOXO1 [208]. Other sirtuins also activate autophagy to varying degrees [209]. Importantly, NAD$^+$ declines during the aging process, because PARP (poly(ADP-ribose) polymerase), an NADase, is activated and consumes NAD$^+$ to repair DNA damage caused by the aging process. NAD$^+$ supplementation is important because NAD$^+$ deficiency leads to a vicious cycle of decreased autophagy activity and subsequent aggravation of intracellular oxidative stress, which exacerbates DNA damage and further NAD$^+$ deficiency [210]. Indeed, supplementation with NAD$^+$ precursors extends lifespan and improves health span in *C. elegans*, Drosophila, and mice [210–212]. NAD$^+$ precursors also prevent memory loss in *C. elegans* and mouse models of Alzheimer disease in a mitophagy-dependent manner [68]. Based on these findings, a number of clinical trials on NAD$^+$ precursor replacement therapy for human age-related diseases have been conducted in recent years. Currently, the safety and pharmacokinetics of NAD$^+$ precursors have been demonstrated, as well as the efficacy of NAD$^+$ precursors in chemoprevention of non-melanoma skin cancer [213], restoration of insulin resistance in prediabetic women [214], and inhibition of neuroinflammation in Parkinson disease patients [215].

## 26    Spermidine

Spermidine is a natural polyamine present in all living organisms and acts as an autophagy inducer [216]. As for the mechanisms of autophagy activation, first, spermidine acts as a substrate for the hypsinylation reaction, and activates autophagy by activating the translation of TFEB, ATG3 and other factors through hypsinylation of the translation factor EIF5A (eukaryotic translation initiation factor 5A) [217, 218]. Second, it activates autophagy by promoting deacetylation of autophagy factors through inhibition of acetyltransferases such as EP300 (E1A binding protein p300) [219]. Spermidine extends lifespan by activating autophagy in yeast, *C. elegans*, and Drosophila [220]. Furthermore, dietary spermidine is known to suppress age-related memory impairment in an autophagy-dependent manner in Drosophila [221]. Spermidine also prolongs lifespan, suppresses cardiac and kidney aging, and maintains cognitive function by activating autophagy and mitophagy in mouse models [222–224]. Consistently, human clinical studies have shown that higher dietary spermidine intake is associated with lower blood pressure and lower incidence of cardiovascular disease in humans [222]. The decline in spermidine level with aging is a common feature in a variety of organisms from yeast to humans. Indeed, in humans, spermidine levels are decreased in lymphocytes from elderly individuals, which correlates with decreased autophagy and impaired lymphocyte function [217, 225]. Importantly, spermidine supplementation in the elderly restores autophagy activity in lymphocytes to the level of younger individuals, and lymphocyte function is completely restored [217, 225]. Based on these results, a randomized controlled trial of spermidine administration has been initiated and has already reported improvements in certain cognitive functions and markers [226].

## 27    Urolithin A

Clinically promising mitophagy inducers include urolithin A, a metabolite of elagitannin derived from the intestinal microbiota [227]. Urolithin A prolongs the healthy life expectancy and lifespan of *C. elegans*, which is dependent on autophagy and mitophagy [228]. Urolithin A also improves locomotor performance in a mouse model of frailty [228]. Furthermore, urolithin A also prevents memory loss in a mitophagy-dependent manner in an Alzheimer disease model of *C. elegans* and mice [68]. The safety of urolithin A has been confirmed in phase I clinical trials [229], and phase III clinical trials have shown that uloritin A improves muscle strength and lowers CRP (C-reactive protein) [230].

## 28 Others

Metformin and trehalose activate autophagy primarily via the AMPK pathway in an MTOR-independent manner, and these drugs have demonstrated autophagy activation, lifespan extension, and protective effects against neurodegenerative diseases in animal models [231]. Resveratrol, a natural polyphenol, extends the lifespan of *C. elegans*, Drosophila and mice by activating the NAD$^+$-dependent deacetylase SIRT1 [232, 233]. Furthermore, an autophagy-inducing peptide (TAT-Beclin 1) based on BECN1 has been developed [234] and reported to improve various age-related diseases including neurodegenerative diseases, heart failure, and chronic kidney disease [235], therefore it is desired to verify the effect of this peptide on lifespan extension in animal models.

## 29 Conclusions and Future Perspectives

Previous studies have shown that (1) autophagy declines with age, (2) activation of autophagy is essential for lifespan extension in various longevity paradigms, and (3) activation of autophagy extends lifespan and healthy life expectancy while impairment of autophagy shortens lifespan and healthy life expectancy. There is no doubt that therapeutic approaches that regulate autophagy are promising for achieving longevity and healthy life expectancy. Therefore, research on the clinical application of autophagy modulators is being conducted worldwide, however, there are several issues to be addressed in the clinical application of autophagy modulators.

The first challenge for clinical application of autophagy modulators is the lack of non-invasive methods for monitoring autophagic activity. Since it has been reported in recent years that excessive induction of autophagy could accelerate aging [39, 49], autophagy should be appropriately maintained within a range that is not excessive. It should also be noted that the decrease in autophagy activity in the elderly is not simply due to a decrease in autophagosome biogenesis, rather, a decrease in autophagosome degradation is an additional important factor. Previous reports have shown that enhancing autophagosome biogenesis alone under situations where autophagosome degradation is impaired could lead to exacerbation of aging and age-related diseases [58]. Therefore, for clinical application of autophagy drugs, it is necessary to pay attention to autophagy status and decide the treatment according to each stage of autophagy disturbance. Furthermore, since there are large biological individual differences in the speed of aging [236], it is inferred that there are also large individual differences in autophagy status. Therefore, the need to monitor autophagy activity in each individual is obvious. Collectively, it is necessary to develop a non-invasive method for monitoring autophagy activity, and to maintain appropriate autophagy activity while monitoring the autophagy status of each individual for the clinical application of autophagy drugs safely without side effects.

The second challenge is that most current autophagy modulators are non-selective autophagy modulators, which is in contrast to the recent accumulation of findings that the autophagy that regulates aging in vivo is mediated by selective autophagy. In the future, it will be necessary to further develop the current research to identify selective substrates of autophagy that directly regulate lifespan and health span in vivo, and to develop methods to activate selective autophagy in vivo. In this regard, a method to induce selective autophagy in vitro has recently been developed [237], holding promise for future application in autophagy modulators.

The third challenge for autophagy modulators is to understand how autophagy regulates the aging process in different organs. In other words, it is necessary to verify whether autophagy in each organ works to maintain homeostasis in each organ and suppresses aging, or whether autophagy in a specific organ regulates aging in the whole individual. Furthermore, while autophagy activity declines with age in most organs, it has been reported that autophagy activity increases with age in adipose tissue, and further promotion of autophagy leads to lypodystrophy and ectopic fat accumulation [238]. Thus, it is necessary to verify whether the activation of autophagy in the elderly might have an adverse effect on certain organs. If organ-specific autophagy that regulates aging in the whole individual is identified, the development of methods to activate autophagy in a organ-specific manner could lead to therapies with fewer side effects.

The fourth challenge is the need to re-examine when autophagy modulators should be used during the life stages. It is generally thought that since autophagy activity declines with age, it is better to use autophagy modulators in old age when autophagy activity declines; however, there are reports that activation of autophagy in old age does not affect lifespan [239] or rather shortens lifespan [16, 53], although autophagy activation in young age prolongs lifespan [239]. Furthermore, there is a report that restoration of autophagy activity after 4 months of autophagy failure in mice results in a marked increase in malignancies, although aging and age-related diseases are suppressed [35]. These reports indicate that the use of autophagy modulators in elderly patients should be carefully monitored for adverse effects on life expectancy and the development of malignant tumors, and future studies on the optimal timing of autophagy activation should be thoroughly examined.

As described above, there are various challenges in the clinical application of autophagy modulators. That is, at which stage of autophagy to activate, which selective autophagy to activate, which tissue autophagy to activate, and at which timing of life stages to activate? Future research to resolve these issues will pave the way for clinical application of autophagy modulators that is safe and has few side effects.

# 30　Introduction to Cardinal Articles

## 30.1　Introduction to the 2 Articles

### 30.1.1　"Suppression of Autophagic Activity by RUBCN Is a Signature of Aging [18]"

Autophagy is a conserved mechanism across species that suppresses aging. Autophagy activity is known to decrease with aging; however, the molecular mechanism underlying the decrease is unknown. In this study, the expression of RUBCN, a negative regulator of autophagy, increases with age in *C. elegans*, Drosophila, and the liver and kidney of mice. The loss of RUBCN extends the lifespan and healthy life expectancy of *C. elegans* and Drosophila, and improves models of age-related diseases such as kidney fibrosis and Parkinson disease model in mice.

- *How did they move the field forward?*

Previously, it was thought that the age-related decline in autophagy activity is caused by impaired autophagosome formation due to a decrease in autophagy-related proteins. However, detailed subsequent analysis revealed that autophagy activity is decreased in aged tissues, accompanied by the accumulation of autophagosomes. This finding suggested that the age-related decline in autophagy activity impairs the process of autophagosome removal, although the underlying molecular mechanisms were unknown. This study advances this field because it reveals that the molecular mechanism by which autophagosome removal is impaired with aging is due to impaired autophagosome maturation associated with increased RUBCN expression.

- *Which questions emerged?*

This study revealed that the increase in RUBCN plays an important role in the decrease in autophagy activity with aging; however, the details of the regulatory mechanism of the increase in RUBCN with aging remain unclear.

Furthermore, in order to verify the organ-specific autophagy mechanism of aging prevention, RUBCN of *C. elegans* was specifically inhibited in each tissue (nerve, intestine, muscle, and hypodermis), and the effect on lifespan extension was verified in this study. The results showed that neuron-specific RUBCN inhibition extends lifespan most efficiently. However, the mechanism by which autophagy activated by RUBCN inhibition in neurons suppresses individual aging remains unclear.

- *What deserves attention*

Most of the conventional studies on autophagy and aging have examined whether lifespan can be extended by artificially increasing the expression of autophagy-related proteins and forcibly promoting the biosynthesis of autophagosomes, however, this approach is non-physiological, and it has been reported that sometimes autophagy is excessive, which rather exacerbates aging pathologies. Conversely, this research identified the molecular basis for the decline in autophagy activity with aging, and by specifically intervening in the molecular basis, the aging-associated autophagy status was "rejuvenated" to that of a youthful autophagy status, which is a physiological and innovative approach to suppressing aging.

### 30.1.2 "Upregulation of RUBCN Promotes Autosis during Myocardial Ischemia/Reperfusion Injury [58]"

Autophagy is generally protective against a variety of diseases, however, uncontrolled or excessive autophagic activation could be detrimental. In this study, myocardial cell death with depletion of intracellular membranes was observed in a model of cardiac ischemia-reperfusion injury, accompanied by accumulation of excess autophagosomes, increased RUBCN, and decreased autophagic flux. RUBCN inhibition rescues cardiac ischemia-reperfusion injury by ameliorating the accumulation of excess autophagosomes, whereas an autophagosome biogenesis-inducing peptide, TAT-Beclin 1, treatment exacerbates cardiac ischemia-reperfusion injury by aggravating the accumulation of excess autophagosomes. Subsequent analysis revealed that during myocardial ischemia-reperfusion injury, autophagosomal membranes produced massively from intracellular membranes are not degraded by lysosomes due to increased RUBCN, impairing autophagosomal membrane recycling, leading to organelle damage due to intracellular membrane depletion, and subsequent deterioration of myocardial ischemia-reperfusion injury.

- *How did they move the field forward?*

It has long been known that impaired autophagy activity is closely associated with aging and the development of age-related diseases, while excessive autophagy could also be detrimental to individual aging under certain circumstances. However, it was unclear which circumstances would make autophagy detrimental. In this study, it was demonstrated that the excessive enhancement of autophagosome biosynthesis is detrimental under conditions where autophagosome maturation is impaired due to increased RUBCN. This mechanism is that autophagosomes formed from intracellular membranes cannot fuse with lysosomes due to increased RUBCN and accumulate intracellularly, resulting in depletion of the intracellular membrane, which induces organelle damage and subsequent cellular damage. Therefore, this study has widely demonstrated that in conditions with reduced autophagosome degradation and recycling due to increased RUBCN or lysosomal dysfunction, interventions that promote autophagosome biosynthesis should be considered with particular caution because they may exacerbate intracellular membrane depletion due to autophagosome accumulation and exacerbate the pathology. This means that autophagy-based therapeutic interventions should not be uniformly applied to promote autophagosome biosynthesis; instead, attention should always be paid to autophagy status and molecular mechanism-specific interventions should be applied according to the stage of autophagy disturbance. In summary, this study advances our knowledge of the use on autophagy-activating drugs for pathological conditions.

- *Which questions emerged?*

This study demonstrated the risk of therapies that promote autophagosome biosynthesis during cardiac ischemia and reperfusion when RUBCN is increased. Recently, it has been reported that RUBCN increases in various organs in the elderly, and that the increase of RUBCN is closely related to age-related diseases [18]. Therefore, it is important to note that therapies that promote autophagosome biosynthesis carry the risk of exacerbating the limited condition of myocardial infarction as well as various age-related diseases in which RUBCN is increased. Recently, research on autophagy drug discovery to promote autophagosome biosynthesis for age-related diseases is currently being driven worldwide, based on the recent results that autophagy combats aging and age-related diseases. It is necessary to carefully examine whether these autophagy-activating drugs could exacerbate age-related diseases through excessive accumulation of autophagosomes, especially when used to diseases with increasing RUBCN.

- *What deserves attention*

This study is noteworthy as it reveals that in pathological conditions in which autophagosome degradation is impaired, interventions that promote autophagosome biosynthesis exacerbate autophagosome accumulation and potentially exacerbate the disease. Even though it has been reported that autophagy activity is decreased in aging pathologies due to a combination of both decreased autophagosome biosynthesis and impaired autophagosome degradation, most autophagy and aging research to date has focused on extending lifespan by forcibly promoting autophagosome biosynthesis. This study raises the alarm against the conventional treatment of aging by blindly promoting autophagosome biosynthesis.

## 30.2    Take Note Of

Abnormalities in autophagy activity during aging include impaired autophagy activity at the following stages: (1) decreased autophagosome biosynthesis, (2) impaired autophagosome maturation, and (3) lysosome dysfunction. Therefore, when considering autophagy-mediated therapeutic interventions for aging and age-related diseases, it is necessary to first understand the autophagy status of each individual and tailor specific therapeutic interventions to the impairment at each stage of autophagy.

## 30.3  Markers and Tools Used

The autophagy process can be divided into the following three stages (Fig. 2): (1) Autophagosome biosynthesis process. In this process, phagophores appear in the cytoplasm, which expand while taking in cytoplasmic components and organelles, and finally autophagosomes are formed. (2) Autophagosome maturation process. (3) Autolysosome formation process. In this process, autophagosomes fuse with lysosomes to form autolysosomes, which degrade their contents. The incorporation of cytoplasmic and organelle components into autophagosomes requires receptor proteins including SQSTM1/p62, which are degraded when autophagy is fully accomplished. Therefore, the expression level of receptor proteins is reduced in the autophagy-activated state.

In this chapter, you will learn how to identify which step of autophagy is impaired by using three markers: SQSTM1/p62 protein level, LC3-II (a marker of autophagosomes) protein level, and lysosomal enzyme activity. First, when autophagosome biosynthesis is decreased, SQSTM1/p62 increases because autophagy activity is reduced. Because autophagosomes do not form, LC3-II is reduced. Lysosomal enzyme activity is invariant as it is not affected by autophagosome biosynthesis. Next, with respect to the case of impaired autophagosome maturation, SQSTM1/p62 increases because autophagy activity is reduced. Autophagosomes form yet fail to mature and accumulate, therefore LC3-II is

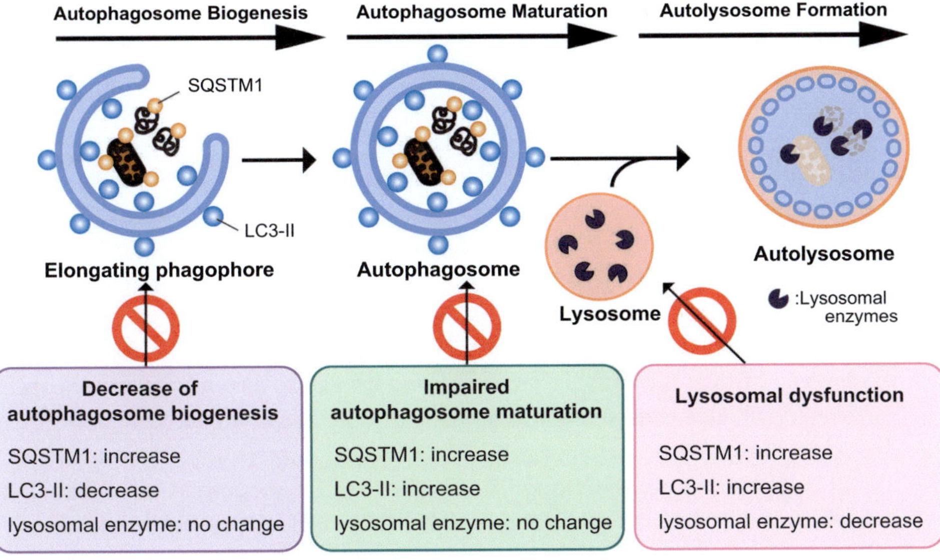



**Fig. 2** The process of autophagy can be divided into three stages: autophagosome biosynthesis, autophagosome maturation, and autolysosome formation. To verify which step of autophagy is impaired, it is possible to differentiate by measuring the protein level of SQSTM1/p62, a receptor protein for autophagy, the protein level of LC3-II, a typical marker of autophagosomes, and lysosomal enzyme activity

increased. Lysosomal enzyme activity is invariant as it is not affected by autophagosome maturation. Finally, when lysosomal damage occurs, SQSTM1/p62 increases because autophagy activity is reduced. Autophagosomes are not degraded by lysosomes and accumulate, resulting in increased LC3-II. Decreased lysosomal enzyme activity is the distinguishing factor between impairment of autophagosomal maturation.

**Take-Home Message**

Autophagy is closely associated with aging: dysregulation of autophagy is a hallmark of aging, whereas activation of autophagy suppresses aging. However, excessive activation of autophagy is also known to exacerbate aging. Based on the above studies, it is considered that a physiological autophagy-mediated approach to suppress aging is to identify the molecular mechanism for dysregulation of autophagy and "rejuvenate" the autophagy status of aging to that of youth by specifically intervening in the molecular mechanism. Because it has been demonstrated that increased RUBCN plays an important role in autophagy dysregulation, especially in aging and age-related diseases, the suppression of RUBCN is a promising therapeutic target to suppress aging and age-related diseases. Conversely, in the pathology of aging, dysregulation of autophagy occurs in various stages of autophagy, such as decreased autophagosome biosynthesis and lysosome dysfunction, as well as increased RUBCN. Therefore, in considering therapeutic interventions based on autophagy, it is necessary to understand the status of autophagy in each individual and condition, and to tailor molecular mechanism-specific therapeutic interventions to each step of autophagy disturbance.

**Answer Guide to Questions**

Questions: Lymphocytes from young and elderly (populations A-C) were collected and evaluated for the protein levels of LC3 and SQSTM1/p62 by western blotting and lysosomal enzyme activity as shown in Fig. 3. Determine the autophagy status of these patients.

Answers: In elderly A, the amount of SQSTM1/p62 protein is increased, the amount of LC3-II protein is decreased, and lysosomal enzyme activity is preserved, which is a decrease in autophagy activity due to impaired autophagosome biosynthesis. In elderly B, SQSTM1/p62 protein level is increased, LC3-II protein level is increased, and lysosomal enzyme activity is decreased, which is a decrease in autophagy activity associated with lysosomal damage. In elderly C, SQSTM1/p62 protein level is increased, LC3-II protein level is increased, lysosomal enzyme activity is preserved, and autophagy activity is decreased due to impaired autophagosome maturation.

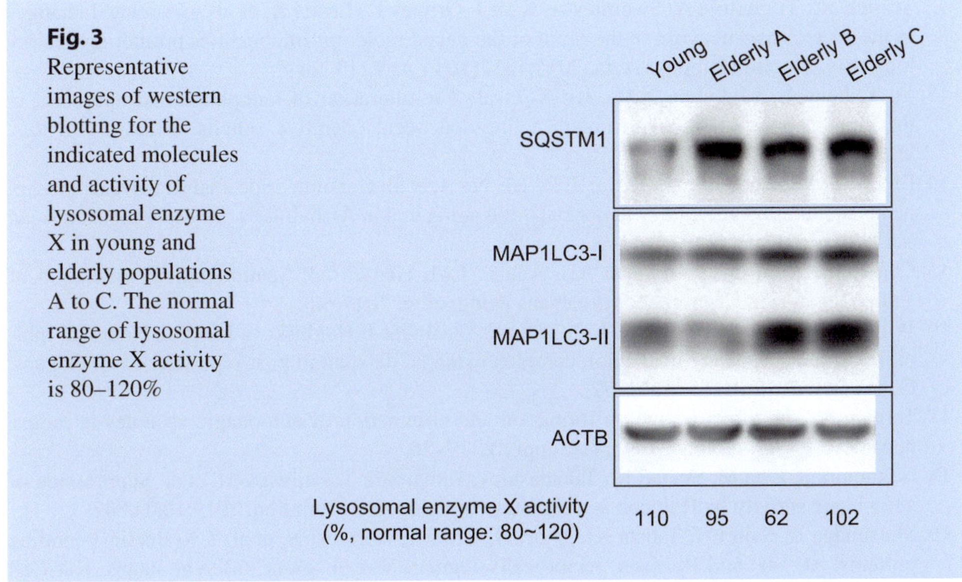



**Fig. 3** Representative images of western blotting for the indicated molecules and activity of lysosomal enzyme X in young and elderly populations A to C. The normal range of lysosomal enzyme X activity is 80–120%

# References


1. Campisi J, Kapahi P, Lithgow GJ, Melov S, Newman JC, Verdin E. From discoveries in ageing research to therapeutics for healthy ageing. Nature. 2019;571(7764):183–92.
2. Partridge L, Deelen J, Slagboom PE. Facing up to the global challenges of ageing. Nature. 2018;561(7721):45–56.
3. López-Otín C, Blasco MA, Partridge L, Serrano M, Kroemer G. Hallmarks of aging: an expanding universe. Cell. 2023;186(2):243–78.
4. Yamamoto H, Zhang S, Mizushima N. Autophagy genes in biology and disease. Nat Rev Genet. 2023:1–19.
5. Vargas JNS, Hamasaki M, Kawabata T, Youle RJ, Yoshimori T. The mechanisms and roles of selective autophagy in mammals. Nat Rev Mol Cell Biol. 2023;24(3):167–85.
6. Leidal AM, Levine B, Debnath J. Autophagy and the cell biology of age-related disease. Nat Cell Biol. 2018;20(12):1338–48.
7. Simonsen A, Cumming RC, Brech A, Isakson P, Schubert DR, Finley KD. Promoting basal levels of autophagy in the nervous system enhances longevity and oxidant resistance in adult Drosophila. Autophagy. 2008;4(2):176–84.
8. Demontis F, Perrimon N. FOXO/4E-BP signaling in Drosophila muscles regulates organism-wide proteostasis during aging. Cell. 2010;143(5):813–25.
9. Bai H, Kang P, Hernandez AM, Tatar M. Activin signaling targeted by insulin/dFOXO regulates aging and muscle proteostasis in Drosophila. PLoS Genet. 2013;9(11):e1003941.
10. Kaushik S, Arias E, Kwon H, Lopez NM, Athonvarangkul D, Sahu S, et al. Loss of autophagy in hypothalamic POMC neurons impairs lipolysis. EMBO Rep. 2012;13(3):258–65.
11. Carnio S, LoVerso F, Baraibar MA, Longa E, Khan MM, Maffei M, et al. Autophagy impairment in muscle induces neuromuscular junction degeneration and precocious aging. Cell Rep. 2014;8(5):1509–21.

12. Triplett JC, Tramutola A, Swomley A, Kirk J, Grimes K, Lewis K, et al. Age-related changes in the proteostasis network in the brain of the naked mole-rat: implications promoting healthy longevity. Biochim Biophys Acta. 2015;1852(10 Pt A):2213–24.

13. Yu Y, Feng L, Li J, Lan XAL, Lv X, et al. The alteration of autophagy and apoptosis in the hippocampus of rats with natural aging-dependent cognitive deficits. Behav Brain Res. 2017;334:155–62.

14. Lipinski MM, Zheng B, Lu T, Yan Z, Py BF, Ng A, et al. Genome-wide analysis reveals mechanisms modulating autophagy in normal brain aging and in Alzheimer's disease. Proc Natl Acad Sci USA. 2010;107(32):14164–9.

15. Chang JT, Kumsta C, Hellman AB, Adams LM, Hansen M. Spatiotemporal regulation of autophagy during Caenorhabditis elegans aging. elife. 2017;6:6.

16. Wilhelm T, Byrne J, Medina R, Kolundžić E, Geisinger J, Hajduskova M, et al. Neuronal inhibition of the autophagy nucleation complex extends life span in post-reproductive C. elegans. Genes Dev. 2017;31(15):1561–72.

17. Terman A. The effect of age on formation and elimination of autophagic vacuoles in mouse hepatocytes. Gerontology. 1995;41(Suppl 2):319–26.

18. Nakamura S, Oba M, Suzuki M, Takahashi A, Yamamuro T, Fujiwara M, et al. Suppression of autophagic activity by Rubicon is a signature of aging. Nat Commun. 2019;10(1):847.

19. Matsunaga K, Saitoh T, Tabata K, Omori H, Satoh T, Kurotori N, et al. Two Beclin 1-binding proteins, Atg14L and Rubicon, reciprocally regulate autophagy at different stages. Nat Cell Biol. 2009;11(4):385–96.

20. Zhong Y, Wang QJ, Li X, Yan Y, Backer JM, Chait BT, et al. Distinct regulation of autophagic activity by Atg14L and Rubicon associated with Beclin 1-phosphatidylinositol-3-kinase complex. Nat Cell Biol. 2009;11(4):468–76.

21. Sun Y, Li M, Zhao D, Li X, Yang C, Wang X. Lysosome activity is modulated by multiple longevity pathways and is important for lifespan extension in C. elegans. elife. 2020;9:9.

22. Donati A, Cavallini G, Paradiso C, Vittorini S, Pollera M, Gori Z, et al. Age-related changes in the regulation of autophagic proteolysis in rat isolated hepatocytes. J Gerontol A Biol Sci Med Sci. 2001;56(7):B288–93.

23. Hughes AL, Gottschling DE. An early age increase in vacuolar pH limits mitochondrial function and lifespan in yeast. Nature. 2012;492(7428):261–5.

24. Menzies FM, Fleming A, Rubinsztein DC. Compromised autophagy and neurodegenerative diseases. Nat Rev Neurosci. 2015;16(6):345–57.

25. Lou G, Palikaras K, Lautrup S, Scheibye-Knudsen M, Tavernarakis N, Fang EF. Mitophagy and neuroprotection. Trends Mol Med. 2020;26(1):8–20.

26. Kenyon CJ. The genetics of ageing. Nature. 2010;464(7288):504–12.

27. Lapierre LR, Kumsta C, Sandri M, Ballabio A, Hansen M. Transcriptional and epigenetic regulation of autophagy in aging. Autophagy. 2015;11(6):867–80.

28. Nakamura S, Yoshimori T. Autophagy and longevity. Mol Cells. 2018;41(1):65–72.

29. Hansen M, Rubinsztein DC, Walker DW. Autophagy as a promoter of longevity: insights from model organisms. Nat Rev Mol Cell Biol. 2018;19(9):579–93.

30. Xiao FH, Chen XQ, Yu Q, Ye Y, Liu YW, Yan D, et al. Transcriptome evidence reveals enhanced autophagy-lysosomal function in centenarians. Genome Res. 2018;28(11):1601–10.

31. Raz Y, Guerrero-Ros I, Maier A, Slagboom PE, Atzmon G, Barzilai N, et al. Activation-induced autophagy is preserved in CD4+ T-cells in familial longevity. J Gerontol A Biol Sci Med Sci. 2017;72(9):1201–6.

32. Lin XX, Sen I, Janssens GE, Zhou X, Fonslow BR, Edgar D, et al. DAF-16/FOXO and HLH-30/TFEB function as combinatorial transcription factors to promote stress resistance and longevity. Nat Commun. 2018;9(1):4400.

33. Kuma A, Komatsu M, Mizushima N. Autophagy-monitoring and autophagy-deficient mice. Autophagy. 2017;13(10):1619–28.

34. Kuma A, Hatano M, Matsui M, Yamamoto A, Nakaya H, Yoshimori T, et al. The role of autophagy during the early neonatal starvation period. Nature. 2004;432(7020):1032–6.

35. Cassidy LD, Young ARJ, Young CNJ, Soilleux EJ, Fielder E, Weigand BM, et al. Temporal inhibition of autophagy reveals segmental reversal of ageing with increased cancer risk. Nat Commun. 2020;11(1):307.

36. Klionsky DJ, Petroni G, Amaravadi RK, Baehrecke EH, Ballabio A, Boya P, et al. Autophagy in major human diseases. EMBO J. 2021;40:e108863.

37. Rubinsztein DC, Marino G, Kroemer G. Autophagy and aging. Cell. 2011;146(5):682–95.

38. Lapierre LR, De Magalhaes Filho CD, McQuary PR, Chu CC, Visvikis O, Chang JT, et al. The TFEB orthologue HLH-30 regulates autophagy and modulates longevity in Caenorhabditis elegans. Nat Commun. 2013;4:2267.

39. Bjedov I, Cocheme HM, Foley A, Wieser D, Woodling NS, Castillo-Quan JI, et al. Fine-tuning autophagy maximises lifespan and is associated with changes in mitochondrial gene expression in Drosophila. PLoS Genet. 2020;16(11):e1009083.

40. Silvestrini MJ, Johnson JR, Kumar AV, Thakurta TG, Blais K, Neill ZA, et al. Nuclear export inhibition enhances HLH-30/TFEB activity, autophagy, and lifespan. Cell Rep. 2018;23(7):1915–21.

41. Ulgherait M, Rana A, Rera M, Graniel J, Walker DW. AMPK modulates tissue and organismal aging in a non-cell-autonomous manner. Cell Rep. 2014;8(6):1767–80.

42. Pyo JO, Yoo SM, Ahn HH, Nah J, Hong SH, Kam TI, et al. Overexpression of Atg5 in mice activates autophagy and extends lifespan. Nat Commun. 2013;4:2300.

43. Fernández ÁF, Sebti S, Wei Y, Zou Z, Shi M, McMillan KL, et al. Disruption of the beclin 1-BCL2 autophagy regulatory complex promotes longevity in mice. Nature. 2018;558(7708):136–40.

44. Wang C, Haas M, Yeo SK, Sebti S, Fernandez AF, Zou Z, et al. Enhanced autophagy in Becn1(F121A/F121A) knockin mice counteracts aging-related neural stem cell exhaustion and dysfunction. Autophagy. 2021;1–14:409–22.

45. Rocchi A, Yamamoto S, Ting T, Fan Y, Sadleir K, Wang Y, et al. A Becn1 mutation mediates hyperactive autophagic sequestration of amyloid oligomers and improved cognition in Alzheimer's disease. PLoS Genet. 2017;13(8):e1006962.

46. Yoshida G, Kawabata T, Takamatsu H, Saita S, Nakamura S, Nishikawa K, et al. Degradation of the NOTCH intracellular domain by elevated autophagy in osteoblasts promotes osteoblast differentiation and alleviates osteoporosis. Autophagy. 2022;18(10):2323–32.

47. Tanaka S, Hikita H, Tatsumi T, Sakamori R, Nozaki Y, Sakane S, et al. Rubicon inhibits autophagy and accelerates hepatocyte apoptosis and lipid accumulation in nonalcoholic fatty liver disease in mice. Hepatology. 2016;64(6):1994–2014.

48. Ando S, Hashida N, Yamashita D, Kawabata T, Asao K, Kawasaki S, et al. Rubicon regulates A2E-induced autophagy impairment in the retinal pigment epithelium implicated in the pathology of age-related macular degeneration. Biochem Biophys Res Commun. 2021;551:148–54.

49. Zhou B, Kreuzer J, Kumsta C, Wu L, Kamer KJ, Cedillo L, et al. Mitochondrial permeability uncouples elevated autophagy and lifespan extension. Cell. 2019;177(2):299–314 e16.

50. Liu Y, Levine B. Autosis and autophagic cell death: the dark side of autophagy. Cell Death Differ. 2015;22(3):367–76.

51. Yu L, Alva A, Su H, Dutt P, Freundt E, Welsh S, et al. Regulation of an ATG7-beclin 1 program of autophagic cell death by caspase-8. Science. 2004;304(5676):1500–2.

52. Shimizu S, Kanaseki T, Mizushima N, Mizuta T, Arakawa-Kobayashi S, Thompson CB, et al. Role of Bcl-2 family proteins in a non-apoptotic programmed cell death dependent on autophagy genes. Nat Cell Biol. 2004;6(12):1221–8.

53. Ezcurra M, Benedetto A, Sornda T, Gilliat AF, Au C, Zhang Q, et al. C. elegans eats its own intestine to make yolk leading to multiple senescent pathologies. Curr Biol. 2018;28(16):2544–56.e5.

54. Xu C, Wang L, Fozouni P, Evjen G, Chandra V, Jiang J, et al. SIRT1 is downregulated by autophagy in senescence and ageing. Nat Cell Biol. 2020;22(10):1170–9.

55. Dou Z, Xu C, Donahue G, Shimi T, Pan JA, Zhu J, et al. Autophagy mediates degradation of nuclear lamina. Nature. 2015;527(7576):105–9.

56. Johmura Y, Yamanaka T, Omori S, Wang TW, Sugiura Y, Matsumoto M, et al. Senolysis by glutaminolysis inhibition ameliorates various age-associated disorders. Science. 2021;371(6526):265–70.

57. Liu Y, Shoji-Kawata S, Sumpter RM Jr, Wei Y, Ginet V, Zhang L, et al. Autosis is a Na+, K+-ATPase-regulated form of cell death triggered by autophagy-inducing peptides, starvation, and hypoxia-ischemia. Proc Natl Acad Sci USA. 2013;110(51):20364–71.

58. Nah J, Zhai P, Huang CY, Fernandez AF, Mareedu S, Levine B, et al. Upregulation of Rubicon promotes autosis during myocardial ischemia/reperfusion injury. J Clin Invest. 2020;130(6):2978–91.

59. Bordi M, Berg MJ, Mohan PS, Peterhoff CM, Alldred MJ, Che S, et al. Autophagy flux in CA1 neurons of Alzheimer hippocampus: increased induction overburdens failing lysosomes to propel neuritic dystrophy. Autophagy. 2016;12(12):2467–83.

60. Lee JH, Yang DS, Goulbourne CN, Im E, Stavrides P, Pensalfini A, et al. Faulty autolysosome acidification in Alzheimer's disease mouse models induces autophagic build-up of Aβ in neurons, yielding senile plaques. Nat Neurosci. 2022;25(6):688–701.

61. Karch J, Schips TG, Maliken BD, Brody MJ, Sargent MA, Kanisicak O, et al. Autophagic cell death is dependent on lysosomal membrane permeability through Bax and Bak. Elife. 2017;6:e30543.

62. Sun N, Youle RJ, Finkel T. The mitochondrial basis of Aging. Mol Cell. 2016;61(5):654–66.

63. Chen G, Kroemer G, Kepp O. Mitophagy: an emerging role in aging and age-associated diseases. Front Cell Dev Biol. 2020;8:200.

64. Palikaras K, Lionaki E, Tavernarakis N. Coordination of mitophagy and mitochondrial biogenesis during ageing in C. elegans. Nature. 2015;521(7553):525–8.

65. Rana A, Oliveira MP, Khamoui AV, Aparicio R, Rera M, Rossiter HB, et al. Promoting Drp1-mediated mitochondrial fission in midlife prolongs healthy lifespan of Drosophila melanogaster. Nat Commun. 2017;8(1):448.

66. Sun N, Yun J, Liu J, Malide D, Liu C, Rovira II, et al. Measuring in vivo mitophagy. Mol Cell. 2015;60(4):685–96.

67. Drummond MJ, Addison O, Brunker L, Hopkins PN, McClain DA, LaStayo PC, et al. Downregulation of E3 ubiquitin ligases and mitophagy-related genes in skeletal muscle of physically inactive, frail older women: a cross-sectional comparison. J Gerontol A Biol Sci Med Sci. 2014;69(8):1040–8.

68. Fang EF, Hou Y, Palikaras K, Adriaanse BA, Kerr JS, Yang B, et al. Mitophagy inhibits amyloid-β and tau pathology and reverses cognitive deficits in models of Alzheimer's disease. Nat Neurosci. 2019;22(3):401–12.

69. Malpartida AB, Williamson M, Narendra DP, Wade-Martins R, Ryan BJ. Mitochondrial dysfunction and mitophagy in Parkinson's disease: from mechanism to therapy. Trends Biochem Sci. 2021;46(4):329–43.

70. Clark IE, Dodson MW, Jiang C, Cao JH, Huh JR, Seol JH, et al. Drosophila pink1 is required for mitochondrial function and interacts genetically with parkin. Nature. 2006;441(7097):1162–6.

71. Greene JC, Whitworth AJ, Kuo I, Andrews LA, Feany MB, Pallanck LJ. Mitochondrial pathology and apoptotic muscle degeneration in Drosophila parkin mutants. Proc Natl Acad Sci USA. 2003;100(7):4078–83.
72. Rana A, Rera M, Walker DW. Parkin overexpression during aging reduces proteotoxicity, alters mitochondrial dynamics, and extends lifespan. Proc Natl Acad Sci USA. 2013;110(21):8638–43.
73. Du F, Yu Q, Yan S, Hu G, Lue LF, Walker DG, et al. PINK1 signalling rescues amyloid pathology and mitochondrial dysfunction in Alzheimer's disease. Brain. 2017;140(12):3233–51.
74. Schiavi A, Maglioni S, Palikaras K, Shaik A, Strappazzon F, Brinkmann V, et al. Iron-starvation-induced mitophagy mediates lifespan extension upon mitochondrial stress in C. elegans. Curr Biol. 2015;25(14):1810–22.
75. West AP, Khoury-Hanold W, Staron M, Tal MC, Pineda CM, Lang SM, et al. Mitochondrial DNA stress primes the antiviral innate immune response. Nature. 2015;520(7548):553–7.
76. Sliter DA, Martinez J, Hao L, Chen X, Sun N, Fischer TD, et al. Parkin and PINK1 mitigate STING-induced inflammation. Nature. 2018;561(7722):258–62.
77. Han S, Schroeder EA, Silva-Garcia CG, Hebestreit K, Mair WB, Brunet A. Mono-unsaturated fatty acids link H3K4me3 modifiers to C. elegans lifespan. Nature. 2017;544(7649):185–90.
78. Qi W, Gutierrez GE, Gao X, Dixon H, McDonough JA, Marini AM, et al. The ω-3 fatty acid α-linolenic acid extends Caenorhabditis elegans lifespan via NHR-49/PPARα and oxidation to oxylipins. Aging Cell. 2017;16(5):1125–35.
79. Chamoli M, Goyala A, Tabrez SS, Siddiqui AA, Singh A, Antebi A, et al. Polyunsaturated fatty acids and p38-MAPK link metabolic reprogramming to cytoprotective gene expression during dietary restriction. Nat Commun. 2020;11(1):4865.
80. Hansen M, Flatt T, Aguilaniu H. Reproduction, fat metabolism, and life span: what is the connection? Cell Metab. 2013;17(1):10–9.
81. Singh R, Kaushik S, Wang Y, Xiang Y, Novak I, Komatsu M, et al. Autophagy regulates lipid metabolism. Nature. 2009;458(7242):1131–5.
82. Folick A, Oakley HD, Yu Y, Armstrong EH, Kumari M, Sanor L, et al. Aging. Lysosomal signaling molecules regulate longevity in Caenorhabditis elegans. Science. 2015;347(6217):83–6.
83. Wang MC, O'Rourke EJ, Ruvkun G. Fat metabolism links germline stem cells and longevity in C. elegans. Science. 2008;322(5903):957–60.
84. Palikaras K, Mari M, Petanidou B, Pasparaki A, Filippidis G, Tavernarakis N. Ectopic fat deposition contributes to age-associated pathology in Caenorhabditis elegans. J Lipid Res. 2017;58(1):72–80.
85. Lapierre LR, Silvestrini MJ, Nuñez L, Ames K, Wong S, Le TT, et al. Autophagy genes are required for normal lipid levels in C. elegans. Autophagy. 2013;9(3):278–86.
86. Shibata M, Yoshimura K, Furuya N, Koike M, Ueno T, Komatsu M, et al. The MAP1-LC3 conjugation system is involved in lipid droplet formation. Biochem Biophys Res Commun. 2009;382(2):419–23.
87. O'Rourke EJ, Ruvkun G. MXL-3 and HLH-30 transcriptionally link lipolysis and autophagy to nutrient availability. Nat Cell Biol. 2013;15(6):668–76.
88. Ramachandran PV, Savini M, Folick AK, Hu K, Masand R, Graham BH, et al. Lysosomal signaling promotes longevity by adjusting mitochondrial activity. Dev Cell. 2019;48(5):685–96.e5.
89. Savini M, Folick A, Lee YT, Jin F, Cuevas A, Tillman MC, et al. Lysosome lipid signalling from the periphery to neurons regulates longevity. Nat Cell Biol. 2022;24(6):906–16.
90. Li Y, Chen B, Zou W, Wang X, Wu Y, Zhao D, et al. The lysosomal membrane protein SCAV-3 maintains lysosome integrity and adult longevity. J Cell Biol. 2016;215(2):167–85.
91. Papadopoulos C, Meyer H. Detection and clearance of damaged lysosomes by the endo-lysosomal damage response and lysophagy. Curr Biol. 2017;27(24):R1330–r41.

92. Hornung V, Bauernfeind F, Halle A, Samstad EO, Kono H, Rock KL, et al. Silica crystals and aluminum salts activate the NALP3 inflammasome through phagosomal destabilization. Nat Immunol. 2008;9(8):847–56.

93. Halle A, Hornung V, Petzold GC, Stewart CR, Monks BG, Reinheckel T, et al. The NALP3 inflammasome is involved in the innate immune response to amyloid-beta. Nat Immunol. 2008;9(8):857–65.

94. Heneka MT, Kummer MP, Stutz A, Delekate A, Schwartz S, Vieira-Saecker A, et al. NLRP3 is activated in Alzheimer's disease and contributes to pathology in APP/PS1 mice. Nature. 2013;493(7434):674–8.

95. Duewell P, Kono H, Rayner KJ, Sirois CM, Vladimer G, Bauernfeind FG, et al. NLRP3 inflammasomes are required for atherogenesis and activated by cholesterol crystals. Nature. 2010;464(7293):1357–61.

96. Bussi C, Heunis T, Pellegrino E, Bernard EM, Bah N, Dos Santos MS, et al. Lysosomal damage drives mitochondrial proteome remodelling and reprograms macrophage immunometabolism. Nat Commun. 2022;13(1):7338.

97. Maejima I, Takahashi A, Omori H, Kimura T, Takabatake Y, Saitoh T, et al. Autophagy sequesters damaged lysosomes to control lysosomal biogenesis and kidney injury. EMBO J. 2013;32(17):2336–47.

98. Papandreou ME, Konstantinidis G, Tavernarakis N. Nucleophagy delays aging and preserves germline immortality. Nat Aging. 2023;3(1):34–46.

99. Medzhitov R. Origin and physiological roles of inflammation. Nature. 2008;454(7203):428–35.

100. Franceschi C, Garagnani P, Parini P, Giuliani C, Santoro A. Inflammaging: a new immune-metabolic viewpoint for age-related diseases. Nat Rev Endocrinol. 2018;14(10):576–90.

101. Desdín-Micó G, Soto-Heredero G, Aranda JF, Oller J, Carrasco E, Gabandé-Rodríguez E, et al. T cells with dysfunctional mitochondria induce multimorbidity and premature senescence. Science. 2020;368(6497):1371–6.

102. Mogilenko DA, Shpynov O, Andhey PS, Arthur L, Swain A, Esaulova E, et al. Comprehensive profiling of an aging immune system reveals clonal GZMK(+) CD8(+) T cells as conserved Hallmark of inflammaging. Immunity. 2021;54(1):99–115.e12.

103. Swanson KV, Deng M, Ting JP. The NLRP3 inflammasome: molecular activation and regulation to therapeutics. Nat Rev Immunol. 2019;19(8):477–89.

104. Latz E, Duewell P. NLRP3 inflammasome activation in inflammaging. Semin Immunol. 2018;40:61–73.

105. Zhou R, Yazdi AS, Menu P, Tschopp J. A role for mitochondria in NLRP3 inflammasome activation. Nature. 2011;469(7329):221–5.

106. Shimada K, Crother TR, Karlin J, Dagvadorj J, Chiba N, Chen S, et al. Oxidized mitochondrial DNA activates the NLRP3 inflammasome during apoptosis. Immunity. 2012;36(3):401–14.

107. Zhong Z, Umemura A, Sanchez-Lopez E, Liang S, Shalapour S, Wong J, et al. NF-kappaB restricts inflammasome activation via elimination of damaged mitochondria. Cell. 2016;164(5):896–910.

108. Shi CS, Shenderov K, Huang NN, Kabat J, Abu-Asab M, Fitzgerald KA, et al. Activation of autophagy by inflammatory signals limits IL-1beta production by targeting ubiquitinated inflammasomes for destruction. Nat Immunol. 2012;13(3):255–63.

109. Kimura T, Jain A, Choi SW, Mandell MA, Schroder K, Johansen T, et al. TRIM-mediated precision autophagy targets cytoplasmic regulators of innate immunity. J Cell Biol. 2015;210(6):973–89.

110. Spalinger MR, Lang S, Gottier C, Dai X, Rawlings DJ, Chan AC, et al. PTPN22 regulates NLRP3-mediated IL1B secretion in an autophagy-dependent manner. Autophagy. 2017;13(9):1590–601.

111. Miller KN, Victorelli SG, Salmonowicz H, Dasgupta N, Liu T, Passos JF, et al. Cytoplasmic DNA: sources, sensing, and role in aging and disease. Cell. 2021;184(22):5506–26.

112. Dou Z, Ghosh K, Vizioli MG, Zhu J, Sen P, Wangensteen KJ, et al. Cytoplasmic chromatin triggers inflammation in senescence and cancer. Nature. 2017;550(7676):402–6.

113. De Cecco M, Ito T, Petrashen AP, Elias AE, Skvir NJ, Criscione SW, et al. L1 drives IFN in senescent cells and promotes age-associated inflammation. Nature. 2019;566(7742):73–8.

114. Zhao M, Wang F, Wu J, Cheng Y, Cao Y, Wu X, et al. CGAS is a micronucleophagy receptor for the clearance of micronuclei. Autophagy. 2021;1-17:3976–91.

115. Han X, Chen H, Gong H, Tang X, Huang N, Xu W, et al. Autolysosomal degradation of cytosolic chromatin fragments antagonizes oxidative stress-induced senescence. J Biol Chem. 2020;295(14):4451–63.

116. Ivanov A, Pawlikowski J, Manoharan I, van Tuyn J, Nelson DM, Rai TS, et al. Lysosome-mediated processing of chromatin in senescence. J Cell Biol. 2013;202(1):129–43.

117. Guo H, Chitiprolu M, Gagnon D, Meng L, Perez-Iratxeta C, Lagace D, et al. Autophagy supports genomic stability by degrading retrotransposon RNA. Nat Commun. 2014;5:5276.

118. Ovadya Y, Landsberger T, Leins H, Vadai E, Gal H, Biran A, et al. Impaired immune surveillance accelerates accumulation of senescent cells and aging. Nat Commun. 2018;9(1):5435.

119. Clarke AJ, Simon AK. Autophagy in the renewal, differentiation and homeostasis of immune cells. Nat Rev Immunol. 2019;19(3):170–83.

120. Wei J, Long L, Yang K, Guy C, Shrestha S, Chen Z, et al. Autophagy enforces functional integrity of regulatory T cells by coupling environmental cues and metabolic homeostasis. Nat Immunol. 2016;17(3):277–85.

121. Bharath LP, Agrawal M, McCambridge G, Nicholas DA, Hasturk H, Liu J, et al. Metformin enhances autophagy and normalizes mitochondrial function to alleviate aging-associated inflammation. Cell Metab. 2020;32(1):44–55 e6.

122. Kabat AM, Harrison OJ, Riffelmacher T, Moghaddam AE, Pearson CF, Laing A, et al. The autophagy gene Atg16l1 differentially regulates Treg and TH2 cells to control intestinal inflammation. elife. 2016;5:e12444.

123. Mannick JB, Del Giudice G, Lattanzi M, Valiante NM, Praestgaard J, Huang B, et al. mTOR inhibition improves immune function in the elderly. Sci Transl Med. 2014;6(268):268ra179.

124. Mannick JB, Morris M, Hockey HP, Roma G, Beibel M, Kulmatycki K, et al. TORC1 inhibition enhances immune function and reduces infections in the elderly. Sci Transl Med. 2018;10(449)

125. Zhang Y, Kim MS, Jia B, Yan J, Zuniga-Hertz JP, Han C, et al. Hypothalamic stem cells control ageing speed partly through exosomal miRNAs. Nature. 2017;548(7665):52–7.

126. Satoh A, Brace CS, Rensing N, Cliften P, Wozniak DF, Herzog ED, et al. Sirt1 extends life span and delays aging in mice through the regulation of Nk2 homeobox 1 in the DMH and LH. Cell Metab. 2013;18(3):416–30.

127. Minnerly J, Zhang J, Parker T, Kaul T, Jia K. The cell non-autonomous function of ATG-18 is essential for neuroendocrine regulation of Caenorhabditis elegans lifespan. PLoS Genet. 2017;13(5):e1006764.

128. Schmid ET, Pyo JH, Walker DW. Neuronal induction of BNIP3-mediated mitophagy slows systemic aging in Drosophila. Nat Aging. 2022;2(6):494–507.

129. Gelino S, Chang JT, Kumsta C, She X, Davis A, Nguyen C, et al. Intestinal autophagy improves Healthspan and longevity in C. elegans during dietary restriction. PLoS Genet. 2016;12(7):e1006135.

130. Lu YX, Regan JC, Eßer J, Drews LF, Weinseis T, Stinn J, et al. A TORC1-histone axis regulates chromatin organisation and non-canonical induction of autophagy to ameliorate ageing. Elife. 2021:10.

131. Rera M, Clark RI, Walker DW. Intestinal barrier dysfunction links metabolic and inflammatory markers of aging to death in Drosophila. Proc Natl Acad Sci USA. 2012;109(52):21528–33.

132. Regan JC, Khericha M, Dobson AJ, Bolukbasi E, Rattanavirotkul N, Partridge L. Sex difference in pathology of the ageing gut mediates the greater response of female lifespan to dietary restriction. Elife. 2016;5:e10956.

133. Fan P, Liu P, Song P, Chen X, Ma X. Moderate dietary protein restriction alters the composition of gut microbiota and improves ileal barrier function in adult pig model. Sci Rep. 2017;7:43412.

134. Hu DJ, Jasper H. Epithelia: understanding the cell biology of intestinal barrier dysfunction. Curr Biol. 2017;27(5):R185–r7.

135. Spalinger MR, Rogler G, Scharl M. Crohn's disease: loss of tolerance or a disorder of autophagy? Dig Dis. 2014;32(4):370–7.

136. Clemmensen C, Müller TD, Woods SC, Berthoud HR, Seeley RJ, Tschöp MH. Gut-brain crosstalk in metabolic control. Cell. 2017;168(5):758–74.

137. Schroeder BO, Bäckhed F. Signals from the gut microbiota to distant organs in physiology and disease. Nat Med. 2016;22(10):1079–89.

138. Priest C, Tontonoz P. Inter-organ cross-talk in metabolic syndrome. Nat Metab. 2019;1(12):1177–88.

139. Demontis F, Piccirillo R, Goldberg AL, Perrimon N. The influence of skeletal muscle on systemic aging and lifespan. Aging Cell. 2013;12(6):943–9.

140. Yousefzadeh MJ, Flores RR, Zhu Y, Schmiechen ZC, Brooks RW, Trussoni CE, et al. An aged immune system drives senescence and ageing of solid organs. Nature. 2021;594(7861):100–5.

141. Ron-Harel N, Notarangelo G, Ghergurovich JM, Paulo JA, Sage PT, Santos D, et al. Defective respiration and one-carbon metabolism contribute to impaired naïve T cell activation in aged mice. Proc Natl Acad Sci USA. 2018;115(52):13347–52.

142. Kumar S, Jia J, Deretic V. Atg8ylation as a general membrane stress and remodeling response. Cell Stress. 2021;5(9):128–42.

143. Deretic V, Lazarou M. A guide to membrane atg8ylation and autophagy with reflections on immunity. J Cell Biol. 2022;221(7)

144. Durgan J, Lystad AH, Sloan K, Carlsson SR, Wilson MI, Marcassa E, et al. Non-canonical autophagy drives alternative ATG8 conjugation to phosphatidylserine. Mol Cell. 2021;81(9):2031–40 e8.

145. Sanjuan MA, Dillon CP, Tait SW, Moshiach S, Dorsey F, Connell S, et al. Toll-like receptor signalling in macrophages links the autophagy pathway to phagocytosis. Nature. 2007;450(7173):1253–7.

146. Heckmann BL, Teubner BJW, Tummers B, Boada-Romero E, Harris L, Yang M, et al. LC3-associated endocytosis facilitates β-amyloid clearance and mitigates neurodegeneration in murine Alzheimer's disease. Cell. 2019;178(3):536–51.e14.

147. Martinez J, Malireddi RK, Lu Q, Cunha LD, Pelletier S, Gingras S, et al. Molecular characterization of LC3-associated phagocytosis reveals distinct roles for Rubicon, NOX2 and autophagy proteins. Nat Cell Biol. 2015;17(7):893–906.

148. Kim JY, Zhao H, Martinez J, Doggett TA, Kolesnikov AV, Tang PH, et al. Noncanonical autophagy promotes the visual cycle. Cell. 2013;154(2):365–76.

149. Cunha LD, Yang M, Carter R, Guy C, Harris L, Crawford JC, et al. LC3-associated phagocytosis in myeloid cells promotes tumor immune tolerance. Cell. 2018;175(2):429–41.e16.

150. Heckmann BL, Boada-Romero E, Cunha LD, Magne J, Green DR. LC3-associated phagocytosis and inflammation. J Mol Biol. 2017;429(23):3561–76.

151. Inomata M, Xu S, Chandra P, Meydani SN, Takemura G, Philips JA, et al. Macrophage LC3-associated phagocytosis is an immune defense against Streptococcus pneumoniae that diminishes with host aging. Proc Natl Acad Sci USA. 2020;117(52):33561–9.

152. Heckmann BL, Teubner BJW, Boada-Romero E, Tummers B, Guy C, Fitzgerald P, et al. Noncanonical function of an autophagy protein prevents spontaneous Alzheimer's disease. Sci Adv. 2020;6(33):eabb9036.

153. Fischer TD, Wang C, Padman BS, Lazarou M, Youle RJ. STING induces LC3B lipidation onto single-membrane vesicles via the V-ATPase and ATG16L1-WD40 domain. J Cell Biol. 2020;219(12)

154. Gui X, Yang H, Li T, Tan X, Shi P, Li M, et al. Autophagy induction via STING trafficking is a primordial function of the cGAS pathway. Nature. 2019;567(7747):262–6.

155. Wan W, Qian C, Wang Q, Li J, Zhang H, Wang L, et al. STING directly recruits WIPI2 for autophagosome formation during STING-induced autophagy. EMBO J. 2023;42(8):e112387.

156. Lee C, Lamech L, Johns E, Overholtzer M. Selective lysosome membrane turnover is induced by nutrient starvation. Dev Cell. 2020;55:289–297.e4.

157. Nakamura S, Shigeyama S, Minami S, Shima T, Akayama S, Matsuda T, et al. LC3 lipidation is essential for TFEB activation during the lysosomal damage response to kidney injury. Nat Cell Biol. 2020;22(10):1252–63.

158. Goodwin JM, Walkup WG, Hooper K, Li T, Kishi-Itakura C, Ng A, et al. GABARAP sequesters the FLCN-FNIP tumor suppressor complex to couple autophagy with lysosomal biogenesis. Sci Adv. 2021;7(40):eabj2485.

159. Murrow L, Malhotra R, Debnath J. ATG12-ATG3 interacts with Alix to promote basal autophagic flux and late endosome function. Nat Cell Biol. 2015;17(3):300–10.

160. Guo H, Chitiprolu M, Roncevic L, Javalet C, Hemming FJ, Trung MT, et al. Atg5 disassociates the V1V0-ATPase to promote exosome production and tumor metastasis independent of canonical macroautophagy. Dev Cell. 2017;43(6):716–30 e7.

161. Keller MD, Ching KL, Liang FX, Dhabaria A, Tam K, Ueberheide BM, et al. Decoy exosomes provide protection against bacterial toxins. Nature. 2020;579(7798):260–4.

162. Leidal AM, Huang HH, Marsh T, Solvik T, Zhang D, Ye J, et al. The LC3-conjugation machinery specifies the loading of RNA-binding proteins into extracellular vesicles. Nat Cell Biol. 2020;22:187–99.

163. Takahashi A, Okada R, Nagao K, Kawamata Y, Hanyu A, Yoshimoto S, et al. Exosomes maintain cellular homeostasis by excreting harmful DNA from cells. Nat Commun. 2017;8:15287.

164. Green CL, Lamming DW, Fontana L. Molecular mechanisms of dietary restriction promoting health and longevity. Nat Rev Mol Cell Biol. 2021;23:56–73.

165. Mattison JA, Colman RJ, Beasley TM, Allison DB, Kemnitz JW, Roth GS, et al. Caloric restriction improves health and survival of rhesus monkeys. Nat Commun. 2017;8:14063.

166. Balasubramanian P, Howell PR, Anderson RM. Aging and caloric restriction research: a biological perspective with translational potential. EBioMedicine. 2017;21:37–44.

167. Spadaro O, Youm Y, Shchukina I, Ryu S, Sidorov S, Ravussin A, et al. Caloric restriction in humans reveals immunometabolic regulators of health span. Science. 2022;375(6581):671–7.

168. Flanagan EW, Most J, Mey JT, Redman LM. Calorie restriction and aging in humans. Annu Rev Nutr. 2020;40:105–33.

169. Prvulovic MR, Milanovic DJ, Vujovic PZ, Jovic MS, Kanazir SD, Todorovic ST, et al. Late-onset calorie restriction worsens cognitive performances and increases frailty level in female Wistar rats. J Gerontol A Biol Sci Med Sci. 2022;77(5):947–55.

170. Sun J, Shen X, Liu H, Lu S, Peng J, Kuang H. Caloric restriction in female reproduction: is it beneficial or detrimental? Reprod Biol Endocrinol. 2021;19(1):1.

171. Sun D, Muthukumar AR, Lawrence RA, Fernandes G. Effects of calorie restriction on polymicrobial peritonitis induced by cecum ligation and puncture in young C57BL/6 mice. Clin Diagn Lab Immunol. 2001;8(5):1003–11.

172. Chung KW, Chung HY. The effects of calorie restriction on autophagy: role on aging intervention. Nutrients. 2019;11(12)

173. Hansen M, Chandra A, Mitic LL, Onken B, Driscoll M, Kenyon C. A role for autophagy in the extension of lifespan by dietary restriction in C. elegans. PLoS Genet. 2008;4(2):e24.

174. Jia K, Levine B. Autophagy is required for dietary restriction-mediated life span extension in C. elegans. Autophagy. 2007;3(6):597–9.

175. Kume S, Uzu T, Horiike K, Chin-Kanasaki M, Isshiki K, Araki S, et al. Calorie restriction enhances cell adaptation to hypoxia through Sirt1-dependent mitochondrial autophagy in mouse aged kidney. J Clin Invest. 2010;120(4):1043–55.

176. Acosta-Rodríguez VA, de Groot MHM, Rijo-Ferreira F, Green CB, Takahashi JS. Mice under caloric restriction self-impose a temporal restriction of food intake as revealed by an automated feeder system. Cell Metab. 2017;26(1):267–77.e2.

177. Mitchell SJ, Bernier M, Mattison JA, Aon MA, Kaiser TA, Anson RM, et al. Daily fasting improves health and survival in male mice independent of diet composition and calories. Cell Metab. 2019;29(1):221–8.e3.

178. Chaix A, Zarrinpar A, Miu P, Panda S. Time-restricted feeding is a preventative and therapeutic intervention against diverse nutritional challenges. Cell Metab. 2014;20(6):991–1005.

179. Hatori M, Vollmers C, Zarrinpar A, DiTacchio L, Bushong EA, Gill S, et al. Time-restricted feeding without reducing caloric intake prevents metabolic diseases in mice fed a high-fat diet. Cell Metab. 2012;15(6):848–60.

180. Pak HH, Haws SA, Green CL, Koller M, Lavarias MT, Richardson NE, et al. Fasting drives the metabolic, molecular and geroprotective effects of a calorie-restricted diet in mice. Nat Metab. 2021;3(10):1327–41.

181. Acosta-Rodríguez V, Rijo-Ferreira F, Izumo M, Xu P, Wight-Carter M, Green CB, et al. Circadian alignment of early onset caloric restriction promotes longevity in male C57BL/6J mice. Science. 2022;376(6598):1192–202.

182. Martinez-Lopez N, Tarabra E, Toledo M, Garcia-Macia M, Sahu S, Coletto L, et al. System-wide benefits of intermeal fasting by autophagy. Cell Metab. 2017;26(6):856–71 e5.

183. Ulgherait M, Midoun AM, Park SJ, Gatto JA, Tener SJ, Siewert J, et al. Circadian autophagy drives iTRF-mediated longevity. Nature. 2021;598(7880):353–8.

184. Yokoyama Y, Onishi K, Hosoda T, Amano H, Otani S, Kurozawa Y, et al. Skipping breakfast and risk of mortality from cancer, circulatory diseases and all causes: findings from the Japan collaborative cohort study. Yonago Acta Med. 2016;59(1):55–60.

185. Uzhova I, Fuster V, Fernández-Ortiz A, Ordovás JM, Sanz J, Fernández-Friera L, et al. The importance of breakfast in atherosclerosis disease: insights from the PESA study. J Am Coll Cardiol. 2017;70(15):1833–42.

186. Blair SN, Kohl HW 3rd, Paffenbarger RS Jr, Clark DG, Cooper KH, Gibbons LW. Physical fitness and all-cause mortality. A prospective study of healthy men and women. JAMA. 1989;262(17):2395–401.

187. Myers J, Prakash M, Froelicher V, Do D, Partington S, Atwood JE. Exercise capacity and mortality among men referred for exercise testing. N Engl J Med. 2002;346(11):793–801.

188. Ruegsegger GN, Booth FW. Health benefits of exercise. Cold Spring Harb Perspect Med. 2018;8(7).

189. Liu Y, Yan T, Chu JM, Chen Y, Dunnett S, Ho YS, et al. The beneficial effects of physical exercise in the brain and related pathophysiological mechanisms in neurodegenerative diseases. Lab Investig. 2019;99(7):943–57.

190. Lira VA, Okutsu M, Zhang M, Greene NP, Laker RC, Breen DS, et al. Autophagy is required for exercise training-induced skeletal muscle adaptation and improvement of physical performance. FASEB J. 2013;27(10):4184–93.

191. He C, Bassik MC, Moresi V, Sun K, Wei Y, Zou Z, et al. Exercise-induced BCL2-regulated autophagy is required for muscle glucose homeostasis. Nature. 2012;481(7382):511–5.
192. Vainshtein A, Tryon LD, Pauly M, Hood DA. Role of PGC-1α during acute exercise-induced autophagy and mitophagy in skeletal muscle. Am J Physiol Cell Physiol. 2015;308(9):C710–9.
193. Schwalm C, Jamart C, Benoit N, Naslain D, Prémont C, Prévet J, et al. Activation of autophagy in human skeletal muscle is dependent on exercise intensity and AMPK activation. FASEB J. 2015;29(8):3515–26.
194. Kuramoto K, Liang H, Hong JH, He C. Exercise-activated hepatic autophagy via the FN1-α5β1 integrin pathway drives metabolic benefits of exercise. Cell Metab. 2023;35(4):620–32.e5.
195. De Miguel Z, Khoury N, Betley MJ, Lehallier B, Willoughby D, Olsson N, et al. Exercise plasma boosts memory and dampens brain inflammation via clusterin. Nature. 2021;600(7889):494–9.
196. Dutta D, Paidi RK, Raha S, Roy A, Chandra S, Pahan K. Treadmill exercise reduces alpha-synuclein spreading via PPARalpha. Cell Rep. 2022;40(2):111058.
197. Almeida MF, Silva CM, Chaves RS, Lima NCR, Almeida RS, Melo KP, et al. Effects of mild running on substantia nigra during early neurodegeneration. J Sports Sci. 2018;36(12):1363–70.
198. Huang J, Wang X, Zhu Y, Li Z, Zhu YT, Wu JC, et al. Exercise activates lysosomal function in the brain through AMPK-SIRT1-TFEB pathway. CNS Neurosci Ther. 2019;25(6):796–807.
199. Saxton RA, Sabatini DM. mTOR signaling in growth, metabolism, and disease. Cell. 2017;168(6):960–76.
200. Song HL, Demirev AV, Kim NY, Kim DH, Yoon SY. Ouabain activates transcription factor EB and exerts neuroprotection in models of Alzheimer's disease. Mol Cell Neurosci. 2019;95:13–24.
201. Kim S, Choi KJ, Cho SJ, Yun SM, Jeon JP, Koh YH, et al. Fisetin stimulates autophagic degradation of phosphorylated tau via the activation of TFEB and Nrf2 transcription factors. Sci Rep. 2016;6:24933.
202. Zhang Z, Chen C, Yang F, Zeng YX, Sun P, Liu P, et al. Itaconate is a lysosomal inducer that promotes antibacterial innate immunity. Mol Cell. 2022;82:2844–2857.e10.
203. Lautrup S, Sinclair DA, Mattson MP, Fang EF. NAD(+) in brain aging and neurodegenerative disorders. Cell Metab. 2019;30(4):630–55.
204. Huang R, Xu Y, Wan W, Shou X, Qian J, You Z, et al. Deacetylation of nuclear LC3 drives autophagy initiation under starvation. Mol Cell. 2015;57(3):456–66.
205. Lee IH, Cao L, Mostoslavsky R, Lombard DB, Liu J, Bruns NE, et al. A role for the NAD-dependent deacetylase Sirt1 in the regulation of autophagy. Proc Natl Acad Sci USA. 2008;105(9):3374–9.
206. Bao J, Zheng L, Zhang Q, Li X, Zhang X, Li Z, et al. Deacetylation of TFEB promotes fibrillar Abeta degradation by upregulating lysosomal biogenesis in microglia. Protein Cell. 2016;7(6):417–33.
207. Hariharan N, Maejima Y, Nakae J, Paik J, Depinho RA, Sadoshima J. Deacetylation of FoxO by Sirt1 plays an essential role in mediating starvation-induced autophagy in cardiac myocytes. Circ Res. 2010;107(12):1470–82.
208. Wan W, Hua F, Fang P, Li C, Deng F, Chen S, et al. Regulation of mitophagy by Sirtuin family proteins: a vital role in aging and age-related diseases. Front Aging Neurosci. 2022;14:845330.
209. Baeken MW. Sirtuins and their influence on autophagy. J Cell Biochem. 2023;125
210. Wilson N, Kataura T, Korsgen ME, Sun C, Sarkar S, Korolchuk VI. The autophagy-NAD axis in longevity and disease. Trends Cell Biol. 2023;33:788–802.
211. Fang EF, Hou Y, Lautrup S, Jensen MB, Yang B, SenGupta T, et al. NAD(+) augmentation restores mitophagy and limits accelerated aging in Werner syndrome. Nat Commun. 2019;10(1):5284.

212. Zhang H, Ryu D, Wu Y, Gariani K, Wang X, Luan P, et al. NAD⁺ repletion improves mitochondrial and stem cell function and enhances life span in mice. Science. 2016;352(6292):1436–43.

213. Chen AC, Martin AJ, Choy B, Fernández-Peñas P, Dalziell RA, McKenzie CA, et al. A phase 3 randomized trial of nicotinamide for skin-cancer chemoprevention. N Engl J Med. 2015;373(17):1618–26.

214. Yoshino M, Yoshino J, Kayser BD, Patti GJ, Franczyk MP, Mills KF, et al. Nicotinamide mononucleotide increases muscle insulin sensitivity in prediabetic women. Science. 2021;372(6547):1224–9.

215. Brakedal B, Dölle C, Riemer F, Ma Y, Nido GS, Skeie GO, et al. The NADPARK study: a randomized phase I trial of nicotinamide riboside supplementation in Parkinson's disease. Cell Metab. 2022;34(3):396–407.e6.

216. Madeo F, Eisenberg T, Pietrocola F, Kroemer G. Spermidine in health and disease. Science. 2018;359:6374.

217. Zhang H, Alsaleh G, Feltham J, Sun Y, Napolitano G, Riffelmacher T, et al. Polyamines control eIF5A hypusination, TFEB translation, and autophagy to reverse B cell senescence. Mol Cell. 2019;76(1):110–25 e9.

218. Lubas M, Harder LM, Kumsta C, Tiessen I, Hansen M, Andersen JS, et al. eIF5A is required for autophagy by mediating ATG3 translation. EMBO Rep. 2018;19(6)

219. Pietrocola F, Lachkar S, Enot DP, Niso-Santano M, Bravo-San Pedro JM, Sica V, et al. Spermidine induces autophagy by inhibiting the acetyltransferase EP300. Cell Death Differ. 2015;22(3):509–16.

220. Eisenberg T, Knauer H, Schauer A, Büttner S, Ruckenstuhl C, Carmona-Gutierrez D, et al. Induction of autophagy by spermidine promotes longevity. Nat Cell Biol. 2009;11(11):1305–14.

221. Gupta VK, Scheunemann L, Eisenberg T, Mertel S, Bhukel A, Koemans TS, et al. Restoring polyamines protects from age-induced memory impairment in an autophagy-dependent manner. Nat Neurosci. 2013;16(10):1453–60.

222. Eisenberg T, Abdellatif M, Schroeder S, Primessnig U, Stekovic S, Pendl T, et al. Cardioprotection and lifespan extension by the natural polyamine spermidine. Nat Med. 2016;22(12):1428–38.

223. De Risi M, Torromino G, Tufano M, Moriceau S, Pignataro A, Rivagorda M, et al. Mechanisms by which autophagy regulates memory capacity in ageing. Aging Cell. 2020;19(9):e13189.

224. Liang W, Yamahara K, Hernando-Erhard C, Lagies S, Wanner N, Liang H, et al. A reciprocal regulation of spermidine and autophagy in podocytes maintains the filtration barrier. Kidney Int. 2020;98:1434–48.

225. Alsaleh G, Panse I, Swadling L, Zhang H, Richter FC, Meyer A, et al. Autophagy in T cells from aged donors is maintained by spermidine and correlates with function and vaccine responses. Elife. 2020;9:9.

226. Schwarz C, Benson GS, Horn N, Wurdack K, Grittner U, Schilling R, et al. Effects of spermidine supplementation on cognition and biomarkers in older adults with subjective cognitive decline: a randomized clinical trial. JAMA Netw Open. 2022;5(5):e2213875.

227. D'Amico D, Andreux PA, Valdés P, Singh A, Rinsch C, Auwerx J. Impact of the natural compound urolithin a on health, disease, and aging. Trends Mol Med. 2021;27(7):687–99.

228. Ryu D, Mouchiroud L, Andreux PA, Katsyuba E, Moullan N, Nicolet-Dit-Felix AA, et al. Urolithin A induces mitophagy and prolongs lifespan in C. elegans and increases muscle function in rodents. Nat Med. 2016;22(8):879–88.

229. Andreux PA, Blanco-Bose W, Ryu D, Burdet F, Ibberson M, Aebischer P, et al. The mitophagy activator urolithin A is safe and induces a molecular signature of improved mitochondrial and cellular health in humans. Nat Metab. 2019;1(6):595–603.

230. Singh A, D'Amico D, Andreux PA, Fouassier AM, Blanco-Bose W, Evans M, et al. Urolithin A improves muscle strength, exercise performance, and biomarkers of mitochondrial health in a randomized trial in middle-aged adults. Cell Rep Med. 2022;3(5):100633.

231. Menzies FM, Fleming A, Caricasole A, Bento CF, Andrews SP, Ashkenazi A, et al. Autophagy and neurodegeneration: pathogenic mechanisms and therapeutic opportunities. Neuron. 2017;93(5):1015–34.

232. Baur JA, Pearson KJ, Price NL, Jamieson HA, Lerin C, Kalra A, et al. Resveratrol improves health and survival of mice on a high-calorie diet. Nature. 2006;444(7117):337–42.

233. Wood JG, Rogina B, Lavu S, Howitz K, Helfand SL, Tatar M, et al. Sirtuin activators mimic caloric restriction and delay ageing in metazoans. Nature. 2004;430(7000):686–9.

234. Shoji-Kawata S, Sumpter R, Leveno M, Campbell GR, Zou Z, Kinch L, et al. Identification of a candidate therapeutic autophagy-inducing peptide. Nature. 2013;494(7436):201–6.

235. He Y, Lu H, Zhao Y. Development of an autophagy activator from class III PI3K complexes, Tat-BECN1 peptide: mechanisms and applications. Front Cell Dev Biol. 2022;10:851166.

236. Elliott ML, Caspi A, Houts RM, Ambler A, Broadbent JM, Hancox RJ, et al. Disparities in the pace of biological aging among midlife adults of the same chronological age have implications for future frailty risk and policy. Nat Aging. 2021;1(3):295–308.

237. Takahashi D, Moriyama J, Nakamura T, Miki E, Takahashi E, Sato A, et al. AUTACs: cargo-specific degraders using selective autophagy. Mol Cell. 2019;76(5):797–810.e10.

238. Yamamuro T, Kawabata T, Fukuhara A, Saita S, Nakamura S, Takeshita H, et al. Age-dependent loss of adipose Rubicon promotes metabolic disorders via excess autophagy. Nat Commun. 2020;11(1):4150.

239. Juricic P, Lu YX, Leech T, Drews LF, Paulitz J, Lu J, et al. Long-lasting geroprotection from brief rapamycin treatment in early adulthood by persistently increased intestinal autophagy. Nat Aging. 2022;2(9):824–36.

# Shear Stress-Dependent Regulation of Autophagy and Metabolism in Kidney Epithelial Cells

Aurore Claude-Taupin, Nicolas Dupont, and Patrice Codogno

**What You Will Learn in This Chapter**
The kidney is the organ consuming the most energy in the body, after the heart. In this chapter, we will define the metabolic needs of kidney epithelial cells to support their principal functions, such as glucose reabsorption and gluconeogenesis. The importance of sensing fluid flow (shear stress) with mechanosensors will be described, with a focus on the primary cilium. Notably, this chapter will describe the molecular machinery regulating autophagy, an important cellular process that transduces mechanical forces in biochemical responses to control the metabolism of renal cells from the proximal tubule.

## 1 Introduction

### 1.1 Metabolism of Proximal Tubule Cells

The kidney is the second most energy-consuming organ in the human body under resting conditions, after the heart [1]. Kidney damage, generated either by acute kidney injury (AKI) or chronic kidney disease (CKD), can be treated by dialysis or kidney transplant in the most severe cases. While AKI is usually linked to slower renal blood flow, provoked by a heart attack, liver failure or during an infection, CKD is a long-term condition. Close to half of CKD cases are patients initially suffering from diabetes, representing over 247,000 patients living with kidney failure in the United States. Kidney diseases are the tenth lead-

A. Claude-Taupin (✉) · N. Dupont · P. Codogno
Université Paris Cité, INSERM UMR-S1151, CNRS UMR-S8253, Institut Necker-Enfants Malades, Paris, France
e-mail: aurore.claude-taupin@inserm.fr; patrice.codogno@inserm.fr

© The Author(s), under exclusive license to Springer Nature Switzerland AG 2025
B. Loos, D. J. Klionsky (eds.), *Autophagy - From Molecular Mechanisms to Flux Control in Health and Disease*, Learning Materials in Biosciences,
https://doi.org/10.1007/978-3-031-88121-3_8

ing cause of death worldwide [2]. This is due to one of the main kidney functions, characterized by the active reabsorption of glucose, ions, and proteins of low molecular weight, driven by ion gradients, essential to control acid-base, electrolyte and water balance and to remove toxins from the blood. Along the nephron, the proximal tubule is by far the most energy-demanding segment, as it is responsible for the reabsorption of close to 70% of water, NaCl, and 100% of glucose and amino acids from the ultrafiltrate, which passed through the glomerulus [3] (Fig. 1a). This explains why cells from the proximal tubule contain more mitochondria than in other parts of the nephron. These mitochondria, located in the basolateral side of proximal tubule cells, produce adenosine triphosphate (ATP) to fuel $Na^+$-$K^+$ pumps, in order to keep intracellular $Na^+$ levels low, thus supporting the passive reabsorption of several ultrafiltrate components, such as glucose, amino acids, and ions (Fig. 1b). In order to produce sufficient amounts of ATP to support reabsorption in physiological conditions, proximal tubule cells preferentially use fatty acids as a mitochondrial fuel. Indeed, beta oxidation of fatty acids is by far the most efficient mechanism to produce ATP in aerobic conditions. For example, the complete oxidation of a molecule of palmitate (16-carbon chain) produces more ATP molecules (129 ATP molecules) than the complete oxidation of a molecule of glucose (38 ATP molecules) [4]. Kidney proximal cells can uptake fatty acids from their apical and basolateral surfaces with SLC27A2/FATP2 and CD36 transporters, respectively [5, 6] (Fig. 1b). Cytosolic fatty acids are then transported into mitochondria, where they undergo beta-oxidation, resulting in the production of acetyl-CoA, which can enter the tricarboxylic acid (TCA) cycle. The electrons derived from NADH and $FADH_2$, synthetized during beta oxidation and the TCA cycle, combined with $O_2$, are processed by the oxidative phosphorylation (OXPHOS) complexes to fuel the ATP synthase [7] (Fig. 1b). Most of the ATP produced in basolateral mitochondria, which is transported in the cytoplasm, is used to fuel the $Na^+$-$K^+$ ATPase, thus promoting reabsorption by maintaining an intracellular $Na^+$ gradient in proximal cells. ATP is also used in proximal cells to stimulate gluconeogenesis [8] (Fig. 1b), in order to contribute to glucose homeostasis. Indeed, intracellular glucose absorbed from the apical surface through the sodium-glucose transporters SLC5A1/SGLT1 and SLC5A2/SGLT2, or generated by gluconeogenesis, reach blood circulation after being transported by basolateral

**Fig. 1** (continued) ultrafiltrate. Most of the water, ions and nutrients from the glomerular ultrafiltrate will be absorbed by the proximal convoluted tubule to form primary urine. Then, the rest of ultrafiltrate is absorbed when reaching the loop of Henle and the distal convoluted tubule, thus producing the urine that will be collected in the collecting duct before going to the bladder. (**b**) Proximal tubule cells produce high levels of energy, thanks to their basolateral mitochondria to fuel the $Na^+$, $K^+$ pumps, which maintain low intracellular $Na^+$ levels. This allows the passive reabsorption of amino acids, ions and nutrients at the apical surface by specific receptors (dependent or not of $Na^+$), from the ultrafiltrate. Gluconeogenesis is also dependent on ATP, in order to maintain glucose homeostasis. ATP production by the mitochondria is fueled essentially by free fatty acids, glutamine and ketone bodies, absorbed from the apical or basolateral sides

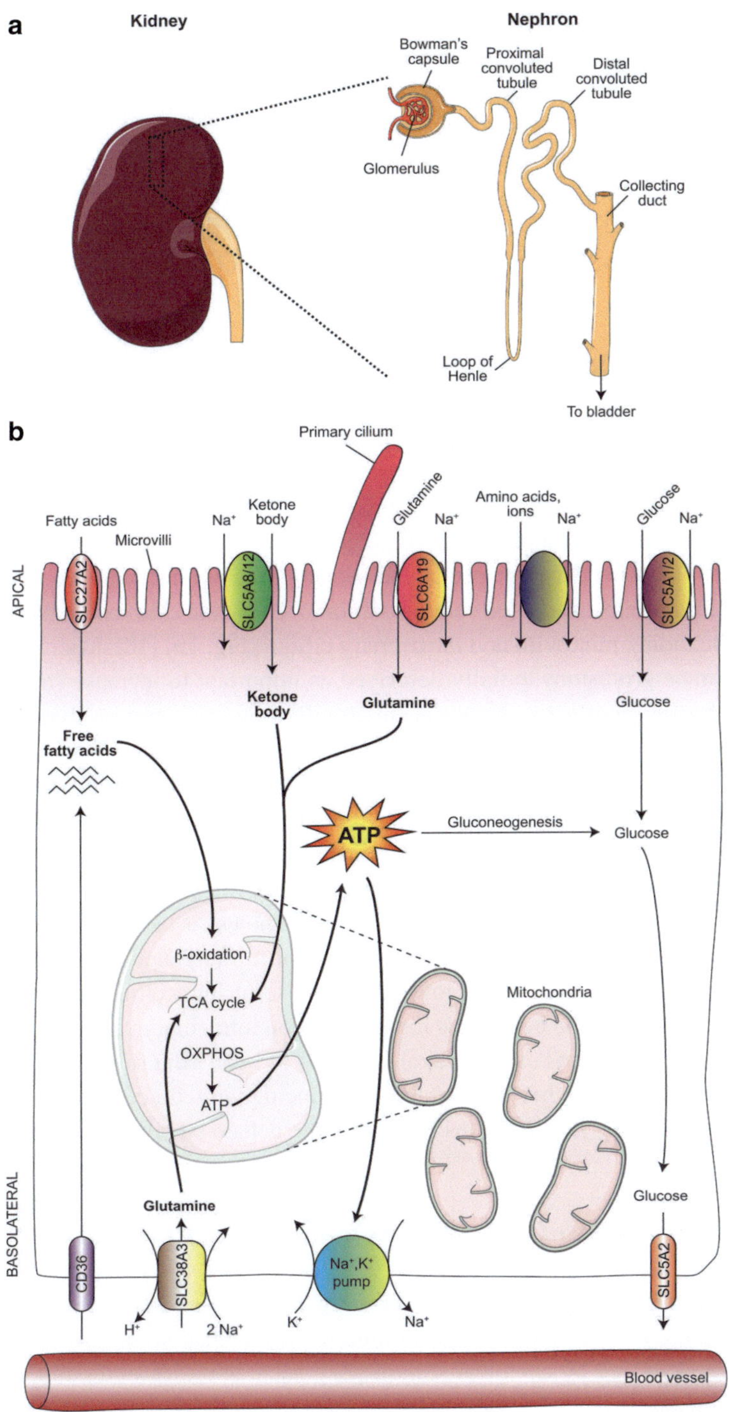



**Fig. 1** Schematic overview of the nephron and summary of metabolic needs of kidney proximal tubule cells. (**a**) The nephron, which is the kidney's functional unit, is composed of different segments. The glomerulus, surrounded by the Bowman's capsule, is filtering the blood to produce the

SLC5A2/GLUT2 [9] (Fig. 1b). While fatty acids are the preferential substrates for ATP synthesis in the proximal tubule, glutamine can also be used in physiological conditions to fuel the TCA cycle, after its entry on the apical and basolateral sides through SLC6A19 and SLC38A3/SNAT3 transporters, respectively [10] (Fig. 1b). ATP production from lactate has also been shown to occur in the proximal tubule [11]. Of note, under starvation, the kidney proximal tubule can use ketones as substrates for ATP synthesis, after their apical reabsorption through the saturable $Na^+$-coupled monocarboxylate transporters SLC5A8/SMCT1 and SLC5A12/SMCT2 (Fig. 1b) [12–14].

## 1.1  Fluid Flow Mechanosensing

The importance of tubular flow rate in regulating the reabsorption levels of proximal cells has been highlighted [15, 16]. Indeed, in order to maintain water and electrolyte balance, tubular cells adapt their reabsorption rate in proportion to the glomerular filtration rate (GFR), named glomerulotubular balance [17]. The regulation of reabsorption in kidney epithelial cells is dependent on specialized sensing machineries, capable of transducing mechanical forces into an adapted cellular response. Proximal cells exhibit several mechanosensors, including microvilli and the primary cilium (Fig. 1b) [18]. Microvilli are actin-based membrane protusions initially described as important to increase surface area and favorize reabsorption efficacy. However, microvilli were also proposed as mechanosensory organelles, as it was shown that fluid-induced microvilli torque is transmitted to the actin cytoskeleton and proportionally modulates $Na^+$ reabsorption in the proximal tubule [19]. In addition, the primary cilium, a microtubule-based organelle protruding from the apical surface of kidney epithelial cells, has been proposed as an important mechanosensory organelle in the kidney [20]. Even though the intraciliary mechanisms activated upon fluid flow (also called shear stress) are still poorly known, it has been shown that shear stress induces a primary cilium-dependent release of intraorganellar calcium within seconds after flow exposure [21–23]. However, this rise in intracellular calcium following shear stress-dependent cilia bending is independent on intraciliary calcium signaling [24]. Nevertheless, recent studies using optical tweezers to bend cilia (mimicking the force of nodal flow), demonstrated that cilia bending is inducing intraciliary calcium influx, which spreads to the cell body [25, 26], thus reopening the debate on ciliary mechanosensation.

## 1.2  Primary Cilium-Dependent Regulation of Intracellular Pathways

As mentioned earlier, the primary cilium has been described as an important mechanosensor in the kidney [20]. In addition to respond to mechanical stimuli, the renal primary cilium can be activated by chemical stimuli [27], such as the vasopressin and dopamine hormones [28, 29], leading to the activation of G-protein-coupled receptors (GPCR) to regulate intraciliary $Ca^{2+}$ signaling. Primary cilia also express the patch receptor for the growth factor family of hedgehog proteins [30] to control the Hedgehog pathway [31–33]. In addition, primary cilia modulate the WNT-CTNNB1/β-catenin signaling, essential during renal development and kidney injury [34–36]. More recently, primary cilia have been described to regulate the autophagy pathway upon starvation [37, 38] but also in kidney epithelial cells submitted to shear stress [39–41] contributing to the regulation of cellular metabolism [42].

## 1.3 The Autophagy Pathway

Macroautophagy (hereafter referred to as autophagy) is an essential pathway regulating cellular homeostasis, playing a key role in maintaining tissue and organ functions, including kidney activity [43]. While this process is essential for innate and adaptive immunity against pathogens, autophagy dysregulation has been associated to many diseases, including cancer, neurodegenerative, hepatic and autoimmune disorders [44]. It involves the formation of a double membrane vesicle, called an autophagosome, which fuses with a lysosome to induce the degradation of its contents (Fig. 2a). Autophagy allows bulk or selective degradation of many intracellular components, such as proteins, lipids or organelles, to provide nutrients, control organellar quality and limit the accumulation of toxic components, such as protein aggregates. For example, the autophagy pathway can regulate cell metabolism by selectively degrading lipid droplets (lipophagy) or mitochondria (mitophagy) [45, 46]. Autophagosomes originate from the phagophore, a selective membrane tightly associated with specialized domains in the endoplasmic reticulum (ER), called omegasomes [47]. The formation of phagophores can be induced by various stimuli (e.g., nutrient starvation, hypoxia and other stress situations including mechanical stress) and is controlled by approximately 20 core autophagy related (ATG) proteins initially discovered in yeast and conserved among eukaryotes [48]. Notably, autophagy initiation is dependent on the activation of the ULK (unc-51 like autophagy activating kinase) complex (or Atg1 complex in yeast), which is negatively controlled by the MTOR (mechanistic target of rapamycin kinase) complex 1 (MTORC1) and positively regulated by the AMP-activated protein kinase (AMPK) [49] (Fig. 2a). Once the ULK complex is activated, it enhances the activity of the class III phosphatidylinositol 3-kinase complex I (PtdIns3K-C1), containing the PIK3C3/VPS34 kinase that synthesizes phosphatidylinositol-3-phosphate (PtdIns3P), allowing the subsequent recruitment and assembly of autophagy related (ATG) proteins at the omegasome [50–52]. Notably, the accumulation of PtdIns3P at these ER subdomains promotes membrane expansion and allows the recruitment of two ubiquitin-like conjugation systems for autophagosome maturation. The first system, called ATG12–ATG5-ATG16L1 complex, which in concert with ATG7 and ATG3, positively regulate the

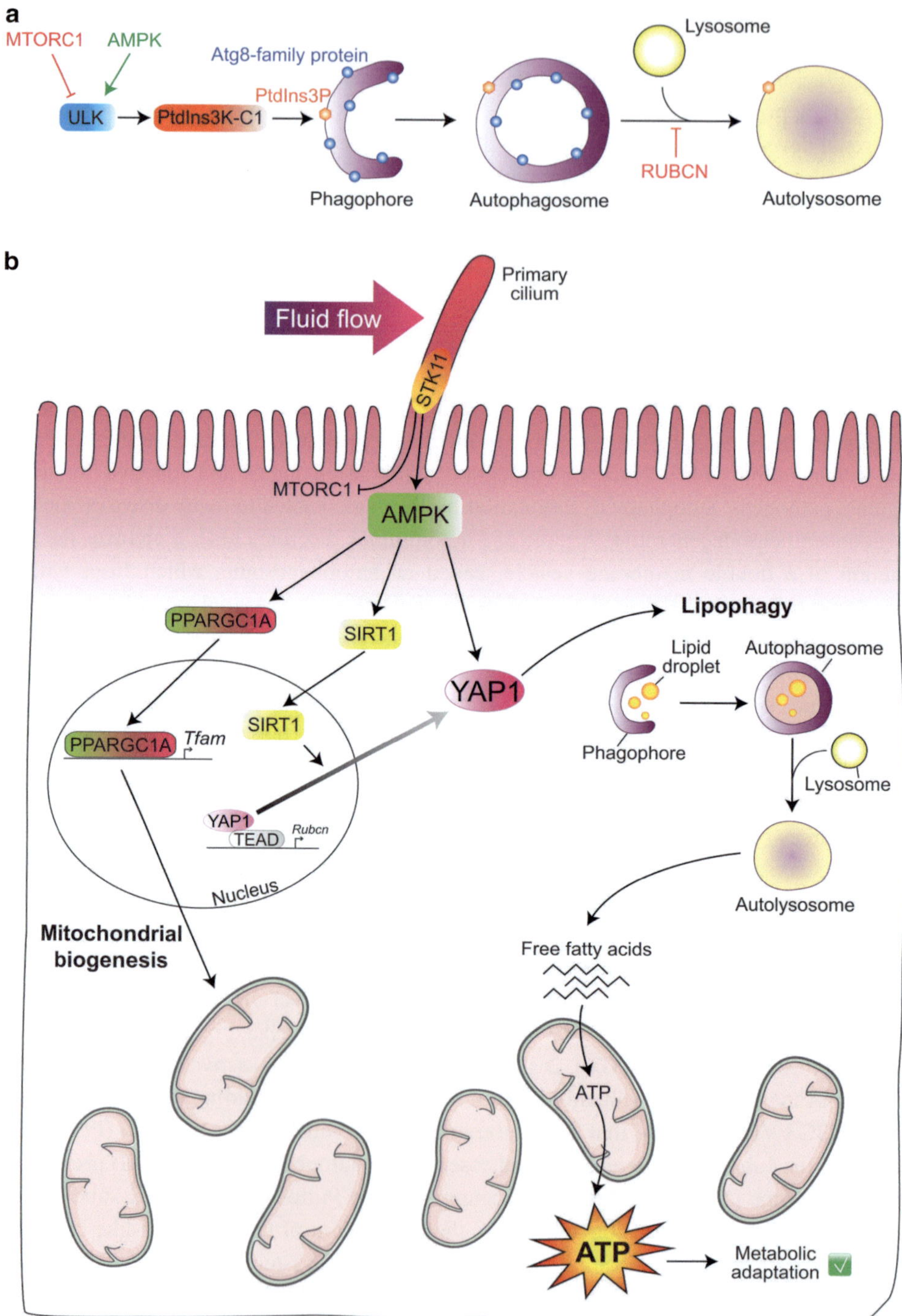



**Fig. 2** Overview of the autophagy pathway and primary cilium-dependent regulation of lypophagy and metabolism in tubular cells. (**a**) The autophagy pathway is controlled by the assembly of the ULK complex, which is positively regulated by AMPK and negatively regulated by MTORC1. ULK

Atg8-family protein lipidation system. Atg8-family proteins are divided in two subfamilies in mammals: the MAP 1LC3/LC3 (microtubule associated protein 1 light chain 3) and the GABARAP (GABA type A receptor-associated protein) family. After the cleavage of Atg8-family proteins by ATG4, they are conjugated to the lipid phosphatidylethanolamine (PE), resulting in their anchoring in the autophagic membranes [53] (Fig. 2a). After sufficient membrane expansion, the phagophore is closed in a process dependent on the endosomal sorting complexes required for transport (ESCRT) machinery, to form an autophagosome [54–56]. Autophagosomes can then fuse with late endosomes/lysosomes, thanks to a soluble N-ethylmaleimide-sensitive-factor attachment protein receptor (SNARE)-dependent process, after a tethering step involving the homotypic vacuole fusion and protein sorting (HOPS) complex [57]. Interestingly, RUBCN (rubicon autophagy regulator), an inhibitor of the fusion between autophagosomes and lysosomes (Fig. 2a), has been recently described as being essential in controlling the metabolism in kidney proximal tubule cells [58].

## 1.4 Primary Cilium-Dependent Regulation of Autophagy

As mentioned earlier, proximal tubule cells can sense shear stress on their apical surface using their primary cilium. The flow of the glomerular ultrafiltrate, inducing the bending of this mechanosensor, has been shown to induce a specific intracellular response in renal cells. Indeed, Boehlke et al. have shown that upon shear stress, the primary cilium regulates the activity of MTORC1 and AMPK, which are respectively negative and positive regulators of autophagy, to regulate cell size [39]. The inactivation of MTORC1 is independent on flow-induced calcium signaling but regulated by STK11/LKB1 (serine/threonine kinase 11) localized in the primary cilium [39]. In parallel, STK11 activates its target

---

**Fig. 2** (continued) activates PtdIns3K-C1, to induce the accumulation of PtdIns3P on the phagophore. This allows phagophore expansion and autophagosome maturation, notably by recruiting the Atg8-family protein conjugation system, which anchors lipidated Atg8-family proteins (LC3 or GABARAP) in the autophagic structures. When the autophagosome is closed, it can fuse with a lysosome to induce the degradation of its contents. This step can be inhibited by RUBCN. (**b**) Kidney epithelial cells are sensing fluid flow with their primary cilium. STK11 activation leads to MTORC1 inhibition and AMPK activation, which controls the metabolism of renal cells by inducing mitochondrial biogenesis and lipophagy. Mitochondrial biogenesis is promoted by the activation of the transcription factor PPARGC1A, which induces the expression of mitochondrial genes, including *Tfam*. To induce lipophagy, AMPK activates the deacetylase SIRT1, which induces the nuclear exit of the co-transcriptional factor YAP1. AMPK also directly phosphorylates YAP1, leading to its sequesration in the cytosol. This thus prevents YAP from binding to transcriptional enhanced associate domain (TEAD) transcription factors and the expression of YAP1 target genes, including the autophagy inhibitor RUBCN The decreased levels of RUBCN promote lipophagy, a type of autophagy degrading lipid droplets. This leads to the release of free fatty acids, which are used as substrates to produce ATP in the basolateral mitochondria, thus promoting the metabolic adaptation of renal cells submitted to physiological shear stress

AMPK in the basal body of primary cilia upon shear stress [39]. This leads to the activation of a signaling cascade resulting in the induction of autophagy [41]. The primary cilium-dependent activation of STK11 and AMPK subsequently activates the deacetylase SIRT1 (sirtuin 1), which induces the nuclear exit of the co-transcriptional factor YAP1 (Yes1 associated transcriptional regulator). The cytoplasmic accumulation of YAP1 inhibits the expression of several YAP1 target genes, including the autophagy inhibitor RUBCN [59], to activate autophagy [41] (Fig. 2b). Autophagy induction by shear stress regulates kidney epithelial cell size [40] and metabolism [42]. To adapt their metabolism, proximal tubule cells submitted to shear stress induce mitochondrial biogenesis (driven by the expression of PPARGC1A/PGC1α) and a selective type of autophagy, to degrade lipid droplets (Fig. 2b). The resulting free fatty acids are used as mitochondrial substrates to generate ATP [42]. This provides sufficient energy to support glucose reabsorption and gluconeogenesis in proximal tubule cells [42] (Fig. 2b). Interestingly, this signal transduction emanating from the primary cilium is dependent on sensing physiological fluid flow. Indeed, increasing fluid flow intensity at the surface of kidney epithelial cells, mimicking the flow observed during chronic kidney disease, has been shown to inhibit AMPK activation at the base of the primary cilium [41]. This is correlated with YAP1 reactivation and an inhibition of autophagy, as observed in kidney biopsies from patients suffering from chronic kidney disease [41]. As this phenomenon seems to be dependent on the presence of a primary cilium, it suggests that this mechanosensor could activate different signaling pathways according to flow intensity. Further studies will be needed to determine what mechanisms are implicated to regulate the downstream pathways according to fluid flow intensity.

## 2 Deciphering the Primary Cilium-Dependent Regulation of Autophagy and Metabolism in Kidney Epithelial Cells

The key findings in the field of primary cilium-dependent regulation of autophagy, as covered in this chapter, are as follows: first, the discovery of cross-talk between the primary cilium and autophagy [37]; second, the role of the primary cilium in regulating renal epithelial cell size in response to shear stress induced by fluid flow [39]; and third, the demonstration that primary cilium-dependent autophagy plays a key role in the regulation of kidney tubular cell size [40]. Thereafter, in order to provide a comprehensive understanding on how fundamental research led to the characterization of a primary cilium-dependent regulation of autophagy and metabolism in kidney epithelial cells, key experiments from two studies [41, 42] will be decrypted in this section. While the first study described the important role of autophagy in adapting the metabolism of kidney epithelial cells submitted to shear stress [42], the second study deciphered the mechanotransduction mechanisms of shear stress to regulate autophagy in proximal tubular cells [41]. The common link between these two studies was the use of a fluidic system mimicking *in vitro* the physiological flow of primary urine at the surface of kidney tubular cells. Overcoming the technical challenges associated with this type of system allowed, for the first time, the characterization of the important role of mechanical forces in regulating autophagy and metabolism of renal cells. In the following paragraphs, we will describe the fundamental aspects to consider when performing shear stress experiments or analyzing autophagy-dependent processes.

## 2.1 Take Note Of

When studying cell response to physiological shear stress, please pay special attention to cell confluency, in order to setup the best conditions allowing cell differentiation. Indeed, kidney tubular cells are physiologically at confluency, meaning that each cell border has to touch another cell. This is important since in these conditions, cells are not proliferating anymore because of contact inhibition. As ciliogenesis and proliferation are mutually exclusive processes, the formation of primary cilia is only possible when cells are totally confluent (Of note, ciliogenesis can be induced by serum starvation in less confluent cells [37, 38]). Thus, only cells confluent since at least 24 h can be used for shear stress experiments, giving cells enough time to generate a functional primary cilium. However, care must be taken when keeping confluent cells in flow-specific chambers, since these slides are usually containing very small volumes of medium. As for example, 500,000 murine kidney epithelial cells (KECs) are seeded in only 150 µL of complete medium, equivalent to 2.5 cm$^2$ of growth area, in slides designed for cell culture under perfusion. In standard culture dishes, this growth area is approximately equivalent to a 24-well plate, where 500 µL of medium can be added. Consequently, the nutrients present in the medium will be consumed much faster in slides for cell perfusion. Thus, medium will have to be renewed at least twice a day to avoid any nutrient starvation, which is a well-known activator of autophagy. Paying attention to these crucial aspects will allow to obtain healthy renal cells, presenting a primary cilium at their apical surface, ready to sense shear stress.

## 2.2 Markers and Tools Used

In order to apply physiological shear stress on renal cells, a pump system allowing long-term cell culture is needed. For example, the Ibidi® Pump system is user-friendly and allows researchers to investigate the effect of shear stress *in vitro*, without requiring an engineering background. The fluidic units available with this system are composed of switching valves, to ensure unidirectional laminar flow inside the chamber for several days. This is necessary to mimic the physiological flow of primary urine at the surface of kidney tubular cells. After subjecting renal cells to at least 24 h of shear stress, cells can be analyzed by several techniques, such as immunofluorescence or western blotting (WB) to study autophagy-related processes. For example, the levels of Atg8-family proteins, such as LC3, are systematically measured to study autophagy flux by WB. In the case of LC3, a membranous marker of phagophores and autophagosomes, it exists in two forms within in cells: a cytosolic form (LC3-I) and a lipidated form (LC3-II), which is associated with autophagic membranes. Even though the LC3-II form is larger in mass, it shows faster mobility in SDS-PAGE gels, probably because of increased hydrophobicity [60]. However, when measuring LC3-II levels by WB, caution has to be given to the rate of autophagic flux, as LC3-II is degraded upon fusion between autophagosomes and lysosomes. Thus, it is important to study the turnover of LC3-II in the presence and absence of a lysosomal inhibitor, such as bafilomycin A1, chloroquine (CQ) or lysosomal enzyme inhibitors. In general, increased LC3-II accumulation in the presence of a lysosomal inhibitor will be indicative of a greater autophagy flux. In parallel to the analysis of autophagy flux using LC3-II, it is important to monitor the degradation of an autophagy substrate, to confirm their proper delivery in the lysosomal compartment. For example, during "bulk" autophagy, decreased levels of SQSTM1/p62, a cytoplasmic protein transported into the lysosome by autophagy, can be observed by WB when autophagy is induced. For the case of lipophagy, the intracellular levels of lipids, such as triglycerides, has to be monitored. Lipid stains, such as BODIPY 493/503, which labels neutral lipids, can be used to visualize lipid droplets by fluorescence microscopy. To further characterize the role of autophagy in the degradation of lipid droplets upon shear stress, it is necessary to perform loss of function experiments, targeting for example *Atg5* by siRNA transfection. This method leads to an acute knockdown when cells are transfected before seeding in fluidic chambers.

Even though siRNA transfection is very efficient in targeting a high percentage of cells, plasmid transfection is more challenging. Indeed, to study the mechanisms regulating autophagy, the expression of protein mutants is recommended to decipher their role. For this, cells have to be transfected with a plasmid coding for the protein of interest. For example, to study the importance of the co-transcription factor YAP1 in regulating lipophagy, YAP1 mutants (constitutively active/inactive) have been transfected in renal cells before seeding them in flow chambers. However, studying autophagy flux by WB can be challenging due to low transfection rates, as LC3-II protein levels reflect the quantity present in the whole cell population. To overcome this, LC3 immunofluorescence can be performed and autophagic flux analyzed by quantification of LC3 puncta (representing autophagic structures) in transfected cells. Cells overexpressing the protein of interest can be detectable by fluorescence microscopy when it is fused with a fluorescent protein, such as GFP, or a protein tag, such as FLAG. Another way to overcome low transfection rates is to perform lentiviral transduction instead of liposome-based transfection protocols. Even though transduction leads to the generation of cell lines stably expressing the protein of interest, please pay special attention to the possible deleterious effects of stably expressing a mutant in cells, as well as the consequences of overexpression on the endogenous protein.

## 2.3    Tasks to Complete

In this section, you will have to analyze data and answer questions to expand and challenge the knowledge you learned along this chapter.

1. Complete the WB below with the following: LC3-I; LC3-II; Static and Shear after analyzing the autophagic flux.

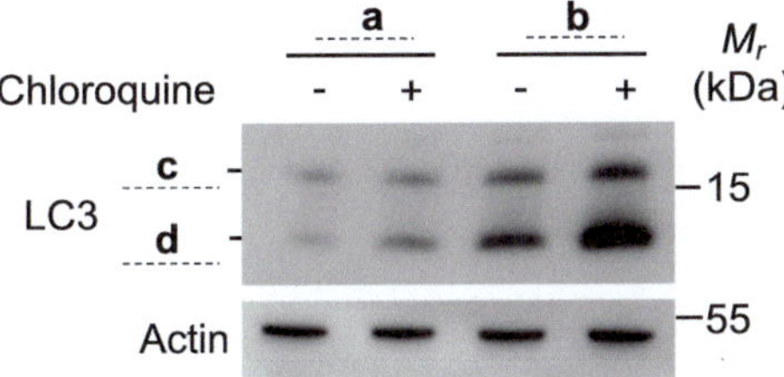

2. According to this WB schematic, which kinase synthetizing phosphatidylinositol-3-phosphate is important for autophagy induction upon shear stress?

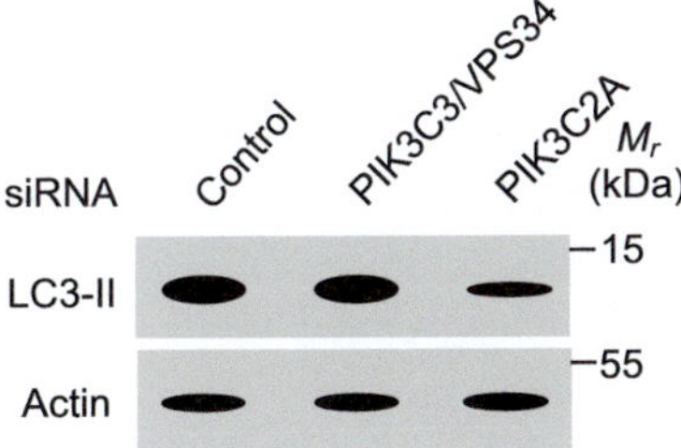

3. Deduce from these BODIPY images which siRNA was transfected into renal cells before submitting them to shear stress: siRNA control or si*Atg5*.

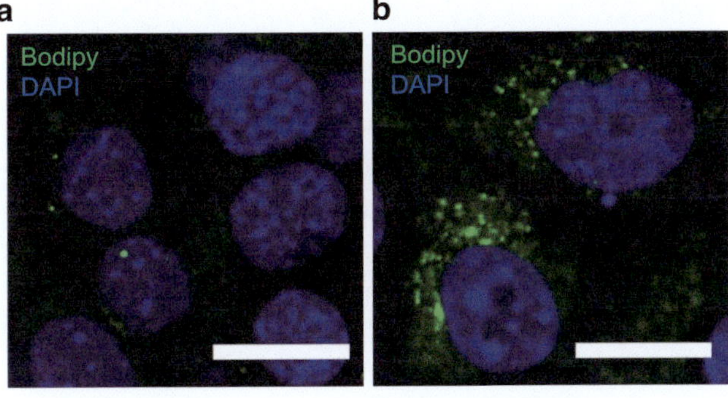

4. Quiz

    (a) The liver and kidney share an important function in glucose metabolism. Which one?

    (b) When sensing physiological flow, the primary cilium activates AMPK, YAP1 and autophagy (Right/Wrong)

    (c) RUBCN is a negative regulator of autophagy (Right/Wrong)

    (d) Why proximal tubule cells use mainly fatty acids instead of glucose to produce energy?

    (e) Upon pathological flow (increased shear stress), autophagy is increased (Right/Wrong)

    (f) Is the accumulation of lipids and mitochondrial damage possible in patients suffering from chronic kidney disease? Why?

**Take Home Message**

In this chapter, we described the mechanisms by which shear stress is regulating autophagy in kidney epithelial cells. The studies that we discussed in Sect. 2 of this chapter [41, 42] have shown that in physiological conditions, fluid flow is sensed by the primary cilium to induce autophagy. The primary cilium activates a signaling pathway involving STK11, AMPK and SIRT1 to inactive YAP1, a co-transcriptional factor responsible for the expression of RUBCN, an autophagy inhibitor. YAP1 inactivation allows the initiation of a specific type of autophagy, called lipophagy, responsible for the degradation of lipid droplets. Lipophagy produces free fatty acids that fuel the synthesis of mitochondrial ATP. The energy produced by lipophagy supports energy-consuming processes, such as glucose reabsorption and neoglucogenesis, essential to maintain the function of kidney proximal cells. However, increasing shear stress at the surface of proximal cells, mimicking the urinary fluid flow detected upon chronic kidney disease, led to a decrease in AMPK activation at the primary cilium, correlated with an inhibition of autophagy [41]. Further studies will be needed to understand why the primary cilium is no longer capable to activate AMPK and if other pathways are activated by this mechanosensor when it is facing pathological shear stress. As several studies highlighted the dysregulation of AMPK, SIRT1, YAP1 and autophagy in chronic kidney disease [61–65], a better understanding of the molecular pathways regulated by the primary cilium could provide new therapeutic strategies to limit the progression of chronic kidney disease.

**Answer Guide to Questions**

1. (a) Static; (b) Shear; (c) LC3-I; (d) LC3-II
2. Decreased LC3-II levels can be observed in cells submitted to shear stress after downregulating PIK3C2A. However, no effect on LC3-II levels are observed upon PIK3C3/VPS34 knockdown. PIK3C3/VPS34 is part of PtdIns3K-C1, regulating PtdIns3P synthesis and autophagy induction upon starvation. These results suggest that autophagy induction during shear stress is PIK3C3/VPS34-independent but rather relies on PIK3C2A. The role of PIK3C2A in regulating autophagy has been described in renal cells submitted to shear stress. For *further reading*, please refer to [66].
3. (a) siRNA control; (b) si*Atg5*. BODIPY 493/503 is labeling neutral lipids. Punctate accumulation of BODIPY 493/503 staining in cells represent lipid droplets. Renal cells submitted to shear stress induce lipophagy, to degrade lipid droplets. In the absence of ATG5, lipophagy is inhibited and leads to the accumulation of lipid droplets. You may also note the cell size difference between siRNA control and si*Atg5*. As discussed in the Sect. 1 of this chapter, autophagy induction during shear stress decreases cell size. For *further reading*, please refer to [40].
4. (a) The liver and kidney are the only organs capable of generating glucose by gluconeogenesis. Even though the liver is responsible for 75% of glucose release in the circulation in the post-absorptive state (50% via glycogenolysis), the liver and kidneys are responsible for the release of equivalent amounts of glucose *via* gluconeogenesis (25% each). For *further reading*, please refer to [67].

    (b) Wrong. Physiological shear stress indeed regulates a primary cilium-dependent activation of AMPK and autophagy. However, YAP1 has to be inactivated to allow autophagy induction, as this co-transcriptional factor is responsible for the expression of RUBCN, an inhibitor of the fusion between autophagosomes and lysosomes [58].

    (c) Right. Conversely, RUBCN can promote a non-canonical form of autophagy, called LC3-associated phagocytosis. For *further reading*, please refer to [68].

    (d) Renal cells from the proximal tubule are producing ATP preferentially from fatty acids because it is far more efficient than with glucose. Moreover, glucose has to be recycled in the blood circulation.

    (e) Wrong. Physiological shear stress induces autophagy through a primary cilium-dependent process. However, mimicking chronic kidney disease, by increasing shear stress on renal cells, leads to the inhibition of the primary cilium-dependent activation of AMPK, with subsequent YAP1 activation and autophagy inhibition [41].

    (f) Yes, the accumulation of lipid droplets is believed to contribute to the progression of chronic kidney diseases, notably by triggering mitochondrial and kidney damage. Several pathways could contribute to these phenotypes, including increased fatty acid uptake into tubular cells and decreased mitochondrial biogenesis. These events could be linked to the dysregulation of the primary cilium-dependent activation of lipophagy and mitochondrial biogenesis. For *further reading*, please refer to [69].

## References


1. Wang Z, Ying Z, Bosy-Westphal A, Zhang J, Schautz B, Later W, et al. Specific metabolic rates of major organs and tissues across adulthood: evaluation by mechanistic model of resting energy expenditure. Am J Clin Nutr. 2010;92(6):1369–77.
2. Kalyesubula R, Conroy AL, Calice-Silva V, Kumar V, Onu U, Batte A, et al. Screening for kidney disease in low- and middle-income countries. Semin Nephrol. 2022;42(5):151315.
3. Curthoys NP, Moe OW. Proximal tubule function and response to acidosis. Clin J Am Soc Nephrol. 2014;9(9):1627–38.
4. Schaub JA, Venkatachalam MA, Weinberg JM. Proximal tubular oxidative metabolism in acute kidney injury and the transition to CKD. Kidney360. 2021;2(2):355–64.
5. Khan S, Cabral PD, Schilling WP, Schmidt ZW, Uddin AN, Gingras A, et al. Kidney proximal tubule lipoapoptosis is regulated by fatty acid transporter-2 (FATP2). J Am Soc Nephrol. 2018;29(1):81–91.
6. Kawakami R, Hanaoka H, Kanai A, Obinata H, Nakano D, Ikeuchi H, et al. Robust capability of renal tubule fatty acid uptake from apical and basolateral membranes in physiology and disease. bioRxiv. 2022:2022.07.04.498762.
7. Bhargava P, Schnellmann RG. Mitochondrial energetics in the kidney. Nat Rev Nephrol. 2017;13(10):629–46.
8. Soltoff SP. ATP and the regulation of renal cell function. Annu Rev Physiol. 1986;48:9–31.
9. Vallon V. Glucose transporters in the kidney in health and disease. Pflugers Arch. 2020;472(9):1345–70.
10. Palmer BF, Clegg DJ. Starvation ketosis and the kidney. Am J Nephrol. 2021;52(6):467–78.
11. Uchida S, Endou H. Substrate specificity to maintain cellular ATP along the mouse nephron. Am J Phys. 1988;255(5 Pt 2):F977–83.
12. Sapir DG, Owen OE. Renal conservation of ketone bodies during starvation. Metabolism. 1975;24(1):23–33.
13. Gopal E, Umapathy NS, Martin PM, Ananth S, Gnana-Prakasam JP, Becker H, et al. Cloning and functional characterization of human SMCT2 (SLC5A12) and expression pattern of the transporter in kidney. Biochim Biophys Acta. 2007;1768(11):2690–7.
14. Gopal E, Fei YJ, Sugawara M, Miyauchi S, Zhuang L, Martin P, et al. Expression of slc5a8 in kidney and its role in Na(+)-coupled transport of lactate. J Biol Chem. 2004;279(43):44522–32.
15. Schnermann J, Wahl M, Liebau G, Fischbach H. Balance between tubular flow rate and net fluid reabsorption in the proximal convolution of the rat kidney. I. Dependency of reabsorptive net fluid flux upon proximal tubular surface area at spontaneous variations of filtration rate. Pflugers Arch. 1968;304(1):90–103.
16. Green R, Moriarty RJ, Giebisch G. Ionic requirements of proximal tubular fluid reabsorption flow dependence of fluid transport. Kidney Int. 1981;20(5):580–7.
17. Vallon V. Tubuloglomerular feedback and the control of glomerular filtration rate. News Physiol Sci. 2003;18:169–74.
18. Verschuren EHJ, Castenmiller C, Peters DJM, Arjona FJ, Bindels RJM, Hoenderop JGJ. Sensing of tubular flow and renal electrolyte transport. Nat Rev Nephrol. 2020;16(6):337–51.
19. Du Z, Duan Y, Yan Q, Weinstein AM, Weinbaum S, Wang T. Mechanosensory function of microvilli of the kidney proximal tubule. Proc Natl Acad Sci USA. 2004;101(35):13068–73.
20. Praetorius HA, Spring KR. The renal cell primary cilium functions as a flow sensor. Curr Opin Nephrol Hypertens. 2003;12(5):517–20.
21. Nauli SM, Alenghat FJ, Luo Y, Williams E, Vassilev P, Li X, et al. Polycystins 1 and 2 mediate mechanosensation in the primary cilium of kidney cells. Nat Genet. 2003;33(2):129–37.

22. Goetz JG, Steed E, Ferreira RR, Roth S, Ramspacher C, Boselli F, et al. Endothelial cilia mediate low flow sensing during zebrafish vascular development. Cell Rep. 2014;6(5):799–808.

23. Raghavan V, Rbaibi Y, Pastor-Soler NM, Carattino MD, Weisz OA. Shear stress-dependent regulation of apical endocytosis in renal proximal tubule cells mediated by primary cilia. Proc Natl Acad Sci USA. 2014;111(23):8506–11.

24. Delling M, Indzhykulian AA, Liu X, Li Y, Xie T, Corey DP, et al. Primary cilia are not calcium-responsive mechanosensors. Nature. 2016;531(7596):656–60.

25. Djenoune L, Mahamdeh M, Truong TV, Nguyen CT, Fraser SE, Brueckner M, et al. Cilia function as calcium-mediated mechanosensors that instruct left-right asymmetry. Science. 2023;379(6627):71–8.

26. Katoh TA, Omori T, Mizuno K, Sai X, Minegishi K, Ikawa Y, et al. Immotile cilia mechanically sense the direction of fluid flow for left-right determination. Science. 2023;379(6627):66–71.

27. Bai Y, Wei C, Li P, Sun X, Cai G, Chen X, et al. Primary cilium in kidney development, function and disease. Front Endocrinol (Lausanne). 2022;13:952055.

28. Jin X, Mohieldin AM, Muntean BS, Green JA, Shah JV, Mykytyn K, et al. Cilioplasm is a cellular compartment for calcium signaling in response to mechanical and chemical stimuli. Cell Mol Life Sci. 2014;71(11):2165–78.

29. Raychowdhury MK, Ramos AJ, Zhang P, McLaughin M, Dai XQ, Chen XZ, et al. Vasopressin receptor-mediated functional signaling pathway in primary cilia of renal epithelial cells. Am J Physiol Renal Physiol. 2009;296(1):F87–97.

30. Ho EK, Stearns T. Hedgehog signaling and the primary cilium: implications for spatial and temporal constraints on signaling. Development. 2021;148(9):dev195552.

31. Hsieh CL, Jerman SJ, Sun Z. Non-cell-autonomous activation of hedgehog signaling contributes to disease progression in a mouse model of renal cystic ciliopathy. Hum Mol Genet. 2022;31(24):4228–40.

32. Tran PV, Talbott GC, Turbe-Doan A, Jacobs DT, Schonfeld MP, Silva LM, et al. Downregulating hedgehog signaling reduces renal cystogenic potential of mouse models. J Am Soc Nephrol. 2014;25(10):2201–12.

33. Silva LM, Jacobs DT, Allard BA, Fields TA, Sharma M, Wallace DP, et al. Inhibition of Hedgehog signaling suppresses proliferation and microcyst formation of human Autosomal Dominant Polycystic Kidney Disease cells. Sci Rep. 2018;8(1):4985.

34. Simons M, Gloy J, Ganner A, Bullerkotte A, Bashkurov M, Kronig C, et al. Inversin, the gene product mutated in nephronophthisis type II, functions as a molecular switch between Wnt signaling pathways. Nat Genet. 2005;37(5):537–43.

35. Saito S, Tampe B, Muller GA, Zeisberg M. Primary cilia modulate balance of canonical and non-canonical Wnt signaling responses in the injured kidney. Fibrogenesis Tissue Repair. 2015;8:6.

36. Lancaster MA, Louie CM, Silhavy JL, Sintasath L, Decambre M, Nigam SK, et al. Impaired Wnt-beta-catenin signaling disrupts adult renal homeostasis and leads to cystic kidney ciliopathy. Nat Med. 2009;15(9):1046–54.

37. Pampliega O, Orhon I, Patel B, Sridhar S, Diaz-Carretero A, Beau I, et al. Functional interaction between autophagy and ciliogenesis. Nature. 2013;502(7470):194–200.

38. Tang Z, Lin MG, Stowe TR, Chen S, Zhu M, Stearns T, et al. Autophagy promotes primary ciliogenesis by removing OFD1 from centriolar satellites. Nature. 2013;502(7470):254–7.

39. Boehlke C, Kotsis F, Patel V, Braeg S, Voelker H, Bredt S, et al. Primary cilia regulate mTORC1 activity and cell size through Lkb1. Nat Cell Biol. 2010;12(11):1115–22.

40. Orhon I, Dupont N, Zaidan M, Boitez V, Burtin M, Schmitt A, et al. Primary-cilium-dependent autophagy controls epithelial cell volume in response to fluid flow. Nat Cell Biol. 2016;18(6):657–67.

41. This reference has to be modified (initially on BioRxiv, now published in Nat. Commun). Here is the new DOI: 10.1038/s41467-023-43775-1 The AMPK-Sirtuin 1-YAP axis is regulated by fluid flow intensity and controls autophagy flux in kidney epithelial cells. Aurore Claude-Taupin, Pierre Isnard, Alessia Bagattin, Nicolas Kuperwasser, Federica Roccio, Biagina Ruscica, Nicolas Goudin, Meriem Garfa-Traoré, Alice Regnier, Lisa Turinsky, Martine Burtin , Marc Foretz, Marco Pontoglio, Etienne Morel, Benoit Viollet, Fabiola Terzi, Patrice Codogno, Nicolas Dupont. Nat Commun 2023Claude-Taupin A, Roccio F, Garfa-Traoré M, Regnier A, Burtin M, Morel E, et al. YAP-dependent autophagy is controlled by AMPK, SIRT1 and flow intensity in kidney epithelial cells. bioRxiv. 2023:2023.01.09.523237.

42. Miceli C, Roccio F, Penalva-Mousset L, Burtin M, Leroy C, Nemazanyy I, et al. The primary cilium and lipophagy translate mechanical forces to direct metabolic adaptation of kidney epithelial cells. Nat Cell Biol. 2020;22(9):1091–102.

43. Tang C, Livingston MJ, Liu Z, Dong Z. Autophagy in kidney homeostasis and disease. Nat Rev Nephrol. 2020;16(9):489–508.

44. Klionsky DJ, Petroni G, Amaravadi RK, Baehrecke EH, Ballabio A, Boya P, et al. Autophagy in major human diseases. EMBO J. 2021;40(19):e108863.

45. Zhang S, Peng X, Yang S, Li X, Huang M, Wei S, et al. The regulation, function, and role of lipophagy, a form of selective autophagy, in metabolic disorders. Cell Death Dis. 2022;13(2):132.

46. Onishi M, Yamano K, Sato M, Matsuda N, Okamoto K. Molecular mechanisms and physiological functions of mitophagy. EMBO J. 2021;40(3):e104705.

47. Zhen Y, Stenmark H. Autophagosome biogenesis. Cells. 2023;12(4):668.

48. Mizushima N, Yoshimori T, Ohsumi Y. The role of Atg proteins in autophagosome formation. Annu Rev Cell Dev Biol. 2011;27:107–32.

49. Licheva M, Raman B, Kraft C, Reggiori F. Phosphoregulation of the autophagy machinery by kinases and phosphatases. Autophagy. 2022;18(1):104–23.

50. Hu Y, Reggiori F. Molecular regulation of autophagosome formation. Biochem Soc Trans. 2022;50(1):55–69.

51. Valverde DP, Yu S, Boggavarapu V, Kumar N, Lees JA, Walz T, et al. ATG2 transports lipids to promote autophagosome biogenesis. J Cell Biol. 2019;218(6):1787–98.

52. Maeda S, Yamamoto H, Kinch LN, Garza CM, Takahashi S, Otomo C, et al. Structure, lipid scrambling activity and role in autophagosome formation of ATG9A. Nat Struct Mol Biol. 2020;27(12):1194–201.

53. Nakatogawa H. Mechanisms governing autophagosome biogenesis. Nat Rev Mol Cell Biol. 2020;21(8):439–58.

54. Zhou F, Wu Z, Zhao M, Murtazina R, Cai J, Zhang A, et al. Rab5-dependent autophagosome closure by ESCRT. J Cell Biol. 2019;218(6):1908–27.

55. Takahashi Y, He H, Tang Z, Hattori T, Liu Y, Young MM, et al. An autophagy assay reveals the ESCRT-III component CHMP2A as a regulator of phagophore closure. Nat Commun. 2018;9(1):2855.

56. Takahashi Y, Liang X, Hattori T, Tang Z, He H, Chen H, et al. VPS37A directs ESCRT recruitment for phagophore closure. J Cell Biol. 2019;218(10):3336–54.

57. Zhao YG, Codogno P, Zhang H. Machinery, regulation and pathophysiological implications of autophagosome maturation. Nat Rev Mol Cell Biol. 2021;22(11):733–50.

58. Matsuda J, Takahashi A, Takabatake Y, Sakai S, Minami S, Yamamoto T, et al. Metabolic effects of RUBCN/Rubicon deficiency in kidney proximal tubular epithelial cells. Autophagy. 2020;16(10):1889–904.

59. Matsunaga K, Saitoh T, Tabata K, Omori H, Satoh T, Kurotori N, et al. Two Beclin 1-binding proteins, Atg14L and Rubicon, reciprocally regulate autophagy at different stages. Nat Cell Biol. 2009;11(4):385–96.

60. Klionsky DJ, Abdel-Aziz AK, Abdelfatah S, Abdellatif M, Abdoli A, Abel S, et al. Guidelines for the use and interpretation of assays for monitoring autophagy (4th edition)(1). Autophagy. 2021;17(1):1–382.
61. Chen J, Wang X, He Q, Bulus N, Fogo AB, Zhang MZ, et al. YAP activation in renal proximal tubule cells drives diabetic renal interstitial fibrogenesis. Diabetes. 2020;69(11):2446–57.
62. Szeto SG, Narimatsu M, Lu M, He X, Sidiqi AM, Tolosa MF, et al. YAP/TAZ are mechanoregulators of TGF-beta-Smad signaling and renal fibrogenesis. J Am Soc Nephrol. 2016;27(10):3117–28.
63. Yan J, Wang J, He JC, Zhong Y. Sirtuin 1 in chronic kidney disease and therapeutic potential of targeting sirtuin 1. Front Endocrinol (Lausanne). 2022;13:917773.
64. Hallows KR, Mount PF, Pastor-Soler NM, Power DA. Role of the energy sensor AMP-activated protein kinase in renal physiology and disease. Am J Physiol Renal Physiol. 2010;298(5):F1067–77.
65. Lin TA, Wu VC, Wang CY. Autophagy in chronic kidney diseases. Cells. 2019;8(1):61.
66. Boukhalfa A, Nascimbeni AC, Ramel D, Dupont N, Hirsch E, Gayral S, et al. PI3KC2alpha-dependent and VPS34-independent generation of PI3P controls primary cilium-mediated autophagy in response to shear stress. Nat Commun. 2020;11(1):294.
67. Gerich JE. Role of the kidney in normal glucose homeostasis and in the hyperglycaemia of diabetes mellitus: therapeutic implications. Diabet Med. 2010;27(2):136–42.
68. Martinez J, Malireddi RK, Lu Q, Cunha LD, Pelletier S, Gingras S, et al. Molecular characterization of LC3-associated phagocytosis reveals distinct roles for Rubicon, NOX2 and autophagy proteins. Nat Cell Biol. 2015;17(7):893–906.
69. Mitrofanova A, Merscher S, Fornoni A. Kidney lipid dysmetabolism and lipid droplet accumulation in chronic kidney disease. Nat Rev Nephrol. 2023;19:629–45.

# Autophagy Assessment in Diagnostic Pathology: Focus on Skeletal Myopathies

Biswarathan Ramani and Marta Margeta

**What You Will Learn This Chapter**

The importance of macroautophagy/autophagy across different cellular functions is demonstrated by the range of human diseases associated with autophagy dysfunction. In this chapter, you will learn about the special role of autophagy in human skeletal muscle homeostasis, delve into the spectrum of muscle diseases tied to dysfunction at various steps of the autophagy-lysosomal pathway, and understand how histopathological assessments of muscle tissues and markers of autophagosomes are used to diagnose these diseases.

## 1    Diseases of Human Skeletal Muscle Caused by Impaired Autophagy

Prior chapters have described and underscored the importance of autophagy in sustaining postmitotic cells throughout an organism's lifespan. This includes brain cells (neurons and glia), skeletal muscles, and cardiac muscles. As such, disruptions in autophagy primarily manifest as neurological, neuromuscular, or cardiac disorders. Specifically, skeletal and cardiac muscles, constituting over a third of the human body mass, endure high metabolic activity and continual mechanical strain from muscle contractions. Consequently, these myocytes rely extensively on autophagy to effectively eliminate damaged proteins, organelles, and structural components, and this homeostatic process is vital for preserving muscle functionality and staving off age-related muscle wasting.

Autophagy plays a pivotal role in skeletal muscle physiology, particularly during exercise-induced stress. For example, acute exercise was shown to induce autophagy in skeletal and cardiac muscles of mice [1]. Preventing this exercise-induced autophagy

B. Ramani · M. Margeta (✉)
School of Medicine, University of California, San Francisco, San Francisco, CA, USA
e-mail: Biswarathan.Ramani@ucsf.edu; Marta.Margeta@ucsf.edu

induction decreases muscle endurance, alters glucose metabolism during acute exercise, and impairs chronic exercise-mediated protection against high-fat-diet-induced glucose intolerance, pointing to a beneficial role of exercise-induced autophagy in potentially combating metabolic syndromes. Additionally, long-term exercise in mice was shown to increase basal autophagy, driven in part by a response to increased reactive oxygen species generated during exercise [2].

The importance of autophagy in skeletal muscle homeostasis is further evidenced through the diseases that result from genetic or pharmacological disruptions of different stages within the autophagy pathway. The detailed steps of autophagy have been discussed in prior chapters but should be briefly reviewed in the context of this chapter, with a specific focus on macroautophagy/degradative autophagy (Fig. 1). Macroautophagy initiates with the formation of the double-membrane phagophore (Step 1), followed by cargo sequestration (Step 2), phagophore closure into an autophagosome (Step 3), autophagosome maturation (Step 4), autophagosome fusion with the single-membrane lysosome (Step 5), and content degradation by lysosomal enzymes (Step 6). Understanding some details of these steps will help illustrate the roles of specific proteins that can be used to diagnostically assay alterations in autophagy in human muscle tissue:

- Step 1 (induction and nucleation): A double-layered membrane is generated de novo, most commonly from the endoplasmic reticulum, and this process is mediated by several ATG (autophagy related) proteins and a protein complex containing ULK1. Of special importance is MAP1LC3/LC3, a member of the Atg8 family, which is critical in both autophagosome formation and, later, the substrate selection. During induction, both the inner and outer membranes are decorated by LC3-II, a lipidated form of cytosolic LC3-I.
- Step 2 (cargo recognition and sequestration): LC3-II plays a key role in selective cargo sequestration by directly interacting either with the cargo itself or, more often, with various autophagy receptors, such as SQSTM1/p62 (sequestosome 1). Along with specific cargo proteins, SQSTM1 can recognize and sequester organelles and larger macromolecular complexes, including mitochondria and stress granules, into the developing and expanding phagophore.
- Step 3 (phagophore closure): The expanding membrane ultimately closes into a double-membrane-bound vesicle called the autophagosome.
- Step 4 (maturation and transport): The autophagosome matures through changes in associated protein machinery, binds to the dynein-dynactin molecular motor, and is transported along the microtubule network to a perinuclear region.
- Step 5 (fusion): At the perinuclear region, the autophagosome fuses with a lysosome to form an autolysosome.
- Step 6 (cargo degradation): The inner membrane of the original autophagosome and all the cargo within it are degraded by lysosomal acid hydrolases, which are only functional at a low pH. Of note, SQSTM1 and LC3-II are degraded along with the targeted substrates within the lysosome.

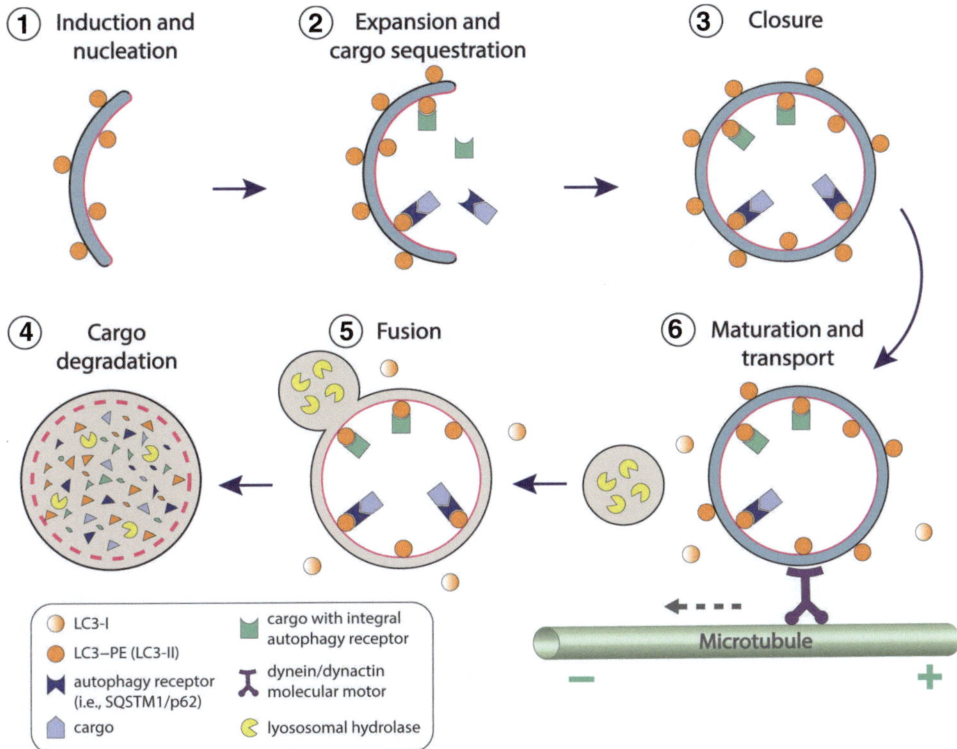


**Fig. 1** Key steps in mammalian macroautophagy. Macroautophagy is a complex process that starts with ① nucleation of the phagophore, a two-layered membrane cisterna that contains phosphatidylethanolamine (PE)-conjugated LC3 (LC3-II; or another mammalian Atg8 ortholog) bound to its inner and outer membranes. ② The phagophore then expands, surrounding the cargo that is being sequestered through its interaction with LC3-II; although LC3-II-cargo binding can be direct, more often it is mediated via an autophagic receptor such as SQSTM1. ③ Eventually, the expanding membrane closes, resulting in the formation of a double-walled vesicle called the autophagosome. ④ The autophagosome then matures; this involves (a) changes in the protein machinery bound to its external surface (including the release of LC3, which can be reused, and binding of various proteins that are required for autophagosome–lysosome fusion) and (b) binding to the dynein-dynactin molecular motor, which then transports the maturing autophagosome centripetally toward the lysosomes located at the minus end of the microtubules (i.e., in the center of the cell). ⑤ The lysosome and autophagosome then fuse, which enables the lysosomal hydrolases to be released into the resulting vesicle (the autolysosome). ⑥ Through the action of these lysosomal enzymes (which requires an acidic intravesicular milieu), the autolysosomal contents are degraded; this degradation includes not only the sequestered cargo but also the inner membrane, LC3-II bound to the inner membrane, and autophagic receptors. Not pictured are two additional steps: (a) kinesin- and FYCO1-mediated centrifugal transport of the phagophore from the perinuclear area (where nucleation occurs) to the periphery of the cell (where cargo is sequestered and the autophagosomes are formed) and (b) fusion of the autophagosome with the late endosome, which sometimes (but not always) occurs prior to the fusion of this combined vesicle (the amphisome) with the lysosome, or direct autophagosome-lysosome fusion. (With the publisher's permission, this figure and its legend are reproduced with modification from [50])

It is worth emphasizing that a static measurement showing an increase in the size of the LC3-II-positive autophagosomal compartment can reflect either autophagy activation (increased on-rate) or autophagy block (decreased off-rate). To appropriately interpret the nature and direction of changes in autophagic flux, it is necessary to assess not only the level of LC3-II but also the level of an autophagy receptor such as SQSTM1; if the SQSTM1 level increases along with the LC3-II level, the findings suggest a block in the late stages of autophagy and an overall decrease in the rate of autophagic flux. An increase in LC3-II with a concomitant decrease in SQSTM1 indicates increased autophagy flux, which can be seen in the setting of nutrient starvation. Examples of these will be discussed in Sect. 2 of the chapter.

Skeletal muscle diseases caused by or associated with dysfunctional autophagy can be classified into three major categories (Fig. 2): diseases due to reduced autophagic flux,

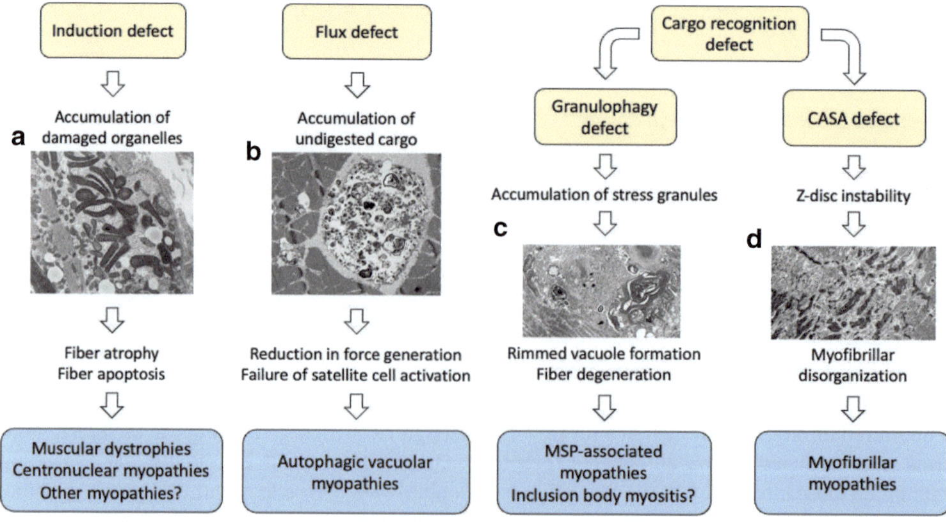


**Fig. 2** The role of autophagic defects in the pathogenesis of skeletal myopathies. In myopathies with autophagy induction defects, the accumulation of damaged cellular organelles activates signaling cascades that contribute to fiber atrophy and apoptosis; however, there is no accumulation of autophagosomes. In contrast, AVMs are caused by defects in the general autophagic flux that lead to a massive accumulation of autophagosomes and undigested cargo, ultimately resulting in a reduction of the mechanical force generated during muscle contraction (i.e., muscle weakness) and a failure of muscle regeneration. Finally, cargo recognition defects lead to two different, defect-specific disease phenotypes: A granulophagy defect results in rimmed vacuole formation and inclusion body myopathy, while a CASA defect causes myofibrillar myopathy by impairing Z-disc stability. The electron micrographs illustrate (**a**) subsarcolemmal accumulation of damaged mitochondria that show marked variation in their size and shape; (**b**) an autophagic vacuole with an accumulation of undigested, electron-dense autophagic cargo; (**c**) a rimmed vacuole with a rim of autophagic material and a center that consists of aggregated proteins; and (**d**) myofibrillar disorganization with streaming of the Z-disc material. (With the publisher's permission, this figure and its legend are reproduced from [50])

diseases due to reduced autophagy induction, and diseases due to impaired recognition and clearance of specific cargo. Diagnosis of these and other muscle diseases often requires muscle biopsy as a primary diagnostic approach; this includes sectioning of the tissue and microscopic evaluation employing specialized stains and antibodies. Hematoxylin and eosin (H&E) are commonly used chemical stains that impart distinct colors to nucleic acids and proteins, thereby revealing characteristic staining patterns that assist in the evaluation of tissue sections. Immunohistochemical stains for LC3 and SQSTM1 can be used to evaluate the quantity, quality, and distribution of autophagosomes/autophagic vacuoles. Finally, ultrastructural evaluation of cellular organelles by electron microscopy can be used to establish the presence and nature of the underlying autophagosomal abnormalities.

## 1.1 Skeletal Muscle Diseases Caused by Decreased Autophagy Flux

Question 1. Why is autophagy considered to be especially important in skeletal muscle, cardiac muscle, and neurons?

Question 2. Which autophagy proteins are involved in the recognition and sequestration of cargo?

Impairments in autophagy frequently underlie human skeletal muscle diseases, with disruptions in bulk autophagic flux, which cause *autophagic vacuolar myopathies* (AVMs), being the most common mechanism. Some AVMs are caused by genetic mutations, while others are a toxic side effect of medications utilized to treat malaria, autoimmune diseases, gout, or cancer. Irrespective of the specific etiology, all diseases in this category are typified by a massive accumulation of autophagic vacuoles (i.e. large sarcoplasmic membrane-bound compartments containing undigested cargo) within skeletal muscle fibers. A few exemplar AVMs are described in more detail below.

### 1.1.1 X-Linked Myopathy with Excessive Autophagy (XMEA)

XMEA, first described in 1988 [3], is caused by mutations in the *VMA21* gene on the X chromosome. *VMA21* encodes a chaperone protein, belonging to a broad class of chaperones that facilitate the proper folding of various proteins. Specifically, the VMA21 protein plays a vital role in orchestrating the proper assembly of a vacuolar ATPase, which uses ATP to pump protons into the lysosome for proper acidification [4]. Reduced expression of functional vacuolar ATPase leads to a failure in lysosomal acidification, effectively obstructing the final step of autophagic cargo degradation (step 6). Individuals with XMEA typically develop progressive weakness within the first or second decade of life, affecting muscles of their upper arms, shoulders, and large muscles of the hip and thighs (collectively this is described as "limb-girdle" weakness). Muscle biopsies in XMEA show the prototypical features of an AVM—large sarcoplasmic vacuoles on light microscopy and frequent enlarged vacuoles containing electron-dense autophagic debris on electron microscopy (Fig. 3) [5].

### 1.1.2 Danon Disease

Described first in 1981, Danon disease is associated clinically with a triad of skeletal myopathy, cardiomyopathy, and intellectual disability [6]. Danon disease is caused by mutations in the X-linked *LAMP2* gene. The LAMP2 protein is localized to lysosomes and late endosomes and is required for autophagosome-lysosome fusion. Therefore, mutations in LAMP2 effectively disrupt the second-to-last step (step 5) of the autophagy lysosomal pathway [7].

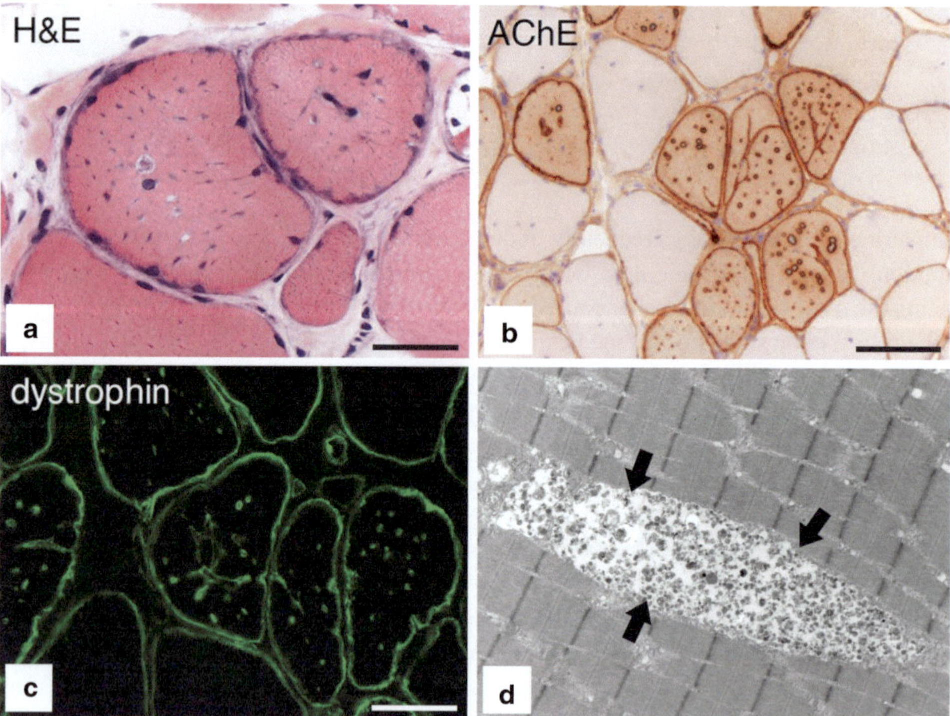


**Fig. 3** Muscle pathology in XMEA. Characteristic features of XMEA that are readily observable by light microscopic analysis of muscle biopsy samples include numerous cytoplasmic autophagic vacuoles best visualized on H&E staining (**a**), acetylcholinesterase (AChE) enzyme histochemistry (**b**), and immunofluorescence for dystrophin (**c**). Electron microscopy (**d**) shows large and small autophagic vacuoles interspersed between myofibrils (black arrows). Scale bars: 50μm (**a** and **c**); 100μm (**b**). (With the publisher's permission, this figure and its legend are reproduced with modifications from [5])

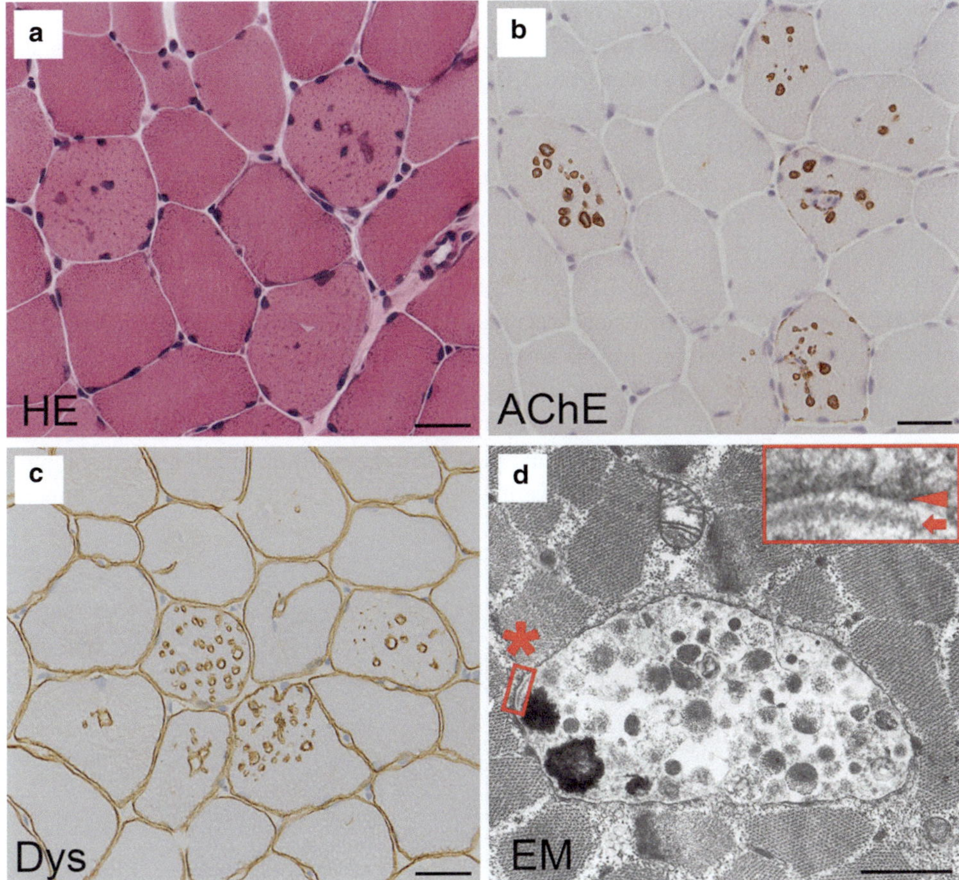

**Fig. 4** Muscle pathology in Danon disease. H&E-stained sections show small dot-like basophilic granular structures in the muscle fibers (**a**), which are acetylcholine esterase (AChE)-positive autophagic vacuoles (**b**). The limiting membranes of autophagic vacuoles have sarcolemmal features, as demonstrated by immunohistochemistry using dystrophin antibodies (**c**, Dys). Electron microscopy (EM) confirms that these vacuoles have autophagic nature by revealing myeloid bodies, electron-dense granular material, and variable cytoplasmic debris; moreover, basal lamina is observed along the inner surface of an autophagic vacuole (**d**, arrow in the inset), providing further evidence that the vacuolar membrane has sarcolemmal features. The inset demonstrates the enlarged view of the area within the square marked with an asterisk. Arrowhead marks the membrane of the autophagic vacuole. Scale bars: 50µm (**a–c**); 1 nm (**d**). (With the publisher's permission, this figure and its legend are reproduced from [8])

As with XMEA, histopathological and ultrastructural evaluation of skeletal muscle from the patients with Danon disease demonstrates characteristic AVM findings (Fig. 4). Intriguingly, Danon disease affects cardiac muscle in addition to skeletal muscle, contrasting it to XMEA, which primarily affects skeletal muscle; the reason for this difference is unclear. Cardiomyocytes in individuals with Danon disease display a similar accumulation of autophagic vacuoles. In mice, removing the *LAMP2* gene replicates several aspects of Danon disease, including the accumulation of autophagic vacuoles in both cardiac and skeletal muscles (although cardiac muscle is less affected in mice than in humans) [8, 9].

### 1.1.3 Pompe Disease

Pompe disease was first described in the 1930s and is caused by loss-of-function mutations in the *GAA* gene, encoding the lysosomal hydrolase acid alpha-glucosidase (GAA) protein. The GAA protein, also known as acid maltase, is responsible for glycogen catabolism in the lysosome [10]. The *GAA* mutations that cause Pompe disease lead to an AVM with the accumulation of autophagic vacuoles in skeletal and cardiac muscle, along with an accumulation of glycogen. (Aside from being an AVM, Pompe disease is also categorized as a glycogen storage disease.) Histopathological examination of skeletal muscle from patients with Pompe disease shows lysosomal accumulation of glycogen along with the evidence of disrupted autophagic flux, evident in the build-up of undigested autophagic cargo (Fig. 5). However, the precise mechanism by which GAA deficiency impairs

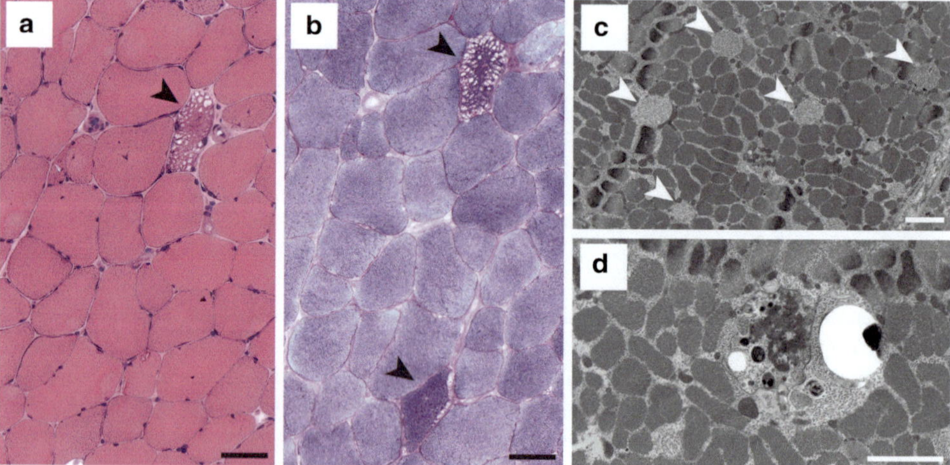

**Fig. 5** Muscle pathology in late-onset Pompe disease. A representative H&E-stained cryosection (**a**) shows largely unremarkable skeletal muscle with a single highly vacuolated muscle fiber in this microscopic field (black arrowhead). The same microscopic field from a PAS (periodic acid-Schiff)-stained cryosection (**b**) shows two vacuolated muscle fibers with mild glycogen accumulation (black arrowheads); the less vacuolated fiber at the bottom of the image shows no obvious changes on the corresponding H&E-stained section (**a**). Electron microscopy shows an increase in the lysosomal (membrane-bound) glycogen (white arrowheads in **c**) as well as accumulation of free sarcoplasmic glycogen accompanied by electron dense autophagic debris (**d**). The focal nature of these pathological findings is typical for late-onset Pompe disease, although the degree of muscle fiber involvement varies from case to case. Scale bars: 60μm (**a** and **b**); 2μm (**c** and **d**). (This figure and its legend are reproduced from [51]; this use is permitted under the Creative Commons Attribution 4.0 International License)

autophagic flux remains unclear; it could involve defects in lysosomal acidification and/or the fusion between autophagosomes and lysosomes (steps 5 or 6).

### 1.1.4    Chloroquine and Hydroxychloroquine Toxicity

Chloroquine (CQ) and its analog hydroxychloroquine (HCQ) were developed in the mid-1950s to treat malaria. Today, they are predominantly used to treat autoimmune diseases, such as systemic lupus erythematosus and rheumatoid arthritis. CQ and HCQ have long been recognized as lysosomotropic weak bases that accumulate in the acidic lysosomal compartment and increase its pH, effectively reducing the ability to degrade cargo by hydrolysis (step 6). There is also some evidence that CQ/HCQ can additionally block autophagosome fusion with the lysosome (step 5) [11]. Given these properties, CQ is often used to experimentally block autophagic flux in cells and animal models. While generally safe and well tolerated, a small subset of patients treated with CQ or HCQ develops skeletal myopathy and cardiomyopathy with the histopathology, not surprisingly, showing characteristic AVM features (Fig. 6a, b). In contrast to the genetic causes of AVMs, which are rare, the widespread use of CQ and HCQ in treatment of autoimmune diseases makes this the most frequent cause of AVM. The effect of CQ/HCQ on skeletal muscle integrity and autophagic marker expression will be discussed further in Sect. 2.

### 1.1.5    Colchicine and Vincristine Toxicity

Colchicine is used in the treatment and prevention of gout and familial Mediterranean fever, while vincristine is a chemotherapeutic agent used to treat some types of cancer. Both drugs bind to tubulin and prevent microtubule polymerization, which is thought to block autophagic flux by preventing autophagosome transport along the microtubule network, a step required for autophagosome maturation (step 4). Colchicine can be acutely toxic leading to multisystem organ failure and death, most likely due to cardiotoxicity. However, chronic colchicine and vincristine toxicity leads to skeletal muscle and peripheral nerve injury, with muscle biopsies that show typical AVM features along with disruption of myofibrils (thick and thin filaments that comprise the contractile apparatus; Fig. 6c, d) [12, 13].

> Question 3. Which steps of autophagy, when inhibited, lead to an autophagic vacuolar myopathy?
>
> Question 4. What X-linked genetic mutations cause an autophagic vacuolar myopathy phenotype?

## 1.2    Skeletal Muscle Diseases Caused by Decreased Autophagy Induction

Diseases caused by decreased autophagic induction (step 1) are rare, reflecting the fundamental importance of autophagy in cellular viability. In mice, knockout of the genes

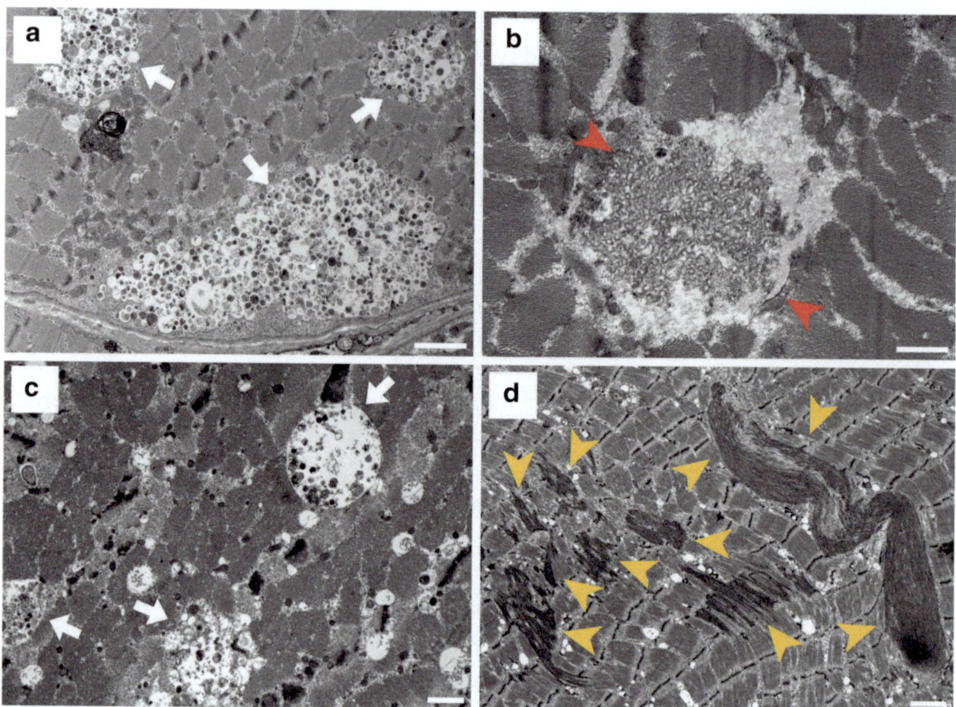

**Fig. 6** Ultrastructural muscle pathology in toxic (drug-induced) AVMs. (**a** and **b**) In HCQ-induced AVM, the key ultrastructural findings include accumulation of autophagic vacuoles containing electron-dense cellular debris (**a**, white arrows) and curvilinear bodies (B, red arrowheads). Aside from CQ/HCQ-induced myopathy, curvilinear bodies (the intralysosomal accumulation of ATP5MC/ subunit c of mitochondrial ATP synthase) are seen only in Batten disease (neuronal ceroid lipofuscinosis), which is a group of neurodegenerative diseases of childhood caused by genetic impairments in the autophagy-lysosomal pathway. (**c** and **d**) Colchicine-induced AVM again shows accumulation of autophagic vacuoles (**c**, white arrows) accompanied by myofibrillar disarray (**d**, yellow arrowheads) that resembles the disarray seen in myofibrillar myopathies (Fig. 8). Scale bars: 2μm (**a** and **d**); 1μm (**b** and **c**)

encoding core proteins involved in autophagy induction, such as ATG7 or ATG5, is perinatal lethal. To circumvent early lethality, muscle-specific deletion of *Atg5* or *Atg7* in mice has been examined and shown to result in muscle fiber atrophy accompanied by accumulation of SQSTM1-positive protein aggregates and abnormal organelles along with impaired muscle function [14, 15]. When autophagy induction is impaired, autophagosomes do not form and cannot accumulate; this results in diseases that lack a distinct "autophagic phenotype" that is seen in myopathies caused by impaired autophagic flux. Contrasting these findings will be the focus of Sect. 2 of this chapter.

### 1.2.1 Disease Caused by Mutations in Autophagic Induction Machinery

In 2016, the first human mutations in an *ATG* gene were described, with two siblings shown to harbor homozygous missense mutations in *ATG5* [16]. They both presented with ataxia, mental retardation, and developmental delay, with cells cultured from these subjects showing autophagic impairment. So far, the histopathological consequences of ATG5 mutations in human skeletal muscle have not been described, but have been studied in mouse models. Raben et al. initially showed that *atg5* knockout in skeletal muscle leads to atrophy of some muscle fiber types and an altered distribution of lysosomes within muscle fibers [15]. As one would expect, there was no accumulation of enlarged autophagosomes as seen in AVMs. Interestingly, the early lethality of *atg5* deletion in mice seems to be driven predominantly by neuronal dysfunction, as Yoshii and colleagues were able to rescue lethality in *atg5*-null mice by virally overexpressing *Atg5* in the central nervous system [17, 18]. The muscles from the rescued mice were severely atrophic and had smaller fibers, increased nuclei (suggesting possibly repeated cycles of fiber degeneration and regeneration), and an accumulation of ubiquitinated proteins and SQSTM1.

In 2021, five families with deleterious mutations in *ATG7* were identified; the affected family members show complex neurodevelopmental phenotypes that include ataxia, developmental delay, and musculoskeletal abnormalities [19, 20], mirroring several of the features of ATG5 mutations. The findings from muscle biopsies from these patients will be discussed in detail in Sect. 2.

### 1.2.2 Muscular Dystrophies and Centronuclear Myopathies

Muscular dystrophies are a large, diverse, and heterogeneous group of diseases that are characterized by progressive degeneration of muscle fibers. Dystrophies are typically caused by genetic mutations in muscle structural elements that make them vulnerable to mechanical stress, with repeated cycles of fiber degeneration and regeneration leading to characteristic histopathological changes.

Among the various mutations that can cause muscular dystrophy, mutations in COL6 (collagen type VI), an important component of the extracellular matrix that also regulates intercellular interactions, have been identified as rare causes of Bethlem myopathy and Ullrich congenital muscular dystrophy. *col6* knockout mice, the mouse model of this disease, were used to demonstrate that impaired autophagic induction could cause a muscle disease [21]. Skeletal muscle from these mice, similarly to skeletal muscle from humans with COL6-associated congenital muscular dystrophy, showed accumulation of abnormal mitochondria and distension of the sarcoplasmic reticulum, mirroring features seen in families with ATG7 mutations. *col6* knockout mice have also been used to demonstrate the importance of autophagy induction during exercise and in skeletal muscle homeostasis [22]. The mechanisms by which COL6 or other extracellular matrix components regulate autophagy in muscle remain unclear. Nonetheless, these findings could indicate that impaired autophagy induction also plays a role in other muscular dystrophies. For example, subsequent studies revealed a partial defect in the initiation of autophagy in mdx mutant mice, a model resembling Duchenne muscular dystrophy [23].

Along with muscular dystrophies, congenital myopathies have also been linked to failure in autophagy induction. Congenital myopathies are a diverse group of genetic muscle diseases caused by mutations that predominantly affect the muscle contractile function (in contrast to dystrophies, which are primarily associated with progressive muscle fiber degeneration). One subtype of congenital myopathies is centronuclear myopathy, which is pathologically characterized by abnormal "central" positioning of muscle fiber nuclei and other cellular organelles. Within the centronuclear myopathy spectrum, three genes—MTM1 (myotubularin 1), DNM2/DYN2 (dynamin 2), and BIN1/amphiphysin 2 (bridging integrator 1)—are exclusively linked to this disease phenotype and play roles in regulating phosphoinositide metabolism and intracellular vesicle transport. MTM1 encodes a phosphoinositide phosphatase, dephosphorylating phosphatidylinositol-3-phosphate (PtdIns3P) and phosphatidylinositol-3,5-bisphosphate, while DNM2 and BIN1 encode phosphoinositide-binding proteins involved in membrane trafficking [24]. The initiation of the phagophore is influenced by the class III PtdIns 3-kinase complex I (PtdIns3K-C1) and localized PtdIns3P production. Therefore, mutations in these genes might affect autophagy regulation, evident in studies where the deletion of MTMR14/Jumpy (myotubularin-related protein 14)—an MTM1-related phosphoinositide phosphatase—results in hyperactive autophagy in cell cultures, contributing to a centronuclear myopathy-like phenotype when combined with another mutation. In mouse models reflecting MTM1-associated centronuclear myopathy, initial stages show evidence of heightened autophagy, but the later stages mirror human disease, featuring defective autophagy induction linked to fiber atrophy and the accumulation of abnormal mitochondria and ubiquitinated protein aggregates [25, 26]. Remarkably, inhibiting MTOR (mechanistic target of rapamycin kinase), which activates autophagy, improved the muscle phenotype in these mice.

Further insights emerged from studies involving mice expressing common human DNM2 mutations, where heterozygous mice displayed abnormalities of mitochondria and sarcoplasmic reticulum, while homozygous littermates demonstrated inhibited autophagy induction, leading to perinatal mortality reminiscent of mice lacking *Atg* genes [27–29]. Collectively, these findings suggest that defects in autophagy induction may contribute to the pathogenesis of several non-autophagic skeletal myopathies. However, the impact of this pathway and the therapeutic potential of manipulating it are likely different across various muscle diseases, necessitating individualized investigations for each condition.

> Question 5. For which *ATG* genes have mutations in human families been identified as a cause of disease?
>
> Question 6. Which genes have been linked to centronuclear myopathy and how do they affect autophagy?

## 1.3 Skeletal Muscle Diseases Caused by Defects in Clearance of Specific Autophagy Cargo

While (macro)autophagy is the most common pathway for clearance of bulk portions of the cytoplasm, including large protein aggregates and aged/dysfunctional organelles, the autophagic machinery is also adapted for the selective recognition and clearance of different substrates. The inability or inefficiency of the autophagic machinery to recognize specific substrates (step 2) can lead to distinct skeletal muscle diseases. In contrast to AVMs that show a large buildup of autophagic vacuoles, these diseases are characterized by protein aggregates accumulated within the sarcoplasm. The two disease subtypes for which the underlying autophagy defect is relatively well understood are discussed in the next section; however, many other diseases may involve related mechanisms.

### 1.3.1 Multisystem Proteinopathies

Multisystem proteinopathies (MSPs) are a group of pleiotropic degenerative genetic diseases that are variably associated within individuals or families with inclusion body myopathy, Paget disease of bone, and frontotemporal dementia, with inclusion body myopathy being the most common manifestation [30, 31]. MSP is most often caused by a mutation in the *VCP* gene, but can also be caused by mutations in *HNRNPA1*, *HNRPA2B1*, *SQSTM1*, and *MATR3*. The proteins encoded by these genes are involved in the recognition and maintenance of subcellular structures called stress granules, which are membraneless condensates of protein and RNA that form during specific stress conditions to slow down translation of proteins, presumably to reduce energy expenditure. The clearance of stress granules is mediated by autophagy (i.e. granulophagy) and critically requires VCP [32–34]. Consequently, mutations in VCP lead to a selective deficiency in granulophagy. Additional

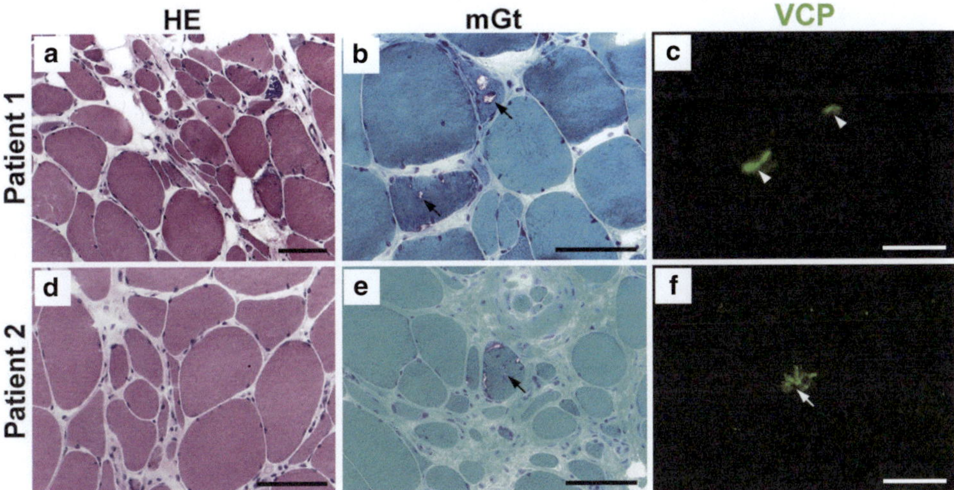

**Fig. 7** Muscle pathology in two patients with MSP1. (**a** and **d**) H&E staining shows groups of atrophic fibers in both patients. (**b** and **e**) Modified Gomori trichrome staining reveals the presence of rimmed vacuoles (black arrows). (**c** and **f**) Immunostaining for VCP (valosin containing protein) demonstrates VCP-positive inclusions. Scale bars: 50 μm. (With the publisher's permission, this figure and its legend are reproduced with modifications from [52])

studies have indicated that VCP is involved in the health of lysosomes themselves (i.e. lysophagy), thereby also potentially affecting autophagic flux [35]. Mutations in other MSP genes may reflect a deficiency in the recognition of stress granules by autophagic machinery, including mutations in the SQSTM1 protein (which is a receptor for a wide range of cargo). The muscles of individuals with MSP are notable for "rimmed vacuoles," (Fig. 7) with protein aggregates containing components of stress granules and other aggregation-prone proteins.

### 1.3.2 Myofibrillar Myopathies

The mechanical stress of skeletal muscle's contractile machinery requires regular maintenance and turnover of its structural elements, including the "Z-disc" of the muscle sarcomere. This muscle maintenance process requires a specialized form of autophagy termed chaperone-assisted selective autophagy (CASA), which utilizes different molecular chaperones to recognize specific substrates for autophagic clearance and initiate phagophore formation around those substrates. (Note that this is a distinct process from chaperone-mediated autophagy, discussed in other chapters, which uses similar chaperones to direct substrates to lysosomes that have already formed.) Some of the core components of the CASA machinery include the chaperones HSPA/HSP70, HSPB8, and BAG3 [36].

Myofibrillar myopathies are characterized by structural disarray of myofibrils, the basic contractile units of muscle fibers. Importantly, most myofibrillar myopathies show the presence of aggregated Z-disc components and increased autophagic vacuoles within the sarcoplasm (Fig. 8); in many cases, true rimmed vacuoles are also present [37].

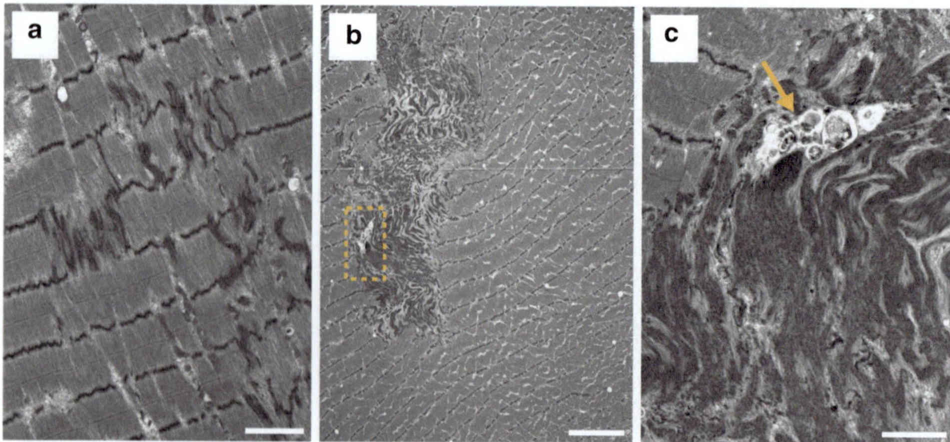

**Fig. 8** Ultrastructural muscle pathology in myofibrillar myopathies. Regardless of the underlying genetic abnormality, all myofibrillar myopathies show widespread disruptions of the Z line (called "Z-band streaming"; panel **a**) accompanied by formation of large protein inclusions (**b**) and scattered autophagic vacuoles (arrow in **c**). Panel **c** is the high magnification image of the boxed area in panel **b**. Scale bars: 2µm (**a** and **c**); 6µm (**b**)

Mutations in genes encoding proteins of Z-disc, which likely decrease Z-disc stability and accelerate its breakdown, lead to a myofibrillar myopathy phenotype. Importantly, mutations in components of the CASA machinery, including HSPB8 and BAG3, also cause myofibrillar myopathy, supporting their role in clearing Z-disc proteins through CASA [38, 39]. Similarly, mutations in DNAJB4 and DNAJB6, which are co-chaperones that modulate the activity of HSPA/HSP70 chaperones, are also causes of myofibrillar myopathies [40–42].

### 1.3.3 Skeletal Muscle Diseases in Which Autophagy May Be Secondarily Impaired

Along with the skeletal muscle diseases described above, in which specific aspects of autophagy are directly impaired, there are a growing number of disorders for which there is clear evidence that autophagy is disrupted, although mechanisms leading to disruption are secondary to a different underlying process. For example, sporadic inclusion body myositis (sIBM) is the most frequent skeletal myopathy in people over the age of 50, with the main symptoms including weakness of the quadriceps muscles and deep finger flexor muscles. The histopathology of sIBM is characterized by, among other features, rimmed vacuoles and accumulation of LC3- and SQSTM1-positive protein aggregates and stress granule proteins. The etiology of sIBM remains poorly understood, but histopathological similarities with MSPs, discussed above, raise the possibility of shared underlying molecular mechanisms [43, 44].

> Question 7. What skeletal muscle material has been shown to preferentially accumulate in myofibrillar myopathies? Why does this material accumulate?
>
> Question 8. What is the characteristic histological manifestation of a defect in stress granule clearance in skeletal muscle?

## 2 Introduction to Cardinal Articles

## 2.1 Assessing Autophagy Defects in Skeletal Muscle Disease

Along with establishing the general importance of autophagy impairment in skeletal muscles, the spectrum and range of different "autophagic" disease phenotypes showcases that the specific manner by which autophagy is impaired can lead to unique clinicopathological manifestations. In other words, different types of autophagic inhibition lead to different disease phenotypes. As such, assessing whether autophagy is impaired at an early step or a late step can help establish a diagnosis. The goal of this section is to discuss two cardinal articles that established the divergence in human muscle disease phenotypes caused by the block of autophagy induction and the block of the bulk autophagic flux, and to showcase the tools that are used for these assessments in human tissue samples, where the autophagic flux cannot be experimentally manipulated.

### 2.1.1    Identifying Defects in Autophagic Flux

The myotoxic effect of CQ was first described in the 1970s [45, 46], with an accumulation of autophagic vacuoles in patient muscle demonstrated by electron microscopy. This led to the increasing recognition that CQ and HCQ inhibit autophagic flux; since then, these two drugs, along with bafilomycin $A_1$, have been used as experimental tools to perturb and study autophagy in cellular and animal models. Similarly, colchicine blocks the maturation of autophagosomes by blocking microtubule transport. Diagnosis of AVM caused by either of these drugs requires evaluation by electron microscopy (Fig. 6), which is expensive and not widely available.

To streamline the diagnostic process, we retrospectively evaluated muscle biopsies from a series of patients treated by HCQ or colchicine [47]. The study consisted of three experimental groups: the normal control group, the drug-treated control group (muscle biopsies from patients treated with HCQ or colchicine but without evidence of significant autophagic build-up on electron microscopy), and the AVM group (muscle biopsies from patients treated with HCQ or colchicine that showed classic AVM features on electron microscopy). In normal control samples, there was no sarcoplasmic staining with either LC3 or SQSTM1 antibodies. In samples from the drug-treated control group, LC3 and SQSTM1 positivity was seen in rare fibers that typically showed sparse, randomly distributed fine puncta. In contrast, samples from the AVM group demonstrated moderate to frequent LC3- and SQSTM1-positive fibers with coarse punctate staining that tended to coalesce, resulting in zones of increased staining running linearly along the longitudinal axis of the fiber and most often located in the fiber center (Fig. 9). The percentage of LC3-positive fibers was significantly higher in the AVM group (median 57.5%, SD 30.8%) compared to either the normal control group (median 0.0%, SD 0.0%; $p < 0.001$) or the drug-treated control group (median 1.8%, SD 4.5%; $p < 0.05$). Similar results were seen with SQSTM1 staining: the percentage of SQSTM1-positive fibers was significantly higher in the AVM group (median 59.0%, SD 32.2%) compared to either the normal control group (median 0.0%, SD 0.3%; $p < 0.001$) or the drug-treated control group (median 2.0%, SD 3.6%; $p < 0.05$).

To confirm that increased LC3 staining in AVM samples was due to a block of the autophagic flux, immunoblotting for LC3 and SQSTM1 was performed on a representative subset of muscle biopsies, with $ATG7^{+/+}$ and $ATG7^{-/-}$ mouse embryonic fibroblasts (MEFs) used as a positive control. As expected, CQ treatment of $ATG7^{+/+}$ MEFs resulted in an increase in LC3-II and SQSTM1 protein levels; in contrast, autophagy-deficient $ATG7^{-/-}$ MEFs showed no LC3-II formation and a large increase in the level of SQSTM1 protein, both of which were independent of CQ treatment. Human AVM samples resembled CQ-treated wild-type ($ATG7^{+/+}$) MEFs, with LC3-II and SQSTM1 protein levels higher than in either normal control or drug-treated control samples (Fig. 10).

In summary, this study demonstrated that immunohistochemical staining for LC3 and SQSTM1 can be used to visualize the buildup of autophagic vacuoles (autophagosomes and autolysosomes) in human muscle fibers, providing a more convenient and cost-effective approach to AVM diagnostics than traditionally used electron microscopy. A

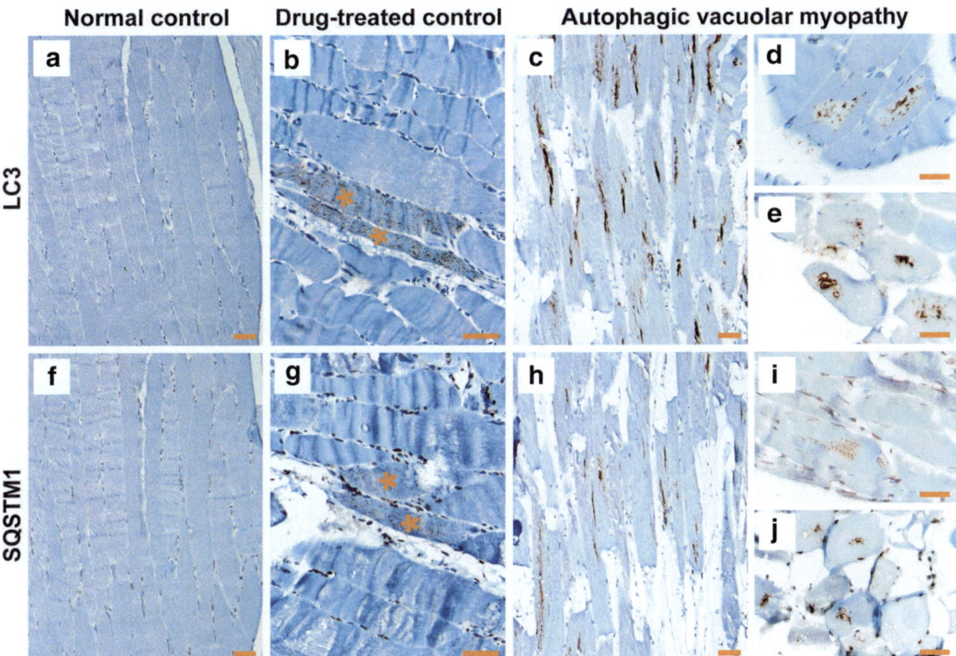


**Fig. 9** LC3 and SQSTM1 immunohistochemical staining patterns in toxic (drug-induced) AVMs. (**a** and **f**) Lack of sarcoplasmic staining in a normal control subject. With SQSTM1, there was background nuclear positivity. (**b** and **g**) Rare fibers (asterisks) with finely punctate staining in a colchicine-treated control subject. (**c–e** and **h–j**) Moderate to frequent LC3- and SQSTM1-positive fibers in AVM subjects. On longitudinal sections (**c** and **h**; HCQ-treated subject), the staining was linear, aligned with the longitudinal axis of the fiber, and often centrally located. On cross sections, the staining was coarsely punctate (**d** and **i**; colchicine-treated subject) or confluent, often lining vacuole rims (**e** and **j**; HCQ-treated subject). Scale bars: 50μm. (This figure and its legend are reproduced with modification from [47]; this use is permitted under the Creative Commons Attribution 4.0 International License)

similar immunohistochemical approach can be used to diagnose CQ- or HCQ-induced cardiomyopathy, a rare but potentially life-threatening condition caused by the blockade of autophagic flux in cardiac myocytes [48].

## 2.1.2 Identifying Defects in Autophagic Induction

As already mentioned, deletion of core autophagy proteins leads to perinatal lethality in murine models; not surprisingly, mutations affecting components of the core autophagy machinery are exceedingly rare in humans. However, in a recent landmark paper, Collier et al. described five unrelated families with recessive biallelic mutations in ATG7, a key autophagy regulatory protein that is required for LC3 lipidation and phagophore induction [19]. The affected family members show a complex neurodevelopmental phenotype characterized by ataxia, developmental delay, musculoskeletal abnormalities, and facial dysmorphism; these clinical features resemble those seen in patients with congenital

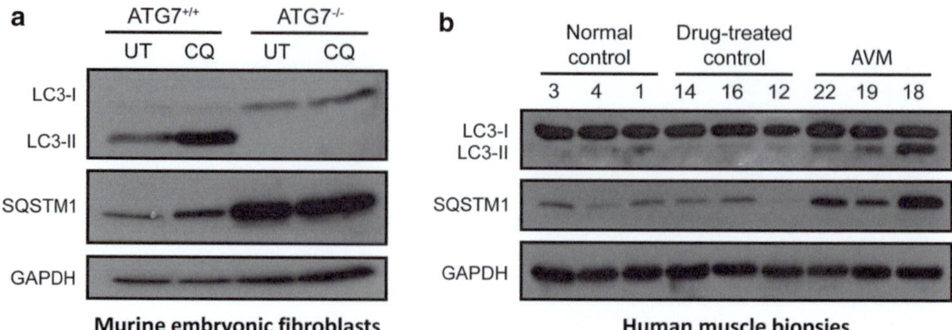



**Fig. 10** Immunoblotting findings in toxic (drug-induced) AVMs. (**a**) WT (ATG7+/+) and autophagy-deficient (ATG7−/−) mouse embryonic fibroblasts (MEFs) were used as positive control. In WT MEFs, 8-h treatment with 30µM chloroquine (CQ) increased the level of LC3-II and SQSTM1 proteins compared to the untreated control (UT); LC3-I was barely detectable in either sample. In autophagy-deficient MEFs, the level of LC3-I and SQSTM1 was high at baseline and did not change following CQ treatment; LC3-II was undetectable in both samples. GAPDH was used as a loading control. (**b**) In subjects from the AVM group, LC3-II and SQSTM1 protein levels were increased relative to subjects from either normal or drug-treated control groups. LC3-I protein level was equally high in all samples, suggesting that this isoform is not detected by LC3 immunohistochemistry. Each lane contains a sample from a different study subject, with subject ID numbers indicated on top (HCQ treatment, subjects 12 and 14; colchicine treatment, subjects 16, 18, 19 and 22). LC3-II and SQSTM1 protein levels were lower in samples #19 and #22 (which showed LC3 and SQSTM1-positive fibers in 13.0–22.5% range) than in sample #18 (which had 64.0% LC3-positive and 86.0% SQSTM1-positive fibers), indicative of a good concordance between the immunoblotting and immunostaining results. GAPDH was used as a loading control. (This figure and its legend are reproduced with modification from [47]; this use is permitted under the Creative Commons Attribution 4.0 International License)

mitochondrial disorders. The muscle biopsies of the individuals with ATG7 deficiency showed subsarcolemmal SQSTM1 aggregates and mild mitochondrial abnormalities (Fig. 11). These myopathological findings resemble those seen in the mice with a muscle-specific conditional knockout of *Atg7* [14] and are distinct from the findings in human (or murine) AVMs described above: without autophagy induction, autophagosomes cannot form and therefore cannot accumulate into LC3-positive vesicles. Conversely, without autophagy induction, the substrates for autophagy cannot be degraded and therefore accumulate in tissues, albeit at lower levels than seen in AVMs; this includes SQSTM1 and organelles that are normally turned over by autophagy (such as aged/abnormal mitochondria). In ATG7-deficient patients, western blot confirmed diminished levels of LC3-II and increased levels of SQSTM1, mimicking the findings seen in ATG7-deficient MEFs (Fig. 10a) and ATG7-deficient mice. In the follow-up functional complementation experiments, Collier et al. demonstrated that the patients' ATG7 mutations were indeed responsible for the observed autophagy deficits; in cultured fibroblasts derived from ATG7-deficient patients, complementation with wild-type ATG7 led to successful LC3 lipidation. Furthermore, expression of patient-associated ATG7 variants in *atg7*-knockout

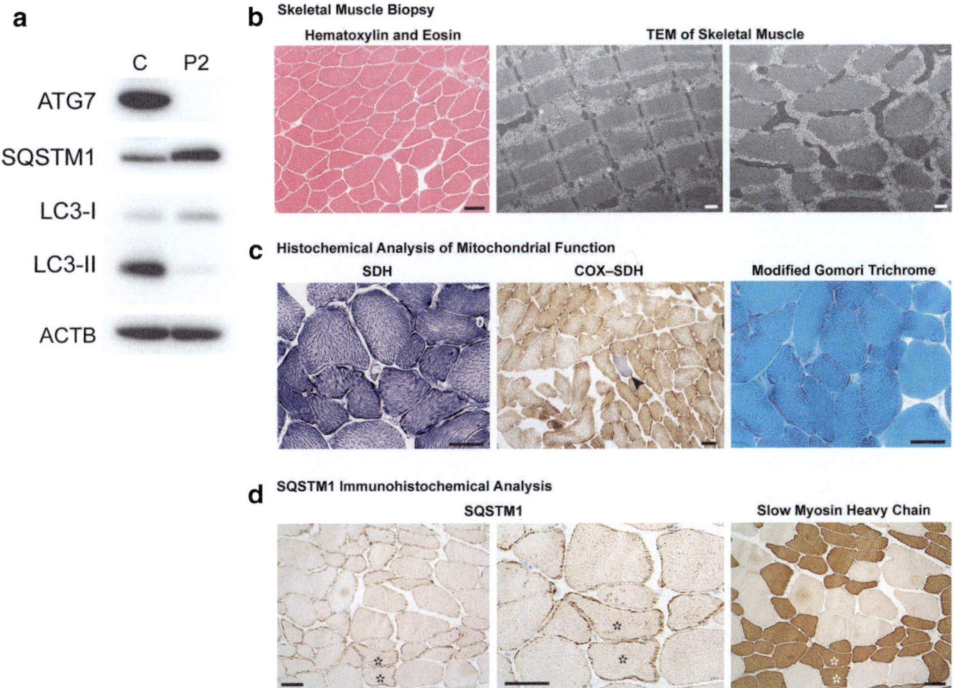

**Fig. 11** Muscle pathology in ATG7-deficient patients. (**a**) Immunoblot analysis of steady-state levels of autophagy proteins ATG7, SQSTM1, and LC3 in myoblasts from an ATG7-deficient patient relative to a control (**c**) cell line. (B) H&E staining and transmission electron microscopy (TEM) showed no vacuoles or internalized nuclei, with normal muscle fiber architecture and slightly enlarged, hyperbranched mitochondria. (**c**) Histochemical reaction for mitochondrial SDH (succinate dehydrogenase) activity was largely normal, but sequential histochemical reaction for cytochrome c oxidase (COX)–SDH activities revealed a fiber (arrowhead) with abnormal mitochondria that lack COX (respiratory chain complex IV) activity. Modified Gomori trichrome staining did not show marked subsarcolemmal mitochondrial aggregation. (**d**) Immunohistochemical analysis with an antibody reactive to SQSTM1 showed abnormal aggregation of SQSTM1 within the subsarcolemmal region of muscle. An accentuation of these accumulations within MYH7/slow myosin heavy chain–positive (type I) muscle fibers can be seen in serial sections (highlighted with asterisks). Scale bars: 50μm. (From [19], page 2412. Copyright © 2021 Massachusetts Medical Society. Reprinted with modification with permission from Massachusetts Medical Society)

MEFs did not lead to the same degree of LC3 lipidation (LC3-II formation) as expression of wild-type ATG7, confirming that ATG7 mutants were functionally deficient.

In summary, patients with genetic alterations in ATG7 show a disease phenotype that is highly reminiscent of the phenotype seen in mice with the conditional knockout of the *Atg7* gene; however, despite a nearly complete loss of ATG7 protein expression, a low level of basal autophagy is preserved in these patients, suggesting that humans have additional autophagy proteins that can compensate for ATG7 deficiency at a level suffi-

Question 9. A patient with weakness undergoes a biopsy of their quadriceps muscle. The histopathological evaluation shows vacuolated fibers, while immunostaining shows frequent and enlarged LC3- and SQSTM1-positive puncta in the fiber centers. What is the most likely defect in the autophagic pathway causing this patient's symptoms?

Question 10. A patient and several of their family members present with weakness and developmental disability. Sequencing showed a mutation in an *ATG* gene that is required for phagophore formation. What results might one expect to see on muscle histopathology and by western blotting for LC3-II and SQSTM1?

cient to enable survival. While defects of the core autophagy machinery are a rare cause of disease in humans, secondary impairments in autophagy induction likely play a role in the pathogenesis of many more common muscle diseases and would be expected to result in similar downstream pathogenetic mechanisms and disease manifestations.

## 2.2    Take Note Of

Remember that an increase in the number of autophagosomes may be due to a reduced autophagic flux or an increase in autophagy induction, while a decrease in the number of autophagosomes may be due to an increased autophagic flux or a reduced autophagy induction. In experimental settings using cultured cells or model animals, these possibilities can be differentiated by using genetic or pharmacological tools to manipulate and measure the rate of flux over a period of time [49]. In contrast, evaluation of human tissue pathology is a snapshot at a specific time; human tissue samples cannot be evaluated longitudinally to measure the true rate of autophagic flux. However, assessments of LC3-II and SQSTM1 protein levels and of LC3-II and SQSTM1 immunostaining patterns in human tissues can serve as reasonable proxy measures for the flux measurements: if LC3-II and SQSTM1 accumulate in parallel, the disease involves impairment of one or more of the late (post-induction) autophagy steps. In contrast, if there is SQSTM1 accumulation in the absence of significant LC3-II accumulation, the disease likely involves a defect in autophagy induction.

Also note the wide spectrum of human diseases caused by reduced autophagy function, often with distinct or characteristic clinicopathological manifestations. This underscores not only that autophagy plays different roles in different tissue and cell types, but also that specific disease phenotypes heavily depend on the exact autophagy steps that are impaired by the disease process.

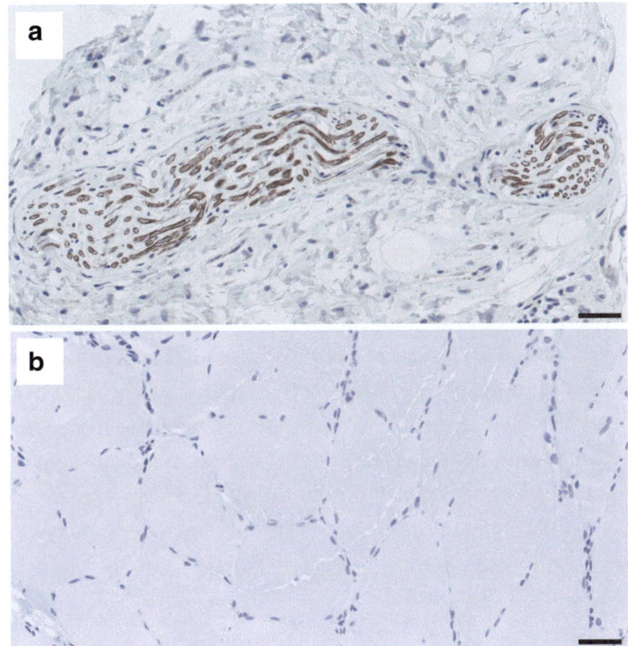

**Fig. 12** Baseline LC3 expression level is cell and tissue specific. (**a**) Using the LC3 immunohistochemistry conditions validated in our diagnostic pathology laboratory, strong basal LC3 staining is seen in the myelin sheaths of large nerve fibers in a normal peripheral nerve twig. (**b**) Under the same conditions, no significant basal LC3 expression is detectable in normal muscle fibers. Scale bars: 50μm

## 2.3  Markers and Tools

The key diagnostic tools in the evaluation of muscle biopsies for autophagic impairment are electron microscopy and immunohistochemical stains for LC3 and SQSTM1. However, the use of these immunostains for diagnosis requires careful antibody titration. For one, the basal levels of LC3 and SQSTM1 can vary significantly between tissues, as seen by the markedly higher baseline amount of LC3 in peripheral nerves in comparison to muscle fibers (Fig. 12). In addition, the antibody concentration must be titrated such that increased levels of LC3- and SQSTM1-positive autophagosomes can be readily recognized; this could require titrating antibody dilution to a point that basal levels of autophagosomes are not visualized in samples from a healthy individual.

In addition to immunohistochemical stains, muscle biopsies (and other human tissue samples) can be evaluated by western blotting to assess LC3-I, LC3-II, and SQSTM1 levels. However, western blotting can only be performed on fresh or frozen muscle tissue, whereas immunohistochemistry is typically done after formalin fixation, a chemical process that crosslinks proteins and greatly improves histopathological visualization by microscopy but precludes further assessment by immunoblotting. Therefore, biopsied skeletal muscle must be portioned as the outset between parts that will be fixed by formalin and parts that will be frozen for possible western blotting (among other assessments).

**Take Home Message**

- Skeletal muscle is a postmitotic tissue with high metabolic rate, low cellular turnover, and constant exposure to mechanical stress, making it highly dependent on autophagy for maintenance.
- Autophagic impairment that leads to skeletal muscle diseases may involve decreased autophagic flux, decreased autophagy induction, or failure to recognize and/or degrade specific autophagic cargo.
- Blocks in autophagic flux cause autophagic vacuolar myopathies, which can be due to genetic or toxic/pharmacological causes, and lead to a buildup of SQSTM1- and LC3-containing autophagosomes and other undigested cargo.
- Autophagy that is ineffective in recognizing and clearing specific proteins or subcellular structures can lead to aggregation of proteins into "rimmed vacuoles".
- Mutations in ATG5 and ATG7 in humans cause a defect in autophagy induction, leading to the accumulation of SQSTM1-positive aggregates and abnormal cellular organelles, but without an accompanying increase in LC3-positive autophagosomes.

**Answers to Questions**

Question 1: Why is autophagy considered to be particularly important in skeletal muscle, cardiac muscle, and neurons?

Answer: Skeletal muscle, cardiac muscle, and neurons are postmitotic and highly metabolically active, requiring autophagy for homeostasis by clearance of misfolding proteins and damaged organelles for the entire lifespan of an organism.

Question 2. Which autophagy proteins are involved in the recognition and sequestration of cargo?

Answer: The Atg8-family protein LC3 and various autophagy receptors, the most common one being SQSTM1, are important in cargo recognition.

Question 3. Which steps of autophagy, when inhibited, lead to an autophagic vacuolar myopathy?

Answer: Inhibition of the last three steps of autophagy (lysosomal cargo degradation, autophagosome/lysosome fusion, and autophagosome maturation) causes a defect in autophagic flux, leading to an autophagic vacuolar myopathy.

Question 4. What X-linked genetic mutations cause autophagic vacuolar myopathy phenotype?

Answer: Mutations in LAMP2 cause Danon disease, whereas mutations in VMA21 cause XMEA.

(continued)

(continued)

Question 5. For which *ATG* genes have mutations in human families been identified as a cause of disease?

Answer: To date, mutations in ATG5 and ATG7 have been identified as causes of human disease.

Question 6. Which genes have been linked to centronuclear myopathy and how do they affect autophagy?

Answer: Three genes that have been linked to centronuclear myopathies are *MTM1*, *DNM2*, and *BIN1*. The proteins encoded by these genes regulate phosphoinositide metabolism and intracellular vesicle transport, which are important in autophagy induction.

Question 7. What skeletal muscle material has been shown to preferentially accumulate in myofibrillar myopathies? Why does this material accumulate?

Answer: Defects in the clearance of damaged/misfolded Z-disc material lead to its accumulation in myofibrillar myopathies. This material either accumulates in excess due to mutations in Z-disc proteins or is improperly cleared due to defects in the chaperone-assisted selective autophagy (CASA) pathway, which is responsible for helping maintain the health and turnover of Z-disc components (which are under constant mechanical stress).

Question 8. What is the characteristic histological manifestation of a defect in stress granule clearance in skeletal muscle?

Answer: Rimmed vacuoles that contain various aggregated proteins, including components of stress granules along with autophagic proteins (LC3 and SQSTM1).

Question 9. A patient with weakness undergoes a biopsy of their quadriceps muscle. The histopathological evaluation shows vacuolated fibers, while immunostaining shows frequent and enlarged LC3- and SQSTM1-positive puncta in the fiber centers. What are the possible defects in the autophagic pathway causing this patient's symptoms?

Answer: The findings are strongly suggestive of an autophagic vacuolar myopathy and are likely linked to a deficit in autophagic flux, which could be due to a defect in lysosomal cargo degradation, autophagosome-lysosome fusion, or maturation of the autophagosome.

Question 10. A patient and several of their family members present with weakness and developmental disability. Sequencing reveals a loss-of-function mutation in an *ATG* gene that is known to be required for phagophore formation. What results might one expect to see on muscle histopathology and by western blotting for LC3-II and SQSTM1?

Answer: Reduced autophagy induction is expected to result in global abnormalities in protein and organelle homeostasis, with accumulation of misfolded proteins and damaged, abnormal mitochondria in skeletal muscle (and other organs). By western blot, there will be increased levels of SQSTM1 but low levels of LC3-II.

## References


1. He C, Bassik MC, Moresi V, Sun K, Wei Y, Zou Z, et al. Exercise-induced BCL2-regulated autophagy is required for muscle glucose homeostasis. Nature. 2012;481(7382):511–5.
2. Lira VA, Okutsu M, Zhang M, Greene NP, Laker RC, Breen DS, et al. Autophagy is required for exercise training-induced skeletal muscle adaptation and improvement of physical performance. FASEB J Off Publ Fed Am Soc Exp Biol. 2013;27(10):4184–93.
3. Kalimo H, Savontaus ML, Lang H, Paljärvi L, Sonninen V, Dean PB, et al. X-linked myopathy with excessive autophagy: a new hereditary muscle disease. Ann Neurol. 1988;23(3):258–65.
4. Ramachandran N, Munteanu I, Wang P, Ruggieri A, Rilstone JJ, Israelian N, et al. VMA21 deficiency prevents vacuolar ATPase assembly and causes autophagic vacuolar myopathy. Acta Neuropathol (Berl). 2013;125(3):439–57.
5. Dowling JJ, Moore SA, Kalimo H, Minassian BA. X-linked myopathy with excessive autophagy: a failure of self-eating. Acta Neuropathol (Berl). 2015;129(3):383–90.
6. Danon MJ, Oh SJ, DiMauro S, Manaligod JR, Eastwood A, Naidu S, et al. Lysosomal glycogen storage disease with normal acid maltase. Neurology. 1981;31(1):51–7.
7. Nascimbeni AC, Fanin M, Angelini C, Sandri M. Autophagy dysregulation in Danon disease. Cell Death Dis. 2017;8(1):e2565.
8. Endo Y, Furuta A, Nishino I. Danon disease: a phenotypic expression of LAMP-2 deficiency. Acta Neuropathol (Berl). 2015;129(3):391–8.
9. Rowland TJ, Sweet ME, Mestroni L, Taylor MRG. Danon disease – dysregulation of autophagy in a multisystem disorder with cardiomyopathy. J Cell Sci. 2016;129(11):2135–43.
10. Fukuda T, Roberts A, Plotz PH, Raben N. Acid alpha-glucosidase deficiency (Pompe disease). Curr Neurol Neurosci Rep. 2007;7(1):71–7.
11. Mauthe M, Orhon I, Rocchi C, Zhou X, Luhr M, Hijlkema KJ, et al. Chloroquine inhibits autophagic flux by decreasing autophagosome-lysosome fusion. Autophagy. 2018;14(8):1435–55.
12. Ching JK, Ju JS, Pittman SK, Margeta M, Weihl CC. Increased autophagy accelerates colchicine-induced muscle toxicity. Autophagy. 2013;9(12):2115–25.
13. Ju JS, Varadhachary AS, Miller SE, Weihl CC. Quantitation of "autophagic flux" in mature skeletal muscle. Autophagy. 2010;6(7):929–35.
14. Masiero E, Agatea L, Mammucari C, Blaauw B, Loro E, Komatsu M, et al. Autophagy is required to maintain muscle mass. Cell Metab. 2009;10(6):507–15.
15. Raben N, Hill V, Shea L, Takikita S, Baum R, Mizushima N, et al. Suppression of autophagy in skeletal muscle uncovers the accumulation of ubiquitinated proteins and their potential role in muscle damage in Pompe disease. Hum Mol Genet. 2008;17(24):3897–908.
16. Kim M, Sandford E, Gatica D, Qiu Y, Liu X, Zheng Y, et al. Mutation in ATG5 reduces autophagy and leads to ataxia with developmental delay. eLife. 2016;5:e12245.
17. Yoshii SR, Kuma A, Akashi T, Hara T, Yamamoto A, Kurikawa Y, et al. Systemic analysis of Atg5-null mice rescued from neonatal lethality by transgenic ATG5 expression in neurons. Dev Cell. 2016;39(1):116–30.
18. Yoshii SR, Kuma A, Mizushima N. Transgenic rescue of Atg5-null mice from neonatal lethality with neuron-specific expression of ATG5: systemic analysis of adult Atg5-deficient mice. Autophagy. 2017;13(4):763–4.
19. Collier JJ, Guissart C, Oláhová M, Sasorith S, Piron-Prunier F, Suomi F, et al. Developmental consequences of defective ATG7-mediated autophagy in humans. N Engl J Med. 2021;384(25):2406–17.
20. Collier JJ, Suomi F, Oláhová M, McWilliams TG, Taylor RW. Emerging roles of ATG7 in human health and disease. EMBO Mol Med. 2021;13(12):e14824.

21. Grumati P, Coletto L, Sabatelli P, Cescon M, Angelin A, Bertaggia E, et al. Autophagy is defective in collagen VI muscular dystrophies, and its reactivation rescues myofiber degeneration. Nat Med. 2010;16(11):1313–20.

22. Nair U, Klionsky DJ. Activation of autophagy is required for muscle homeostasis during physical exercise. Autophagy. 2011;7(12):1405–6.

23. De Palma C, Morisi F, Cheli S, Pambianco S, Cappello V, Vezzoli M, et al. Autophagy as a new therapeutic target in Duchenne muscular dystrophy. Cell Death Dis. 2012;3(11):e418.

24. Jungbluth H, Gautel M. Pathogenic mechanisms in centronuclear myopathies. Front Aging Neurosci. 2014;6:339.

25. Al-Qusairi L, Prokic I, Amoasii L, Kretz C, Messaddeq N, Mandel JL, et al. Lack of myotubularin (MTM1) leads to muscle hypotrophy through unbalanced regulation of the autophagy and ubiquitin-proteasome pathways. FASEB J Off Publ Fed Am Soc Exp Biol. 2013;27(8):3384–94.

26. Fetalvero KM, Yu Y, Goetschkes M, Liang G, Valdez RA, Gould T, et al. Defective autophagy and mTORC1 signaling in myotubularin null mice. Mol Cell Biol. 2013;33(1):98–110.

27. Cann GM, Guignabert C, Ying L, Deshpande N, Bekker JM, Wang L, et al. Developmental expression of LC3alpha and beta: absence of fibronectin or autophagy phenotype in LC3beta knockout mice. Dev Dyn Off Publ Am Assoc Anat. 2008;237(1):187–95.

28. Komatsu M, Waguri S, Ueno T, Iwata J, Murata S, Tanida I, et al. Impairment of starvation-induced and constitutive autophagy in Atg7-deficient mice. J Cell Biol. 2005;169(3):425–34.

29. Kuma A, Komatsu M, Mizushima N. Autophagy-monitoring and autophagy-deficient mice. Autophagy. 2017;13(10):1619–28.

30. Pfeffer G, Lee G, Pontifex CS, Fanganiello RD, Peck A, Weihl CC, et al. Multisystem proteinopathy due to VCP mutations: a review of clinical heterogeneity and genetic diagnosis. Genes. 2022;13(6):963.

31. Taylor JP. Multisystem proteinopathy: intersecting genetics in muscle, bone, and brain degeneration. Neurology. 2015;85(8):658–60.

32. Buchan JR, Kolaitis RM, Taylor JP, Parker R. Eukaryotic stress granules are cleared by autophagy and Cdc48/VCP function. Cell. 2013;153(7):1461–74.

33. Johnson AE, Shu H, Hauswirth AG, Tong A, Davis GW. VCP-dependent muscle degeneration is linked to defects in a dynamic tubular lysosomal network in vivo. elife. 2015;4:e07366.

34. Papadopoulos C, Kirchner P, Bug M, Grum D, Koerver L, Schulze N, et al. VCP/p97 cooperates with YOD1, UBXD1 and PLAA to drive clearance of ruptured lysosomes by autophagy. EMBO J. 2017;36(2):135–50.

35. Arhzaouy K, Papadopoulos C, Schulze N, Pittman SK, Meyer H, Weihl CC. VCP maintains lysosomal homeostasis and TFEB activity in differentiated skeletal muscle. Autophagy. 2019;15(6):1082–99.

36. Tedesco B, Vendredy L, Timmerman V, Poletti A. The chaperone-assisted selective autophagy complex dynamics and dysfunctions. Autophagy. 2023;19(6):1619–41.

37. de Siqueira Carvalho AA, Lacene E, Brochier G, Labasse C, Madelaine A, da Silva VG, et al. Genetic mutations and demographic, clinical, and morphological aspects of myofibrillar myopathy in a French cohort. Genet Test Mol Biomark. 2018;22(6):374–83.

38. Ghaoui R, Palmio J, Brewer J, Lek M, Needham M, Evilä A, et al. Mutations in HSPB8 causing a new phenotype of distal myopathy and motor neuropathy. Neurology. 2016;86(4):391–8.

39. Selcen D, Muntoni F, Burton BK, Pegoraro E, Sewry C, Bite AV, et al. Mutation in BAG3 causes severe dominant childhood muscular dystrophy. Ann Neurol. 2009;65(1):83–9.

40. Inoue M, Noguchi S, Inoue YU, Iida A, Ogawa M, Bengoechea R, et al. Distinctive chaperonopathy in skeletal muscle associated with the dominant variant in DNAJB4. Acta Neuropathol (Berl). 2023;145(2):235–55.

41. Ruggieri A, Brancati F, Zanotti S, Maggi L, Pasanisi MB, Saredi S, et al. Complete loss of the DNAJB6 G/F domain and novel missense mutations cause distal-onset DNAJB6 myopathy. Acta Neuropathol Commun. 2015;3:44.
42. Weihl CC, Töpf A, Bengoechea R, Duff J, Charlton R, Garcia SK, et al. Loss of function variants in DNAJB4 cause a myopathy with early respiratory failure. Acta Neuropathol (Berl). 2023;145(1):127–43.
43. Askanas V, Engel WK, Nogalska A. Pathogenic considerations in sporadic inclusion-body myositis, a degenerative muscle disease associated with aging and abnormalities of myoproteostasis. J Neuropathol Exp Neurol. 2012;71(8):680–93.
44. Nogalska A, D'Agostino C, Terracciano C, Engel WK, Askanas V. Impaired autophagy in sporadic inclusion-body myositis and in endoplasmic reticulum stress-provoked cultured human muscle fibers. Am J Pathol. 2010;177(3):1377–87.
45. Macdonald RD, Engel AG. Experimental chloroquine myopathy. J Neuropathol Exp Neurol. 1970;29(3):479–99.
46. Neville HE, Maunder-Sewry CA, McDougall J, Sewell JR, Dubowitz V. Chloroquine-induced cytosomes with curvilinear profiles in muscle. Muscle Nerve. 1979;2(5):376–81.
47. Lee HS, Daniels BH, Salas E, Bollen AW, Debnath J, Margeta M. Clinical utility of LC3 and p62 immunohistochemistry in diagnosis of drug-induced autophagic vacuolar myopathies: a case-control study. PLoS One. 2012;7(4):e36221.
48. Daniels BH, McComb RD, Mobley BC, Gultekin SH, Lee HS, Margeta M. LC3 and p62 as diagnostic markers of drug-induced autophagic vacuolar cardiomyopathy: a study of 3 cases. Am J Surg Pathol. 2013;37(7):1014–21.
49. Loos B, du Toit A, Hofmeyr JHS. Defining and measuring autophagosome flux – concept and reality. Autophagy. 2014;10(11):2087–96.
50. Margeta M. Autophagy defects in skeletal myopathies. Annu Rev Pathol. 2020;15:261–85.
51. Margeta M. Neuromuscular disease: 2022 update. Free Neuropathol. 2022;3:3–5.
52. Inoue M, Iida A, Hayashi S, Mori-Yoshimura M, Nagaoka A, Yoshimura S, et al. Two novel VCP missense variants identified in Japanese patients with multisystem proteinopathy. Hum Genome Var. 2018;5:9.

## Further Reading

Franco-Romero A, Sandri M. Role of autophagy in muscle disease. Mol Asp Med. 2021;82:101041. https://doi.org/10.1016/j.mam.2021.101041.
Margeta M. Autophagy defects in skeletal myopathies. Annu Rev Pathol. 2020;15:261–85. https://doi.org/10.1146/annurev-pathmechdis-012419-032618.
Xia Q, Huang X, Huang J, Zheng Y, March ME, Li J, Wei Y. The role of autophagy in skeletal muscle diseases. Front Physiol. 2021;12:638983. https://doi.org/10.3389/fphys.2021.638983.

# The Role of Autophagy and Mitophagy in Cardiomyocyte Ischemic Injury

Roberta A. Gottlieb and Somayeh Pourpirali

**What You Will Learn in This Chapter**
You will be introduced to autophagy/mitophagy machinery in the heart, including the key proteins and regulators of the process. Next you'll learn about the stressors that can invoke autophagy/mitophagy, especially those relevant to the heart. The next section will cover methodology for studying autophagy, including biochemical approaches as well as microscopy. Additional methods for studying autophagy and autophagic flux in cells and mice will be presented. Finally, this chapter includes a discussion on the role of autophagy in the context of heart disease.

## 1 Overview of Autophagy and Mitophagy in the Heart

### 1.1 Autophagy

Once the components and processes of autophagy were largely worked out in yeast and simple organisms, investigation of autophagy in mammals and in specific tissues followed rapidly, particularly after the introduction of GFP-MAP1LC3/LC3 (microtubule associated protein 1 light chain 3) as a reporter of autophagosome formation [1]. Macroautophagy (delivery of organelles, protein aggregates, and other bulk cytosol, and henceforth referred to as autophagy) is the focus of this chapter, although there are other mechanisms for delivering cellular material to the lysosome for degradation, including microautophagy, mitochondrial-derived vesicles, and chaperone-mediated autophagy.

<u>Initiation</u> of autophagy is mediated by ULK1 (unc-51 like autophagy activating kinase 1), a protein kinase which itself can be phosphorylated and activated by AMP-activated

R. A. Gottlieb (✉)
Department of Cardiology, Cedars-Sinai Medical Center, Los Angeles, CA, USA

S. Pourpirali
Department of Biological Science and Technology, Islamic Azad University,
Najafabad, Isfahan, Iran

© The Author(s), under exclusive license to Springer Nature Switzerland AG 2025
B. Loos, D. J. Klionsky (eds.), *Autophagy - From Molecular Mechanisms to Flux Control in Health and Disease*, Learning Materials in Biosciences,
https://doi.org/10.1007/978-3-031-88121-3_10

protein kinase (AMPK) in response to nutrient deprivation or other energy-related cues [2]. ULK1 (or its homolog ULK2) form a large multiprotein complex needed for initiation of autophagy, a process requiring the activation of ubiquitin ligase-like enzymes responsible for conjugating ATG12 (autophagy related 12) onto ATG5 and the amine group of phosphatidylethanolamine (PE) onto an acceptor lysine on LC3. This links LC3 to the concave and convex faces of the developing autophagosomal membrane, or phagophore. Conjugation of LC3 onto PE-containing membranes is thought to be nonspecific, and is regulated by ATG4 (autophagy related 4 cysteine peptidase), that both cleaves LC3 to expose an N-terminal glycine and removes LC3 from its PE membrane anchor [3]. Interestingly, because ATG4 is redox sensitive, removal of LC3 from membranes is suppressed where high local reactive oxygen species (ROS) are present, such as dysfunctional mitochondria, providing an important mechanism for tagging and eliminating damaged mitochondria [4].

Engulfment occurs when the cup-shaped membrane elongates around its cargo assisted by receptor proteins such as SQSTM1/p62 (sequestosome 1), a bifunctional protein that binds ubiquitinated proteins and LC3. Upon closure of the expanding ends of the phagophore, creating a double-membrane structure with cargo in the lumen, the autophagosome is associated with microtubules and is routed to a lysosome. Delivery to the lysosome depends upon trafficking along microtubules; therefore, lysosomal blockade can also be achieved with microtubule poisons such as nocodazole [5].

Degradation begins after the outer membrane of the autophagosome fuses with the lysosomal membrane, thereby introducing degradative enzymes (proteases, lipases, and polysaccharidases) to the cargo. Fusion between the autophagosome and lysosome and subsequent degradation require acidification of the compartments (to an impressive pH of 4.5–5.0). As will be presented in a subsequent section, we can take advantage of the requirement for acidification to block autophagosome-lysosome fusion and subsequent cargo degradation, allowing for interrogation of autophagic activity over time, also referred to as autophagic flux. Acidification is accomplished through the vacuolar-type proton-translocating ATPase (V-ATPase), a protein complex similar to the mitochondrial $F_0F_1$ ATPase, but in this case, it hydrolyzes ATP in order to pump protons against the gradient to acidify the lysosome. Bafilomycin $A_1$ is a fungal product that potently (in the nM concentration range) inhibits the V-ATPase and can be used to block autophagic flux [6]. Weak bases such as $NH_4Cl$ or chloroquine accumulate in the lysosome and neutralize its pH to block flux.

Cessation of autophagy occurs when amino acids (especially branched chain amino acids) are exported from the lysosome and are sensed by MTOR (mechanistic target of rapamycin kinase), a master regulator, which inactivates ULK1 to suppress further autophagosome initiation. Inhibition of MTOR (e.g., by rapamycin) can potently induce autophagy, as can activation of AMPK.

It is important to note that autophagy is positively and negatively regulated by AMPK and MTOR, respectively, closely linking autophagy to nutrient status. Moreover, this dynamic process follows a circadian pattern related to food intake; nutritional excess, especially prior to sleep, can prevent the normal activation of autophagy and may result in the accumulation of damaged organelles and protein aggregates over time. This may have profound consequences for long-lived cells (cardiomyocytes, neurons, certain immune cells), which depend heavily on autophagy for protein homeostasis to maintain cell function.

The heart is largely comprised of cardiomyocytes, fibroblasts, blood vessels, and tissue-resident macrophages, but the dominant cell type (by mass) is the cardiomyocyte, which contains the actin-myosin contractile machinery and many mitochondria to provide the necessary ATP. Mitochondria make up approximately one-third of the mass of heart tissue, so maintaining mitochondrial function at a high level is essential for both contractile function and electrical conduction needed to achieve rhythmic, organized contraction of the chambers. Autophagy comes into play for selective removal of dysfunctional mitochondria which, if depolarized, consume rather than produce ATP, which can generate damaging ROS, and which can contribute to abnormal electromechanical activity if they fail to handle calcium appropriately [7]. For these reasons, cardiomyocytes possess robust mechanisms for elimination of damaged mitochondria, including mitophagy [8], formation and clearance of mitochondria-derived vesicles/MDVs [9], and the mitochondrial unfolded protein response [10, 11]. Here we will focus on mitophagy, which is initiated in response to nutrient stress, activation of AMPK (high AMP:ATP), activation of sirtuins, mitochondrial ROS, mitochondrial depolarization, and a variety of other stresses.

## 1.2   Mitophagy

Mitophagy is the selective removal of mitochondria through autophagy and depends upon specialized receptors to engage the autophagic machinery. In addition to the ROS- and ATG4-mediated accumulation of LC3 on mitochondrial membranes described above, additional mechanisms exist, notably BNIP3 (BCL2 interacting protein 3) and the PINK1 (PTEN induced kinase 1)-PRKN (parkin RBR E3 ubiquitin protein ligase) pathway. BNIP3 is a pro-apoptotic member of the BCL2 family that possess an LC3-interacting region (LIR); under conditions of mitochondrial dysfunction, BNIP3 associates with the mitochondrial outer membrane and recruits an LC3-studded phagophore to engulf the organelle [12]. However, because BNIP3 has proapoptotic functions including mitochondrial outer membrane permeabilization (MOMP) to release CYCS/cytochrome *c* [8], and additionally to trigger opening of the inner membrane permeability transition pore, cell death and autophagy may be competing ongoing processes, leading to the occurrence of what is termed autophagic cell death. Perhaps the most important mechanism of mitophagy in cardiomyocytes is the PINK1-PRKN pathway. PINK1 is synthesized in the cytosol and inserted into the mitochondrion where under normal conditions it is promptly degraded by proteases including the intermembrane space protease PARL (presenilin associated rhomboid like). However, if mitochondrial membrane potential is lost, protein import ceases and PINK1 accumulates on the mitochondrial outer membrane (OM) where it phosphorylates various OM proteins as well as ubiquitin monomers [13]. PRKN interacts with the OM phosphoproteins and is potently activated by phospho-ubiquitin. PRKN is a ubiquitin ligase and may be unique in its preference for phosphoubiquitin; it ubiquitinates numerous OM proteins, resulting in accumulation of SQSTM1/p62, which as noted above binds ubiquitin chains as well as LC3, thereby promoting engulfment of the decorated mitochondrion.

In the heart, autophagy and mitophagy were recognized to arise in response to ischemic stress [14, 15]. PRKN plays an important role in clearing damaged mitochondria in the setting of ischemia-reperfusion (I/R) injury [16]. Ischemic preconditioning, in which a brief ischemic episode confers protection against a subsequent more prolonged ischemia/ reperfusion insult, depends upon PRKN-mediated mitophagy [17]. Notably, *PRKN* deletion exacerbates post-infarction heart failure, and heart failure is more prevalent among patients diagnosed with Parkinson disease, for which mutation of PRKN is a known genetic risk factor [18]. However, a recent study by an Israel-based group concluded that Parkinson disease was associated with lower risk of heart failure [19]. Some investigators have argued that PRKN is not essential in the heart [20], although the evidence was based on low *Prkn* mRNA expression and a faint signal on western blot in wild-type mice, and lack of heart failure in cardiac-specific knockout of *Prkn* in mice aged up to 20 weeks (the authors did not report longer aging studies in the knockout mice). In contrast, others have reported accelerated cardiac aging in PRKN-deficient mice [21, 22]. Interestingly, although PINK1 is responsible for phosphorylating ubiquitin (phospho-ubiquitin is the preferred substrate for PRKN's ubiquitin ligase activity) PRKN recruitment to mitochondria and subsequent mitophagy can proceed even in the absence of PINK1, suggesting alternative pathways to PRKN activation [23].

Inflammation is linked to impaired mitophagy: failure to clear damaged mitochondria leads to release of mitochondrial DNA (mtDNA) into the cytosol where it activates the NLRP3 (NLR family pyrin domain containing 3) inflammasome [24] and the STING1 (stimulator of interferon response cGAMP interactor 1) pathway [25]. ROS-damaged mitochondria, in which inner membrane cardiolipin is exposed (either through OM disruption or transfer of the lipid to the OM), can serve as a substrate for assembly of the NLRP3 inflammasome; both NLRP3 and CASP1 (caspase 1; another component of the inflammasome) bind to cardiolipin and the resulting complex processes pro-IL1B/IL-1β (interleukin 1 beta) to its mature form. Inflammation is tightly associated with cardiac dysfunction and is a feature of pathologic remodeling after myocardial infarction or pressure overload. Indeed, it was shown that mtDNA from failed autophagy elicits sterile inflammation and heart failure [26]. Thus, damaged mitochondria serve as an inflammatory trigger, while mitophagy suppresses the inflammatory process, by clearing damaged mitochondria [27].

## 1.3    Perspective

While mitochondria perform essential functions for ATP production, calcium handling, and essential biosynthetic functions (e.g., uridine, heme, lipid and prostaglandin synthesis), they also play a central role in programmed cell death and inflammatory signaling. Because mitochondria which have lost membrane potential can actually hydrolyze ATP, eliminating those wasteful mitochondria can improve ATP availability; moreover, mitophagy is linked to mitochondrial biogenesis, so removal of damaged mitochondria is followed by replacement with new, well-functioning organelles. Similarly, elimination of mitochondria that are generating excessive ROS will improve overall homeostasis by attenuating redox stress. However, mitophagy is not always beneficial; in settings where mitophagy is excessive, especially if replacement biogenesis cannot keep pace, or where lysosomal function is impaired, mitophagy can be deleterious.

> **Question**
> 1. What are autophagy steps and what molecules can be used as a marker for each step?
> 2. What cells comprise the majority of the heart mass?
> 3. What is mitophagy and how does it affect the heart?

## 2    Key Concepts and Methodology

Autophagy and mitophagy can be assessed through imaging techniques, western blotting, and additional methodology. A comprehensive and detailed discussion of methodology is presented in Klionsky's Guidelines for the use and interpretation of assays for monitoring autophagy [28]. Here we will present a few of the most common methods employed in the cardiac field.

## 2.1    Immunoblotting

Immunoblotting of cell/tissue lysates is widely used because of convenience, the small amount of material required (10–50 μg), and the wide availability of suitable antibodies for interrogation of autophagy. RIPA buffer is acceptable, and similar results can be obtained with a nonionic detergent such as Triton X-100. LC3 (a homolog of yeast Atg8) is an excellent indicator of autophagy, because lipidation (conjugation to PE) alters its migration in SDS-PAGE, resulting in two bands: the unlipidated LC3-I and the faster-migrating lipidated form, LC3-II. An increase in LC3-II is generally assumed to represent increased autophagosome initiation. However, an increase can also indicate impaired autophagic flux, so care must be taken to keep the latter possibility in mind. A thorough discussion of tips and tricks to improved detection is provided on the Novus Biologicals website [https://www.novusbio.com/support/faqs-autophagy-and-lc3#Are%20there%20 any%20tips]. Wherever possible, a flux experiment should be performed (see below). One can also examine conjugation of ATG12 to ATG5, so a comparison of free ATG5 ($M_r$ ~30 kDa) to the ATG12–ATG5 conjugate ($M_r$ ~50 kDa) can provide supporting evidence of autophagy. An increase in SQSTM1/p62 is also considered a reflection of increased autophagy, but here again, flux must be considered.

## 2.2    Immunohistology/Immunofluorescence

This technique can be used to document presence of autophagy machinery, but because LC3 is constitutively expressed, one must look at the pattern of distribution in the cell: diffuse distribution is seen in quiescent cells, while a punctate pattern reflects autophagosomes. Similarly, cells may be transfected with a construct expressing LC3 fused to a fluorescent reporter such as GFP or mCherry, and transgenic mice expressing GFP-LC3 and mCherry-LC3 are available. Again, it is important to recall that both increased autophagy initiation and decreased flux (due to impaired delivery or lysosomal degradation) can give rise to an increase in the number of puncta.

## 2.3    Flux Assessment

To accurately assess autophagic activity, we need to look into autophagic flux which is defined as the amount of autophagic degradation. This measurement remains complicated, still, efforts have been made to develop new techniques or improve the existing ones for a better flux analysis.

### 2.3.1    ATG8 Turnover Assay

Assessment of flux involves paired samples, whether parallel dishes of cells or pairs of animals. In either case, experimental design should include a control (no intervention except vehicle), a stress intervention (e.g., starvation, I/R, or drug administration), and a second set in which a lysosomal blockade (chloroquine or bafilomycin $A_1$) is applied for a period of time (often 1–4 h, must be empirically determined for a given system), then the four conditions (Control, Control + blockade, Stress, Stress + blockade) are harvested and imaged or analyzed by western blot as described above. Under control conditions, one can expect a modest increase in LC3-II after lysosomal blockade. In the test Stress condition, one would expect to detect increased LC3-II which would be further enhanced with lysosomal blockade. If, however, the test Stress actually causes lysosomal dysfunction, then the abundance of LC3-II would be similar with/without blockade or might show a small increase. An expanded discussion of autophagy assessment and pitfalls is covered in a previous review article [29].

A recent advancement has been reported using cardiac tissue slices (350 µm) from dogs and mice; canine tissue retained viability for a longer period of time in culture and demonstrated increased LC3-II in response to rapamycin (an MTOR inhibitor and activator of autophagy), and in response to lysosomal blockade with chloroquine [30]. It is likely that this approach could be applied to human heart tissue (e.g., endomyocardial biopsies) and might provide additional insights into autophagic function in a variety of cardiac disease processes.

### 2.3.2 Simultaneous Detection of Autophagosomes and Autolysosomes

A recent improvement in analysis of autophagy has been the use of a tandem mCherry-GFP-LC3 construct [31]. GFP is readily quenched and/or degraded in the acidic lysosome whereas mCherry retains its fluorescence in the lysosome for a longer period of time. Thus, a yellow signal (overlay of red + green) reflects autophagosomes while a red-only signal indicates autolysosomes. This allows an estimation of flux, based on the abundance of red-only puncta. Similarly, a construct targeted to mitochondrial inner membrane using the COX8 (cytochrome c oxidase subunit 8) mitochondrial targeting sequence fused to GFP-mCherry can be used to assess mitochondrial flux [32]. Mito-Keima is a mitochondrial-targeted fluorescent protein in which the excitation wavelength is predominantly short wavelength at the pH of mitochondrial matrix (pH 8.0), whereas the excitation wavelength shifts to a longer wavelength in the acidic lysosome (pH 4.5). This probe allows for ratiometric imaging and is compatible with live-cell imaging; a transgenic mouse line has been established [33].

### 2.3.3 Analysis of Autolysosome Size

As mentioned before, at the final stage of autophagy, cytosolic materials are degraded by lysosomal hydrolases inside autolysosomes. A defect in lysosome function increases the number of autolysosomes that are larger than normal [34]. Careful analysis of enlarged autolysosomes shows malfunction of the degradation process in the autolysosome. This can be achieved through analyzing cells expressing mCherry-GFP-LC3 where the diameter or circumference of each spot size of mCherry$^+$ GFP$^+$ (autolysosomes) increases during blockade of lysosome function [35].

## 2.4 Assessment of Mitochondrial Mass

The most reliable indicator of mitophagy is mitochondrial mass analysis. This can be accomplished using immunoblotting, which measures the amount of mitochondrial proteins in relation to levels of traditional loading markers. It is recommended to cover all mitochondrial compartment proteins, including the outer (e.g., TOMM20 [translocase of outer mitochondrial membrane 20), VDAC [voltage dependent anion channel]) and inner mitochondrial membrane proteins (such as COX [cytochrome c oxidase] or SLC25A4/ANT1 [solute carrier family 25 member 4]), intermembrane space (CYCS [cytochrome c, somatic]), and matrix proteins (SOD2/MnSOD [superoxide dismutase 2], HSPD1/HSP60 [heat shock protein family D (Hsp60) member 1]). Furthermore, quantitative PCR can be used to calculate the mitochondrial:nuclear DNA ratio by using *MT-RNR2/16S* (mitochondrially encoded 16S rRNA) rRNA as an index of mtDNA and HK2 (hexokinase 2) as an index of nuclear DNA. However, in order to accurately interpret these findings, mitochondrial biogenesis must also be assessed to verify that it is not compromised. It should also be highlighted that to detect net loss of mitochondrial proteins, large levels of mitophagy may be required [28].

## 2.5 MitoTimer

MitoTimer is a fantastic tool for monitoring mitochondrial turnover in real time. DsRed-E5, or fluorescent timer protein, is a mutant form of the red fluorescent protein DsRed that changes color from green to red as it matures [36]. MitoTimer, a mitochondrial matrix-targeting protein, was developed by fusing the timer protein with the mitochondrial targeting region of COX8A [37, 38]. Subsequently, transgenic mice were generated expressing heart-specific MitoTimer via the cardiac *Myh6* (myosin, heavy polypeptide 6, cardiac muscle, alpha) promoter [1]. Unlike MYH6, which is expressed equally in both atrial and ventricular myocardium, MitoTimer is predominantly expressed in the ventricles and to a lesser extent in the atria. MitoTimer expression in vivo makes its application in experiments easier in investigating mitophagy's involvement in cardiac muscle.

## 2.6 Models of Cardiac Ischemia and Reperfusion

Given the clinical significance of myocardial infarctions, a variety of models have been introduced, including culture of cardiomyocyte-like cell lines subjected to chemical ischemia (and reperfusion), tissue slices, isolated perfused organs, and in vivo animal models including mice, rats, rabbits, and pigs. Models may provide readouts addressing early changes (acute infarct size, gene expression, proteomics, cellular processes) and longer-term consequences (cardiac contractility, fibrosis). The most widely used models will be covered here.

### 2.6.1 Cell Culture

Cell lines include mouse HL-1, rat H9c2, human AC16, and human iPSC-derived cardiomyocytes, whether generated in-house or obtained through commercial suppliers. Primary cell culture using neonatal rodent cardiomyocytes or adult cardiomyocytes obtained by collagenase digestion of the heart are also accepted models, although all cell-based models have their own limitations. Cells are maintained in appropriate medium until the test ischemia is applied, preceded by a brief wash in phosphate-buffered saline to remove residual culture media. Ischemia is a complex stress: in addition to oxygen deprivation, glucose is also unavailable, potassium is increased in the extracellular space, and pH is low. A thoughtful discussion of cell models of I/R injury is presented by Chen and Vunjak-Novakovic [39]. A common model includes 12 mM potassium, 20 mM lactate, pH 6.5 in a physiologic salt solution; hypoxia is achieved in a gas-tight chamber flushed with 95% $N_2$, 5% $CO_2$ (if using a bicarbonate buffer) or with an oxygen depleting system such as the BD GasPak (if using a HEPES-based buffer). Many cell lines (which have been selected for their ability to grow well in cell culture) are adapted to utilize glucose anaerobically; for that reason, glucose deprivation or the use of 2-deoxyglucose, have been introduced. Metabolic poisons such as cyanide or antimycin A have also been used in place of hypoxia, while others have used mineral oil over the top of culture medium to achieve a very gradual depletion of oxygen in the culture medium. Duration of simulated ischemia varies according to cell line and must be empirically determined, with the objective to achieve 50% cell death after ischemia and reperfusion. This is chosen so that one can assess interventions that either exacerbate or mitigate the injury. In this system, autophagy can be measured by western blot for LC3 lipidation, or by fluorescence microscopy as described above. Flux measurements can be performed using parallel dishes treated with bafilomycin $A_1$ or chloroquine.

### 2.6.2 Slice Culture

Thin slices of viable tissue can be obtained using a vibratome. Canine cardiac tissue slices (350 µm) maintained viability for >7 days in culture [40]. Our unpublished experience with mouse cardiac tissue was that thin slices (50 µm) maintained viability for a few hours in culture; thicker slices lost mitochondrial function quickly, possibly due to limited perfusion of oxygen into the center of the slice. More work is needed to establish this model system as suitable for studies of I/R injury.

### 2.6.3 Ex Vivo Organ Perfusion

The isolated perfused heart model was developed by H. Newell Martin (1883) and Oscar Langendorff (1895), using retrograde perfusion (reverse flow into the aorta) with oxygenated buffer. The working heart model was introduced in the 1960s [41] which is more physiological but requires more technical sophistication. In either model, ischemia is achieved by stopping flow of oxygenated buffer, or in certain instances, by perfusing with buffer bubbled with $N_2$ rather than $O_2$. The ex vivo perfused heart model allows for a variety of measurements of cardiac contractility, histological staining (including TUNEL staining for apoptosis), as well as biochemical measurements (LDH release, troponin release, and western blotting of tissue lysates) and functional measurements of respiration and autophagy. However, it requires one animal per experiment compared to slice culture or primary cell culture in which one animal heart can provide enough material for several conditions. Autophagy can be assessed using the above described immunoblot methods, or the tissue can be homogenized and subcellular fractions can be recovered by differential sedimentation (nuclei, crude mitochondrial pellet, crude cytosol including ER and plasma membrane). This enables detection of translocation of cytosolic factors to the heavy membrane compartment, such as PRKN and SQSTM1/p62; flux measurements require parallel heart perfusions with chloroquine or bafilomycin $A_1$ or the use of fluorescent reporter constructs that will survive fixation and sectioning. We used the Langendorff model to conduct polysome profiling before ischemia, and end-ischemia, and during reperfusion; the methods are described in detail [42].

### 2.6.4 Working Heart Model

Here the heart is perfused in anterograde, with oxygenated buffer usually including a small amount of insulin and in many cases, fatty acids and amino acids in addition to glucose. The flow and pressure regulate afterload and preload; functional parameters can be analyzed with a pressure balloon while metabolic information ($O_2$ extraction, lactate, and other metabolites) can be measured in the perfusate. This model represents the most sophisticated *ex vivo* approach.

### 2.6.5 *In Vivo* Models

Rodent models of ischemic injury involve ligating the left coronary artery for a period of 30–60 min followed by reperfusion of 30–120 min. Such models allow for interventions such as ischemic preconditioning or postconditioning, drug administration, and longer reperfusion times. The typical readout is infarct size (measured by triphenyltetrazolium chloride staining of formalin-fixed cross sections) and cardiac contractility (determined by placing a pressure balloon into the LV), collagen deposition (scar formation, indicated by Masson-trichrome staining or picrosirius red), ventricular hypertrophy (LV wall thickness), apoptosis (TUNEL staining), as well as detailed biochemical analysis of tissue lysates or subcellular compartments (nucleus, heavy membrane fraction, crude cytosol). This approach allows for interrogation of ischemic and at-risk tissue (distal to the coronary occlusion) as well as the remote tissue (left ventricle (LV) above the suture line. Most studies dissect the LV separate from RV and atria. Autophagy and mitophagy measurements can be performed on tissue lysates (immunoblotting), subcellular fractions (to assess mitophagy markers in the heavy membrane fraction), or fixed tissue as above. Flux measurements require administration of lysosomal inhibitor (bafilomycin $A_1$ or chloroquine) for 3–4 h in a parallel set of animals.

> **Question**
> 1. How can we use LC3 to detect autophagy?
> 2. What is autophagy flux and why is it important?
> 3. What are the methods for measurement of mitophagy?

# 3 Implication of Autophagy in Cardiovascular Disorders

According to the World Health Organization, cardiovascular disease is the leading cause of mortality worldwide, killing 17.9 million people each year [43]. In fact, almost 50% of the adult population in the US suffers from some type of cardiovascular disease. Therefore, it is of great significance to address this group of disorders and find the underlying mechanisms so as to prevent them or develop therapeutic strategies. In this part, we will discuss several of these disorders to see how autophagy is implicated in their occurrence and/or treatment.

## 3.1 Diabetic Cardiomyopathy (DCM)

Cardiovascular problems are widely recognized as a major source of morbidity and mortality in diabetic patients. In fact, DCM develops as a result of dysregulated glucose and lipid metabolism, which causes increased oxidative stress along with activation of multiple inflammatory pathways that mediate cellular and extracellular injury, pathological cardiac remodeling, and diastolic and systolic dysfunction [44, 45]. Multiple biological processes have been proposed to explain the pathogenesis of DCM [46]. For example, insulin acts as a strong inducer of the nutrient-sensing MTOR pathway, which controls protein translation, cell growth and proliferation, and autophagy as described before [47]. Also, the involvement of AMPK pathway has been shown in this process. This pathway detects energy deficit and modifies cellular metabolism, mainly by activating glucose and fatty acid absorption and oxidation [48]. It has been demonstrated that in a diabetic cardiomyopathy animal model, the downregulated AMPK pathway impaired cardiac autophagy, resulting in cardiac failure. When treated with metformin, a well-characterized AMPK activator, animals showed an enhanced autophagic activity which prevented the development of DCM [49].

Generally, in diabetic and metabolic syndrome models, both cardiac autophagy and mitophagy are suppressed under pathological settings, most likely with negative repercussions [50, 51]. Consistent with this notion, cardiac autophagic flux is impaired in hearts of mice with type II diabetes. Indeed, autophagosome clearance is significantly suppressed in hearts of mice fed with a high-fat diet [52]. As a result, autophagy inducers such as resveratrol and trehalose were shown to improve cardiac function [53, 54].

While high glucose levels suppress autophagy by itself, they can also produce pro-autophagic mitochondrial malfunction and oxidative stress. As a result, depending on the context and cell type, excessive glucose exposure may suppress or activate autophagic flux [55, 56]. Importantly, an examination of the hearts of type II diabetes patients revealed greater levels of the autophagy markers LC3-II and BECN1 (beclin 1) as well as decreased SQSTM1/p62 levels, demonstrating that autophagy is activated by other mechanisms despite the high glucose levels in these patients; alternatively, this may reflect impaired flux [57].

## 3.2 Myocardial Ischemia/Reperfusion Injury

Each year in the United States, about one million people suffer from myocardial infarction. Besides, over 700,000 patients undergo cardiac surgery or circulatory arrest [58]. These patients all show a common feature which is I/R injury which is defined by an ischemic phase during which there is no blood flow to the heart, followed by reperfusion where blood flow is restored. Although both events cause cardiomyocyte stress, they differ in numerous respects, including food supply quantity, levels of ROS buildup, and oxygen availability. Still, it has been shown that cardiomyocyte autophagy contributes to both phases of cardiac injury—ischemia and reperfusion—but with contradicting effects [59]. In fact, one can say that autophagy plays a role of "double-edged sword" in the process of myocardial I/R.

On the one hand, autophagic flow increases significantly during ischemia, with autophagy playing a protective role at this time. AMPK, as an energy sensor, is activated by low ATP levels in this phase which in turn activates autophagy by direct activation of ULK1 or inhibition of MTORC1 [60]. It has been reported that downregulating autophagy genes *Atg5* and *Lamp2* significantly reduce the survival rate of cardiomyocytes subjected to hypoxia/anoxia. Besides, the blockage of autophagy with bafilomycin $A_1$ results in the exacerbation of infarction in a model of permanent coronary artery occlusion [61]. In line with these findings, starvation or treatment with pro-autophagic rapamycin or metformin reduced infarct size [62].

On the other hand, during reperfusion, autophagy is highly activated which results in cardiomyocyte death [60, 63]. As blood flow is restored, AMPK no longer is active, instead, BECN1 plays the vital role in this phase. *Becn1* heterozygous knockout mice demonstrate a reduced infarct size further pointing at a detrimental effect of autophagy during reperfusion [61].

## 3.3 Cardiac Hypertrophy

Increases in afterload (e.g., hypertension, aortic stenosis) as well as various other disease-related stressors cause hypertrophic development of the myocardium. This pathological stress-induced hypertrophy is an adaptive response to an increase in heart strain. At first, the increased heart mass serves to regulate wall stress, allowing the heart to operate normally at rest. This type of adaptation is known as compensatory hypertrophy. While some data suggests that cardiac hypertrophy may be advantageous in the short term, it can eventually lead to heart failure. In fact, if the stimulus for pathological hypertrophy is strong enough or long enough, the ventricle dilates, cardiac function declines, and the heart dies [64, 65].

Recent research has focused on the role of cardiomyocyte autophagy in the natural course of this disease. It has been shown that the degree of left ventricular hypertrophy correlates with increased autophagic flux in a mouse model of pressure overload induced by transverse aortic constriction (TAC) surgery [66]. Genetic studies have demonstrated that BECN1 overexpression or its haploinsufficiency, results in amplified or diminished pressure overload-induced heart failure, respectively [60]. Also, blocking autophagy by administration of 3-methyladenine reduced TAC damage [67].

Interestingly, there are studies arguing that some level of autophagy is required for an adaptive response. For example, in one such case, administering rapamycin prior to TAC greatly reduced the load-induced rise in heart weight by 70% [65]. Also, mice lacking the cardiac-specific Drp1 gene and defective mitophagy, showed mitochondrial malfunction, cardiac hypertrophy, and heart failure in response to pressure overload [68]. Mitophagy is also involved in cardiac hypertrophy following a heart attack. Indeed, Kubli et al. found cardiac hypertrophy and higher infarct diameters in PRKN-deficient animals following myocardial infarction [16].

## 3.4    Heart Failure

Heart failure (HF), the most common cardiovascular disorder, is a medical problem caused by a structural and/or functional cardiac defect that results in decreased cardiac output and/or higher intracardiac pressures at rest or under stress [69]. Atypical cardiac tissue remodeling is typically coupled with changes in the architecture of the failing heart. Fibrotic lesions, an increase in the size and shape of cardiomyocytes, and the presence of inflammatory cells are all common morphological changes seen in the myocardium of HF patients [69, 70].

While autophagy can efficiently control the myocardium at the physiological level, its influence on the pathological situation of HF is debatable. One study showed that cardiac-specific deletion of *Dnase2* (deoxyribonuclease 2, lysosomal) can increase mortality and cause severe myocarditis and dilated cardiomyopathy 10 days after aortic constriction to create pressure overload [26].

A recent study has shown that HF is associated with an isoform transition from PRKAA2/AMPK2 (protein kinase AMP-activated catalytic subunit alpha 2) to PRKAA1/ AMPK1. PRKAA2 expression is linked to increased mitophagy and a reduction in HF symptoms, whereas PRKAA1 expression is linked to an increase in HF severity. In cardiomyocytes of the failing heart, overexpression of PRKAA2 reverses the effects of HF and increases mitochondrial activity and mitophagy. These findings point to PRKAA2 as a crucial element in avoiding and alleviating HF symptoms, and they suggest the possibility that PRKAA2 therapy can repair cardiomyocytes during the beginning of HF [70]. In another study where heart tissue of HF patients were studied, autophagic cardiomyocytes are detected [71]. Besides, a mouse model of HF demonstrated upregulated lysosomal markers and autophagosomes containing organelles [72].

Therefore, one could conclude that autophagy plays an adaptive role in progressive heart failure and protects myocardial cells. In the late phase of heart failure, overinduction of autophagy may lead to depletion of mitochondria and other organelles [28].

**Question**

1. What is diabetic cardiomyopathy and how is it developed?
2. What happens during ischemia and how does it affect autophagy?
3. What is the role of autophagy during reperfusion?
4. What are different types of hypertrophy? how is autophagy involved in this process?
5. What is the significance of autophagy in heart failure?

## 4 Modulation of Autophagy in Treatment of Cardiovascular Disorders

As we discussed before, autophagy is involvement in the pathobiology of various heart disorders. Hence, modulating this process can be used as a potential therapeutic approach in preventing or limiting damage to the heart.

Rapamycin, as an MTOR inhibitor and inducer of autophagy, protects against hypertrophy and I/R injury [65, 73]; it can even improve the heart function of diabetic mice [74]. Therefore, this component can be used as a clinically relevant strategy to treat several heart conditions.

Resveratrol is another molecule with a protective potential. As an antioxidant, it activates autophagy pathway through SIRT1 (sirtuin 1) and AMPK which, in turn, improves the cardiac function in diabetic cardiomyopathy [75, 76]. This molecule is shown to inhibit hypertrophy and cardiac remodeling in a rat model [77]. These effects can also be achieved through upregulation of autophagy proteins such as BECN1, SQSTM1/p62 and LC3-II [78].

Metformin, as the first-line medication for type 2 diabetes, not only improves cardiac function in DCM, but also in hypertrophy through activation of autophagy via AMPK [49, 79]. Interestingly, in a mouse model of I/R, metformin induced cardioprotection through inhibition of autophagy through AKT (AKT serine/threonine kinase) signaling pathway in a non-diabetic context [80].

There are other cases where inhibiting autophagy improved cardiac function. For example, using 3-methyladenine to block autophagy in a rat model of overload exercise alleviated the heart injury [81]. Also, in a mouse model of hypertrophy, this reagent reduced fibrosis [67].

HDAC inhibitors are another group of pharmacological agents showing therapeutic potential in treating cardiovascular disorders [82]. In fact, by inducing autophagy, HDAC inhibitors can mitigate the effects of in a rabbit model of I/R injury [83].

**Question**

1. How does rapamycin treatment affect cardiovascular disorders?
2. What is resveratrol and which pathways are affected by this component?
3. What is the effect of metformin on autophagy?

> **Take-Home Messages**
> - Macroautophagy is a cell-autonomous process in which protein aggregates or damaged organelles are engulfed within a double-membrane autophagosome which is then routed to the lysosome.
> - This homeostatic process is particularly important to long-lived cells to preserve optimal cell function and cellular survival.
> - Macroautophagy diminishes with age and its failure is a factor in many age-related diseases.
> - Mitochondria-selective autophagy—mitophagy—is particularly important in cells that are dependent on oxidative phosphorylation for energy production and to support cellular functions, as in cardiomyocytes and neurons.
> - Fasting, metformin, and other small molecules promote autophagy, whereas nutritional overload can suppress autophagy.
> - Autophagic activity is dynamic, necessitating measures of flux.
> - Failure of autophagy can lead to cellular dysfunction and release of proinflammatory signals.

## 5    Introduction to Cardinal Articles

### 5.1    1 Mizushima 2004

This group made tremendous contributions to our understanding of autophagy, initially through studies in yeast, identifying the key proteins involved in the process, and subsequently through studies in mammalian cells and animals. This landmark paper describes transgenic mice expressing the autophagy protein LC3 fused to green fluorescent protein. Importantly, the redistribution of LC3 from diffuse cytosolic distribution to a punctate pattern revealed activation of autophagy. This greatly advanced the study of autophagy *in vivo*.

### 5.2    8 Hamacher-Brady 2006

This, along with the earlier paper by Vatner's group in 2005 [ncbi.nlm.nih.gov/articles/PMC1224362/], pointed to the existence of autophagy in mammalian heart tissue, and implicated it in cardioprotection, where it prevented progression to apoptosis.

### 5.3    13 Narendra 2010

This important study outlined the central role of PINK1 and Parkin in initiating mitochondrial autophagy. PINK1, normally imported to mitochondrial matrix and degraded, is retained on the outer mitochondrial membrane of depolarized mitochondria, where it phosphorylates targets facilitating the recruitment and retention of Parkin, an ubiquitin ligase essential for mitophagy.

## 5.4    17 Huang 2011

Although ischemic and pharmacological preconditioning had been investigated for a number of years, this study was the first to elucidate autophagy as a key mechanism of cardioprotection.

## 5.5    22 Hoshino 2013

This work established a mechanistic link between p53 and inhibition of mitophagy. Doxorubicin, an effective chemotherapeutic agent, exhibits considerable cardiotoxicity by activating p53. In this study the authors showed that p53 bound and sequestered Parkin, interfering with its translocation to damaged mitochondria. This prevented protective mitophagy and led to the accumulation of damaged mitochondria and cardiac injury.

## 5.6    26 Oka 2012

In this study, the authors took advantage of mice lacking the lysosomal enzyme DNase II needed to degrade mtDNA in lysosomes. They showed that in this setting of lysosomal dysfunction, mtDNA was released into the extracellular space where it triggered inflammatory signaling as a Damage-Associated Molecular Pattern. Subsequent studies have expanded the important connection between mitophagy dysfunction and inflammation.

## 5.7    37 Ferree 2013

The novel fluorescent probe consisting of Timer protein targeted to mitochondrial matrix has been used by several groups to interrogate mitochondrial biogenesis and turnover. This study elegantly employs MitoTimer to assess the relative age of mitochondria in neuronal outgrowths vs the soma.

## 5.8    48 Mihaylova 2011

This review article provides a comprehensive discussion of the AMPK signaling pathway, which promotes autophagy and suppresses cell growth. It is responsive to metabolic cues and in turn, regulates downstream metabolic pathways.

## 5.9  75 Wang 2014

Resveratrol, widely recognized for its potential anti-aging effects, activates sirtuins which in turn activate autophagy. In this study, resveratrol is used in the setting of streptozotocin-induced diabetes to protect the heart against hyperglycemia-induced oxidative stress. It is important to note that the authors performed flux measurements and multiple readouts to conclude that diabetes resulted in impaired autophagic flux in the heart, that mitochondria were a frequent cargo in autophagosomes, and that resveratrol restored autophagic flux.

## 5.10  83 Xie 2014

Histone deacetylases function in a wide variety of contexts to regulate gene expression. Here, Joe Hill's group used SAHA, a histone deacetylase inhibitor, in the setting of myocardial ischemia/reperfusion injury. They found that SAHA was protective through upregulation of autophagy, as silencing essential autophagy genes ATG5 or ATG7 abrogated the protective effects of SAHA.

**Answers to Questions**

**Part 1 Questions**

Question 1:

**Steps of Autophagy:**

1. **Initiation:** ULK1 complex activation triggers the formation of the phagophore.
   - **Marker:** ULK1 and phosphorylated ULK1.
2. **Membrane nucleation and expansion:** ATG proteins, including ATG12 and ATG5, drive phagophore expansion; LC3 is conjugated to the membrane.
   - **Marker:** LC3-I (cytosolic) converts to LC3-II (membrane-bound form).
3. **Cargo selection and engulfment:** Damaged organelles and proteins are tagged for engulfment with the help of receptors such as SQSTM1/p62.
   - **Marker:** SQSTM1/p62, ubiquitinated proteins.
4. **Autophagosome formation:** Closure of the phagophore forms a double-membrane autophagosome.
   - **Marker:** LC3-II remains associated with autophagosomal membranes.
5. **Lysosomal fusion and degradation:** The autophagosome fuses with the lysosome, where its contents are degraded by acidic hydrolases.
   - **Marker:** LAMP1 (lysosomal marker), degradation of LC3-II and SQSTM1/p62.

(continued)

(continued)

Question 2:

The majority of the heart's mass is made up of **cardiomyocytes**, which are responsible for contractile function and contain abundant mitochondria. Other contributing cell types include fibroblasts, vascular smooth muscle and endothelial cells of blood vessels, and tissue-resident macrophages.

Question 3:

**Mitophagy** is the selective removal of damaged or dysfunctional mitochondria through autophagy. It helps maintain mitochondrial quality and prevents the accumulation of defective mitochondria, which can:

- Generate excessive reactive oxygen species (ROS), causing oxidative stress.
- Produce ATP inefficiently, reducing energy availability, or even hydrolyze ATP in an effort to maintain mitochondrial membrane potential.
- Release mitochondrial DNA, HSPD1/HSP60, and other factors that trigger inflammation.

In the heart, mitophagy is critical for maintaining the functionality of cardiomyocytes. It plays a role in protecting the heart during stress conditions like ischemia-reperfusion injury and promotes recovery by clearing damaged mitochondria. Dysregulated mitophagy, however, can contribute to inflammation and cardiac dysfunction, including heart failure.

**Part 2 Questions**

Question 1:

**LC3 (microtubule associated protein 1 light chain 3):** A key marker for autophagy, particularly because its lipidation alters its behavior in SDS-PAGE.

- **LC3-I and LC3-II:** LC3 exists in two forms:
  - LC3-I (unlipidated)
  - LC3-II (lipidated), which migrates faster in SDS-PAGE.
- **Increase in LC3-II:** Indicates autophagosome formation but may also reflect impaired autophagic flux.
- **Immunoblotting:** Detects LC3-I and LC3-II bands to monitor autophagic activity.
- **Immunofluorescence:** Differentiates diffuse (inactive) LC3 distribution from punctate (autophagosome-associated) patterns.

(continued)

(continued)

Question 2:

- **Autophagy Flux:** Refers to the dynamic process of autophagy, encompassing the formation, maturation, and degradation of autophagosomes and their cargo.
- **Importance:**
  - Helps differentiate between **increased autophagosome formation** and **impaired autophagic degradation**.
  - Provides a more accurate measure of overall autophagic activity.
  - Essential to avoid misinterpreting an increase in autophagosome markers (e.g., LC3-II) as increased autophagy when it might be due to impaired flux.
- **Flux Measurement Techniques:**
  - **Lysosomal Blockade:** Using inhibitors like chloroquine or bafilomycin $A_1$ to assess changes in LC3-II levels.
  - **Tandem Fluorescent LC3 Constructs:** Distinguish autophagosomes (yellow) from autolysosomes (red-only).

Question 3:
**Mitochondrial Mass Analysis:**

- Immunoblotting for mitochondrial proteins (e.g., TOMM20, COX) relative to loading controls.
- Quantitative PCR to calculate mitochondrial DNA:nuclear DNA ratio.**MitoTimer:**

- A fluorescent protein that changes from green to red over time, allowing real-time monitoring of mitochondrial turnover.
- Specific for cardiac tissue via transgenic mouse models.**Tandem mCherry-GFP Constructs:**

- Targeted to mitochondria using COX8.
- Green fluorescence quenched in acidic lysosomes, while red fluorescence persists, enabling differentiation between mitochondria in autophagosomes and aut olysosomes.**Mito-Keima:**

- A pH-sensitive fluorescent protein with excitation wavelength shifts indicating the transition of mitochondria from the cytosol to the acidic lysosome.
- Suitable for live-cell imaging.

(continued)

(continued)
**Part 3 Questions**
Question 1:
Diabetic cardiomyopathy (DCM) is a cardiac disorder independent of hypertension or coronary artery disease, often associated with diabetes mellitus. It is characterized by structural and functional changes in the heart, including diastolic dysfunction, fibrosis, and left ventricular hypertrophy. DCM develops through dysregulated glucose and lipid metabolism, which increases oxidative stress, mitochondrial dysfunction, and inflammation. These pathological processes impair autophagy, a cellular mechanism critical for removing damaged organelles and proteins, further contributing to the progression of DCM. The downregulation of energy-sensing pathways like AMPK exacerbates these effects, leading to cardiac dysfunction.

Question 2:
Ischemia occurs when blood flow to the heart is restricted, leading to oxygen and nutrient deprivation. During this period, the energy-sensing AMPK pathway is activated, promoting autophagy as a survival mechanism. Autophagy helps remove damaged cellular components, providing essential nutrients and energy to maintain cellular function. However, excessive autophagy during prolonged ischemia may degrade essential cellular structures, exacerbating injury. Thus, autophagy plays a dual role, balancing protective effects and potential cellular damage.

Question 3:
Reperfusion refers to restoring blood flow after ischemia, which paradoxically induces additional injury due to oxidative stress and inflammation. During reperfusion, autophagy initially protects cardiac cells by removing damaged organelles and mitigating oxidative stress. However, overactive or impaired autophagy during this stage may lead to excessive degradation of functional cellular components, exacerbating cell death and contributing to reperfusion injury. Modulating autophagy to balance its protective and harmful effects is a therapeutic target in myocardial ischemia/reperfusion injury.

Question 4:
Cardiac hypertrophy is the enlargement of the heart muscle in response to stress or injury and can be classified into two types:

- **Physiological hypertrophy:** A beneficial adaptation, often seen in athletes, characterized by balanced growth of cardiomyocytes and enhanced cardiac function. Autophagy supports physiological hypertrophy by maintaining cellular quality control.
- **Pathological hypertrophy:** A maladaptive response due to chronic pressure overload, myocardial infarction, or other stressors, leading to fibrosis and cardiac dysfunction. Dysregulated or impaired autophagy contributes to the accumulation of damaged organelles and proteins, exacerbating pathological remodeling and dysfunction.

(continued)

(continued)

Question 5:

In heart failure, autophagy is critical for maintaining cardiomyocyte function by removing damaged cellular components and preventing toxic accumulation. Impaired autophagy contributes to cellular stress, inflammation, and mitochondrial dysfunction, which exacerbate heart failure. While basal autophagy is protective, excessive or dysregulated autophagy can lead to cardiomyocyte death and worsen cardiac function. Targeting autophagy through pharmacological or genetic approaches offers potential therapeutic strategies for managing heart failure.

**Part 4 Questions**

Question 1:

Rapamycin, an MTOR inhibitor, induces autophagy and has protective effects against various cardiovascular disorders. It:

- Protects against cardiac hypertrophy and ischemia/reperfusion (I/R) injury.
- Improves heart function in diabetic mice, making it a clinically relevant therapeutic strategy for several heart conditions.

Question 2:

Resveratrol is an antioxidant with protective potential for cardiovascular health. It:

- Activates the autophagy pathway through **SIRT1 (sirtuin 1)** and **AMPK**.
- Improves cardiac function in diabetic cardiomyopathy.
- Inhibits hypertrophy and cardiac remodeling in rat models.
- Upregulates autophagy-related proteins like **BECN1, SQSTM1/p62**, and **LC3-II** to achieve its effects.

Question 3:

Metformin, a first-line medication for type 2 diabetes, has context-dependent effects on autophagy:

- **Activation of autophagy via AMPK**: Improves cardiac function in diabetic cardiomyopathy (DCM) and hypertrophy.
- **Inhibition of autophagy via AKT signaling**: Provides cardioprotection in a mouse model of ischemia/reperfusion injury in a non-diabetic context.

## References


1. Mizushima N, Yamamoto A, Matsui M, Yoshimori T, Ohsumi Y. In vivo analysis of autophagy in response to nutrient starvation using transgenic mice expressing a fluorescent autophagosome marker. Mol Biol Cell. 2004;15(3):1101–11.
2. Kim J, Kundu M, Viollet B, Guan KL. AMPK and mTOR regulate autophagy through direct phosphorylation of Ulk1. Nat Cell Biol. 2011;13(2):132–41.
3. Kirisako T, Ichimura Y, Okada H, Kabeya Y, Mizushima N, Yoshimori T, et al. The reversible modification regulates the membrane-binding state of Apg8/Aut7 essential for autophagy and the cytoplasm to vacuole targeting pathway. J Cell Biol. 2000;151(2):263–76.
4. Scherz-Shouval R, Shvets E, Fass E, Shorer H, Gil L, Elazar Z. Reactive oxygen species are essential for autophagy and specifically regulate the activity of Atg4. EMBO J. 2007;26(7):1749–60.
5. Matteoni R, Kreis TE. Translocation and clustering of endosomes and lysosomes depends on microtubules. J Cell Biol. 1987;105(3):1253–65.
6. Yamamoto A, Tagawa Y, Yoshimori T, Moriyama Y, Masaki R, Tashiro Y. Bafilomycin A1 prevents maturation of autophagic vacuoles by inhibiting fusion between autophagosomes and lysosomes in rat hepatoma cell line, H-4-II-E cells. Cell Struct Funct. 1998;23(1):33–42.
7. Solhjoo S, O'Rourke B. Mitochondrial instability during regional ischemia-reperfusion underlies arrhythmias in monolayers of cardiomyocytes. J Mol Cell Cardiol. 2015;78:90–9.
8. Hamacher-Brady A, Brady NR, Gottlieb RA, Gustafsson AB. Autophagy as a protective response to Bnip3-mediated apoptotic signaling in the heart. Autophagy. 2006;2(4):307–9.
9. Soubannier V, McLelland GL, Zunino R, Braschi E, Rippstein P, Fon EA, et al. A vesicular transport pathway shuttles cargo from mitochondria to lysosomes. Curr Biol: CB. 2012;22(2):135–41.
10. Zhao Q, Wang J, Levichkin IV, Stasinopoulos S, Ryan MT, Hoogenraad NJ. A mitochondrial specific stress response in mammalian cells. EMBO J. 2002;21(17):4411–9.
11. Liu J, He X, Zheng S, Zhu A, Wang J. The mitochondrial unfolded protein response: a novel protective pathway targeting cardiomyocytes. Oxidative Med Cell Longev. 2022;2022:6430342.
12. Zhang J, Ney PA. Role of BNIP3 and NIX in cell death, autophagy, and mitophagy. Cell Death Differ. 2009;16(7):939–46.
13. Narendra DP, Jin SM, Tanaka A, Suen DF, Gautier CA, Shen J, et al. PINK1 is selectively stabilized on impaired mitochondria to activate Parkin. PLoS Biol. 2010;8(1):e1000298.
14. Hamacher-Brady A, Brady NR, Gottlieb RA. Enhancing macroautophagy protects against ischemia/reperfusion injury in cardiac myocytes. J Biol Chem. 2006;281(40):29776–87.
15. Depre C, Vatner SF. Mechanisms of cell survival in myocardial hibernation. Trends Cardiovasc Med. 2005;15(3):101–10.
16. Kubli DA, Zhang X, Lee Y, Hanna RA, Quinsay MN, Nguyen CK, et al. Parkin protein deficiency exacerbates cardiac injury and reduces survival following myocardial infarction. J Biol Chem. 2013;288(2):915–26.
17. Huang C, Andres AM, Ratliff EP, Hernandez G, Lee P, Gottlieb RA. Preconditioning involves selective mitophagy mediated by Parkin and p62/SQSTM1. PLoS One. 2011;6(6):e20975.
18. Zesiewicz TA, Strom JA, Borenstein AR, Hauser RA, Cimino CR, Fontanet HL, et al. Heart failure in Parkinson's disease: analysis of the United States medicare current beneficiary survey. Parkinsonism Relat Disord. 2004;10(7):417–20.
19. Leshchinski T, Rozani V, Giladi N, Bitan M, Peretz C. Incidence of cardiovascular morbidity among Parkinson's disease patients; a large-scale cohort study in a 16-year time window around disease onset. Parkinsonism Relat Disord. 2023;114:105795.
20. Song M, Gong G, Burelle Y, Gustafsson ÅB, Kitsis RN, Matkovich SJ, et al. Interdependence of Parkin-mediated mitophagy and mitochondrial fission in adult mouse hearts. Circ Res. 2015;117(4):346–51.

21. Kubli DA, Quinsay MN, Gustafsson AB. Parkin deficiency results in accumulation of abnormal mitochondria in aging myocytes. Commun Integr Biol. 2013;6(4):e24511.

22. Hoshino A, Mita Y, Okawa Y, Ariyoshi M, Iwai-Kanai E, Ueyama T, et al. Cytosolic p53 inhibits Parkin-mediated mitophagy and promotes mitochondrial dysfunction in the mouse heart. Nat Commun. 2013;4:2308.

23. Kubli DA, Cortez MQ, Moyzis AG, Najor RH, Lee Y, Gustafsson ÅB. PINK1 is dispensable for mitochondrial recruitment of Parkin and activation of mitophagy in cardiac myocytes. PLoS One. 2015;10(6):e0130707.

24. Zhong Z, Umemura A, Sanchez-Lopez E, Liang S, Shalapour S, Wong J, et al. NF-κB restricts inflammasome activation via elimination of damaged mitochondria. Cell. 2016;164(5):896–910.

25. Sliter DA, Martinez J, Hao L, Chen X, Sun N, Fischer TD, et al. Parkin and PINK1 mitigate STING-induced inflammation. Nature. 2018;561(7722):258–62.

26. Oka T, Hikoso S, Yamaguchi O, Taneike M, Takeda T, Tamai T, et al. Mitochondrial DNA that escapes from autophagy causes inflammation and heart failure. Nature. 2012;485(7397):251–5.

27. Marek-Iannucci S, Ozdemir AB, Moreira D, Gomez AC, Lane M, Porritt RA, et al. Autophagy-mitophagy induction attenuates cardiovascular inflammation in a murine model of Kawasaki disease vasculitis. JCI Insight. 2021;6(18):e151981.

28. Klionsky DJ, Abdel-Aziz AK, Abdelfatah S, Abdellatif M, Abdoli A, Abel S, et al. Guidelines for the use and interpretation of assays for monitoring autophagy (4th edition)(1). Autophagy. 2021;17(1):1–382.

29. Gottlieb RA, Andres AM, Sin J, Taylor DP. Untangling autophagy measurements: all fluxed up. Circ Res. 2015;116(3):504–14.

30. Boukhalfa A, Robinson SR, Meola DM, Robinson NA, Ling LA, LaMastro JN, et al. Using cultured canine cardiac slices to model the autophagic flux with doxorubicin. PLoS One. 2023;18(3):e0282859.

31. Castillo K, Valenzuela V, Oñate M, Hetz C. A molecular reporter for monitoring autophagic flux in nervous system in vivo. Methods Enzymol. 2017;588:109–31.

32. Wang H, Ni HM, Chao X, Ma X, Rodriguez YA, Chavan H, et al. Double deletion of PINK1 and Parkin impairs hepatic mitophagy and exacerbates acetaminophen-induced liver injury in mice. Redox Biol. 2019;22:101148.

33. Sun N, Malide D, Liu J, Rovira II, Combs CA, Finkel T. A fluorescence-based imaging method to measure in vitro and in vivo mitophagy using mt-Keima. Nat Protoc. 2017;12(8):1576–87.

34. Chevrier M, Brakch N, Celine L, Genty D, Ramdani Y, Moll S, et al. Autophagosome maturation is impaired in Fabry disease. Autophagy. 2010;6(5):589–99.

35. Nyfeler B, Bergman P, Wilson CJ, Murphy LO. Quantitative visualization of autophagy induction by mTOR inhibitors. Methods Mol Biol. 2012;821:239–50.

36. Terskikh A, Fradkov A, Ermakova G, Zaraisky A, Tan P, Kajava AV, et al. "Fluorescent timer": protein that changes color with time. Science. 2000;290(5496):1585–8.

37. Ferree AW, Trudeau K, Zik E, Benador IY, Twig G, Gottlieb RA, et al. MitoTimer probe reveals the impact of autophagy, fusion, and motility on subcellular distribution of young and old mitochondrial protein and on relative mitochondrial protein age. Autophagy. 2013;9(11):1887–96.

38. Gottlieb RA, Stotland A. MitoTimer: a novel protein for monitoring mitochondrial turnover in the heart. J Mol Med (Berl). 2015;93(3):271–8.

39. Chen T, Vunjak-Novakovic G. In vitro models of ischemia-reperfusion injury. Regener Eng Transl Med. 2018;4(3):142–53.

40. Meki MH, Miller JM, Mohamed TMA. Heart slices to model cardiac physiology. Front Pharmacol. 2021;12:617922.

41. Neely JR, Liebermeister H, Battersby EJ, Morgan HE. Effect of pressure development on oxygen consumption by isolated rat heart. Am J Phys. 1967;212(4):804–14.

42. Stastna M, Thomas A, Germano J, Pourpirali S, Van Eyk JE, Gottlieb RA. Dynamic proteomic and miRNA analysis of polysomes from isolated mouse heart after Langendorff perfusion. J Vis Exp: JoVE. 2018;138:58079.

43. https://www.who.int/health-topics/cardiovascular-diseases#tab=tab_1.

44. Tan Y, Zhang Z, Zheng C, Wintergerst KA, Keller BB, Cai L. Mechanisms of diabetic cardiomyopathy and potential therapeutic strategies: preclinical and clinical evidence. Nat Rev Cardiol. 2020;17(9):585–607.

45. Jia G, Hill MA, Sowers JR. Diabetic cardiomyopathy: an update of mechanisms contributing to this clinical entity. Circ Res. 2018;122(4):624–38.

46. Al Hroob AM, Abukhalil MH, Hussein OE, Mahmoud AM. Pathophysiological mechanisms of diabetic cardiomyopathy and the therapeutic potential of epigallocatechin-3-gallate. Biomed Pharmacother. 2019;109:2155–72.

47. Jia G, DeMarco VG, Sowers JR. Insulin resistance and hyperinsulinaemia in diabetic cardiomyopathy. Nat Rev Endocrinol. 2016;12(3):144–53.

48. Mihaylova MM, Shaw RJ. The AMPK signalling pathway coordinates cell growth, autophagy and metabolism. Nat Cell Biol. 2011;13(9):1016–23.

49. Xie Z, Lau K, Eby B, Lozano P, He C, Pennington B, et al. Improvement of cardiac functions by chronic metformin treatment is associated with enhanced cardiac autophagy in diabetic OVE26 mice. Diabetes. 2011;60(6):1770–8.

50. Xie Z, He C, Zou M-H. AMP-activated protein kinase modulates cardiac autophagy in diabetic cardiomyopathy. Autophagy. 2011;7(10):1254–5.

51. Li ZL, Woollard JR, Ebrahimi B, Crane JA, Jordan KL, Lerman A, et al. Transition from obesity to metabolic syndrome is associated with altered myocardial autophagy and apoptosis. Arterioscler Thromb Vasc Biol. 2012;32(5):1132–41.

52. Jaishy B, Zhang Q, Chung HS, Riehle C, Soto J, Jenkins S, et al. Lipid-induced NOX2 activation inhibits autophagic flux by impairing lysosomal enzyme activity. J Lipid Res. 2015;56(3):546–61.

53. Sulaiman M, Matta MJ, Sunderesan NR, Gupta MP, Periasamy M, Gupta M. Resveratrol, an activator of SIRT1, upregulates sarcoplasmic calcium ATPase and improves cardiac function in diabetic cardiomyopathy. Am J Phys Heart Circ Phys. 2010;298(3):H833–H43.

54. Liu Y, Wu S, Zhao Q, Yang Z, Yan X, Li C, et al. Trehalose ameliorates diabetic cardiomyopathy: role of the PK2/PKR pathway. Oxidative Med Cell Longev. 2021;2021:6779559.

55. Kim K-A, Shin Y-J, Akram M, Kim E-S, Choi K-W, Suh H, et al. High glucose condition induces autophagy in endothelial progenitor cells contributing to angiogenic impairment. Biol Pharm Bull. 2014;37(7):1248–52.

56. Kobayashi S, Xu X, Chen K, Liang Q. Suppression of autophagy is protective in high glucose-induced cardiomyocyte injury. Autophagy. 2012;8(4):577–92.

57. Munasinghe PE, Riu F, Dixit P, Edamatsu M, Saxena P, Hamer NSJ, et al. Type-2 diabetes increases autophagy in the human heart through promotion of Beclin-1 mediated pathway. Int J Cardiol. 2016;202:13–20.

58. Ebermann L, Wika S, Klumpe I, Hammer E, Klingel K, Lassner D, et al. The mitochondrial respiratory chain has a critical role in the antiviral process in Coxsackievirus B3-induced myocarditis. Lab Investig. 2012;92(1):125–34.

59. Matsui Y, Takagi H, Qu X, Abdellatif M, Sakoda H, Asano T, et al. Distinct roles of autophagy in the heart during ischemia and reperfusion: roles of AMP-activated protein kinase and Beclin 1 in mediating autophagy. Circ Res. 2007;100(6):914–22.

60. Ma X, Liu H, Murphy JT, Foyil SR, Godar RJ, Abuirqeba H, et al. Regulation of the transcription factor EB-PGC1alpha axis by beclin-1 controls mitochondrial quality and cardiomyocyte death under stress. Mol Cell Biol. 2015;35(6):956–76.

61. Kanamori H, Takemura G, Goto K, Maruyama R, Tsujimoto A, Ogino A, et al. The role of autophagy emerging in postinfarction cardiac remodelling. Cardiovasc Res. 2011;91(2):330–9.

62. Aisa Z, Liao GC, Shen XL, Chen J, Li L, Jiang SB. Effect of autophagy on myocardial infarction and its mechanism. Eur Rev Med Pharmacol Sci. 2017;21(16):3705–13.

63. Du J, Li Y, Zhao W. Autophagy and myocardial ischemia. Adv Exp Med Biol. 2020;1207:217–22.

64. Schiattarella GG, Hill JA. Therapeutic targeting of autophagy in cardiovascular disease. J Mol Cell Cardiol. 2016;95:86–93.

65. McMullen JR, Sherwood MC, Tarnavski O, Zhang L, Dorfman AL, Shioi T, Izumo S. Inhibition of mTOR signaling with rapamycin regresses established cardiac hypertrophy induced by pressure overload. Circulation. 2004;109(24):3050–5.

66. Zhu H, Tannous P, Johnstone JL, Kong Y, Shelton JM, Richardson JA, et al. Cardiac autophagy is a maladaptive response to hemodynamic stress. J Clin Invest. 2007;117(7):1782–93.

67. Weng LQ, Zhang WB, Ye Y, Yin PP, Yuan J, Wang XX, et al. Aliskiren ameliorates pressure overload-induced heart hypertrophy and fibrosis in mice. Acta Pharmacol Sin. 2014;35(8):1005–14.

68. Shirakabe A, Zhai P, Ikeda Y, Saito T, Maejima Y, Hsu CP, et al. Drp1-dependent mitochondrial autophagy plays a protective role against pressure overload-induced mitochondrial dysfunction and heart failure. Circulation. 2016;133(13):1249–63.

69. Bielawska M, Warszynska M, Stefanska M, Blyszczuk P. Autophagy in heart failure: insights into mechanisms and therapeutic implications. J Cardiovasc Dev Dis. 2023;10(8):352.

70. Wang R, Tan J, Chen T, Han H, Tian R, Tan Y, et al. ATP13A2 facilitates HDAC6 recruitment to lysosome to promote autophagosome–lysosome fusion. J Cell Biol. 2018;218(1):267–84.

71. Knaapen MW, Davies MJ, De Bie M, Haven AJ, Martinet W, Kockx MM. Apoptotic versus autophagic cell death in heart failure. Cardiovasc Res. 2001;51(2):304–12.

72. Akazawa H, Komazaki S, Shimomura H, Terasaki F, Zou Y, Takano H, et al. Diphtheria toxin-induced autophagic cardiomyocyte death plays a pathogenic role in mouse model of heart failure. J Biol Chem. 2004;279(39):41095–103.

73. Das S, Ferlito M, Kent OA, Fox-Talbot K, Wang R, Liu D, et al. Nuclear miRNA regulates the mitochondrial genome in the heart. Circ Res. 2012;110:1596–603.

74. Das S, Bedja D, Campbell N, Dunkerly B, Chenna V, Maitra A, et al. miR-181c regulates the mitochondrial genome, bioenergetics, and propensity for heart failure in vivo. PLoS One. 2014;9:1–9.

75. Wang B, Yang Q, Sun YY, Xing YF, Wang YB, Lu XT, et al. Resveratrol-enhanced autophagic flux ameliorates myocardial oxidative stress injury in diabetic mice. J Cell Mol Med. 2014;18(8):1599–611.

76. Zordoky BNM, Robertson IM, Dyck JRB. Preclinical and clinical evidence for the role of resveratrol in the treatment of cardiovascular diseases. Biochim Biophys Acta (BBA) – Mol Basis Dis. 2015;1852(6):1155–77.

77. Guan P, Sun ZM, Wang N, Zhou J, Luo LF, Zhao YS, Ji ES. Resveratrol prevents chronic intermittent hypoxia-induced cardiac hypertrophy by targeting the PI3K/AKT/mTOR pathway. Life Sci. 2019;233:116748.

78. Fan S, Hu Y, You Y, Xue W, Chai R, Zhang X, et al. Role of resveratrol in inhibiting pathological cardiac remodeling. Front Pharmacol. 2022;13:924473.

79. Li Y, Chen C, Yao F, Su Q, Liu D, Xue R, et al. AMPK inhibits cardiac hypertrophy by promoting autophagy via mTORC1. Arch Biochem Biophys. 2014;558:79–86.

80. Huang KY, Que JQ, Hu ZS, Yu YW, Zhou YY, Wang L, et al. Metformin suppresses inflammation and apoptosis of myocardiocytes by inhibiting autophagy in a model of ischemia-reperfusion injury. Int J Biol Sci. 2020;16(14):2559–79.

81. Liu H, Lei H, Shi Y, Wang JJ, Chen N, Li ZH, et al. Autophagy inhibitor 3-methyladenine alleviates overload-exercise-induced cardiac injury in rats. Acta Pharmacol Sin. 2017;38(7):990–7.
82. Gillette TG. HDAC inhibition in the heart. Circulation. 2021;143(19):1891–3.
83. Xie M, Kong Y, Tan W, May H, Battiprolu PK, Pedrozo Z, et al. Histone deacetylase inhibition blunts ischemia/reperfusion injury by inducing cardiomyocyte autophagy. Circulation. 2014;129(10):1139–51.

# Chaperone-Mediated Autophagy in Health and Disease

Susmita Kaushik [iD]

## Abbreviations

| | |
|---|---|
| CA | CMA activator |
| CAMK2/Camkii | Calcium/calmodulin dependent protein kinase II |
| CMA | Chaperone-mediated autophagy |
| GBA | Glucosylceramidase beta |
| HSPA8/HSC70 | Heat shock protein family A (Hsp70) member 8 |
| LAMP2A | Lysosomal associated membrane protein 2A |
| LRRK2 | Leucine rich repeat kinase 2 |
| PARK7 | Parkinsonism associated deglycase |
| TARDBP/TDP-43 | TAR DNA binding protein |
| UCHL1 | Ubiquitin C-terminal hydrolase L1 |
| VPS35 | VPS35 retromer complex component |

S. Kaushik (✉)
Department of Developmental and Molecular Biology, Albert Einstein College of Medicine, Bronx, NY, USA

Institute for Aging Research, Albert Einstein College of Medicine, Bronx, NY, USA
e-mail: susmita.kaushik@einsteinmed.edu

**What Will You Learn in This Chapter**

In this chapter, you will learn what chaperone-mediated autophagy (CMA) is, how it is different from other forms of autophagy, and what the CMA substrates are. You will also learn about the CMA effectors and regulators and how they function together to modulate CMA activity under physiological conditions. You will learn how CMA plays an essential part in protein quality control in every cell and tissue in addition to the proteasome and other forms of autophagy. While damaged proteins are degraded by CMA, you will learn that, by degrading functional undamaged proteins, CMA regulates diverse cellular processes in which those proteins participate in. You will learn about the functions of CMA in different tissues, and how malfunction of CMA contributes to specific disease conditions and aging. Finally, you will learn that restoration of CMA in old organisms serves as an anti-aging intervention and a potential therapeutic tool against age-related disorders such as neurodegeneration, retinal degeneration, and atherosclerosis.

# 1    An Introduction to CMA

Autophagy is a cellular recycling program that occurs in lysosomes, the only degradative organelle containing digestive enzymes to breakdown every cellular macromolecule. Chaperone-mediated autophagy (CMA) is one of the three major types of autophagy in mammalian cells in addition to macroautophagy and microautophagy [1] with the latter two processes relying on vesicle formation and entrapment of cargo inside a vesicle [1, 2].

CMA was the first selective form of autophagy discovered—in times when lysosomal degradation was thought to be random and mostly digest cellular components "in bulk". CMA is the degradation in lysosomes of soluble proteins that bear a chaperone-recognized specific sequence tag (KFERQ-like motif) (Fig. 1). The key steps in this pathway are (1) substrate recognition: HSPA8/HSC70 (heat shock protein family A (Hsp70) member 8) binds the KFERQ-tag on the substrate protein; (2) binding: substrate-chaperone complex binds to the essential membrane receptor LAMP2A (lysosomal associated membrane protein 2A) triggering the multimerization of the LAMP2A receptor; (3) translocation: substrate protein unfolds, translocates across the lysosomal membrane assisted by lysosomal HSPA8; (4) degradation: substrate proteins get degraded by the proteases inside the lysosomes; (5) disassembly: multimer translocation complex disassembles into LAMP2A monomer to undergo another round of substrate translocation [3]. The distinguishing features of CMA that make this type of autophagy unique are: (i) selectivity towards its substrates and (ii) one-by-one translocation of the unfolded substrates inside lysosomes across the membrane via the receptor LAMP2A.

Broadly, CMA plays a dual role in maintaining cellular health (Fig. 2). (i) Protein quality control: In addition to the proteasome and other forms of autophagy, CMA is a major

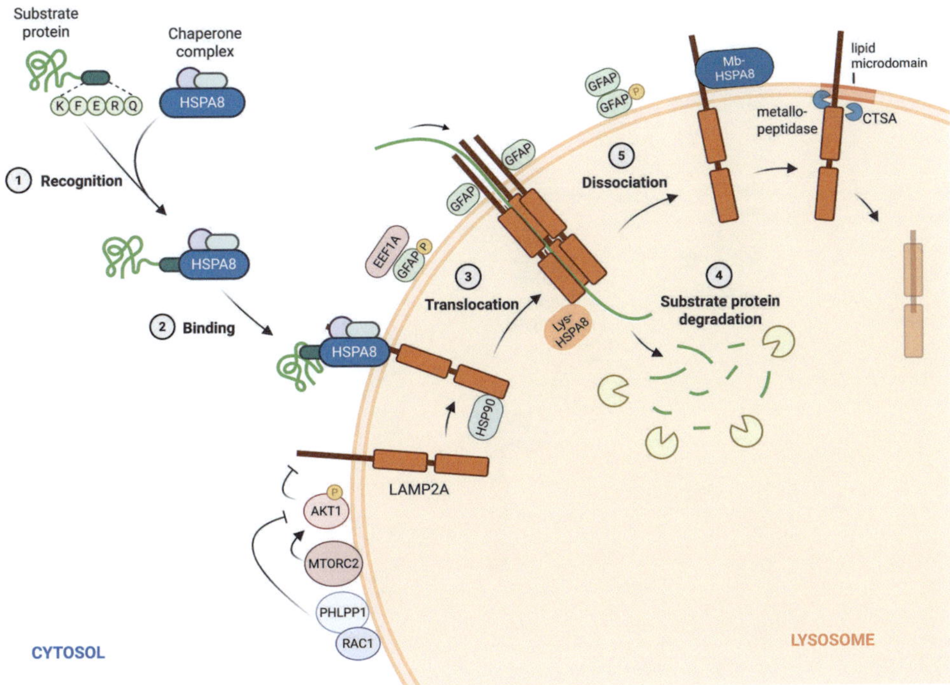


**Fig. 1** Chaperone-mediated autophagy—pathway steps and lysosomal effectors and regulators. Steps of chaperone-mediated autophagy (CMA): (1) Recognition of KFERQ-like motif in the substrate proteins by HSPA8; (2) Binding of the substrate-chaperone complex to LAMP2A, the CMA receptor; (3) Translocation of the substrate protein after unfolding via the CMA translocation complex mediated by lysosomal HSPA8; (4) Substrate protein degradation by abundant lysosomal proteases; (5) Dissociation of LAMP2A from the translocation complex to facilitate new rounds of substrate binding and translocation. Figure shows several lysosomal regulators that affect the dynamic cycles of assembly-disassembly of the CMA translocation complex and the stability of LAMP2A monomer/multimers. See text for details. Turnover of LAMP2A occurs in lysosomal lipid microdomains by the dual action of a metallopeptidase and CTSA. LAMP2A lysosomal associated membrane protein 2A, HSPA8 heat shock protein family A (Hsp70) member 8, Mb-HSPA8 membrane-associated HSPA8, Lys-HSPA8 lysosomal HSPA8, HSP90 heat shock protein 90, EEF1A eukaryotic translation elongation factor 1 alpha, GFAP glial fibrillary acidic protein, AKT1 AKT serine/threonine kinase 1, PHLPP1 PH domain and leucine rich repeat protein phosphatase 1, RAC1 Rac family small GTPase 1, MTORC2 MTOR complex 2, CTSA cathepsin A. (Created with BioRender.com)

component of the proteostasis network. CMA degrades soluble prone-to-aggregate proteins thereby preventing them from aggregation. Cells respond to stress conditions that increase the cellular load of damaged proteins by activating CMA as a first line of defense to degrade/eliminate toxic proteins before they can aggregate and cause further cellular damage. (ii) Proteome remodeling: CMA selectively degrades soluble and functional (undamaged) proteins. By degrading specific proteins at the 'right' time, CMA decreases

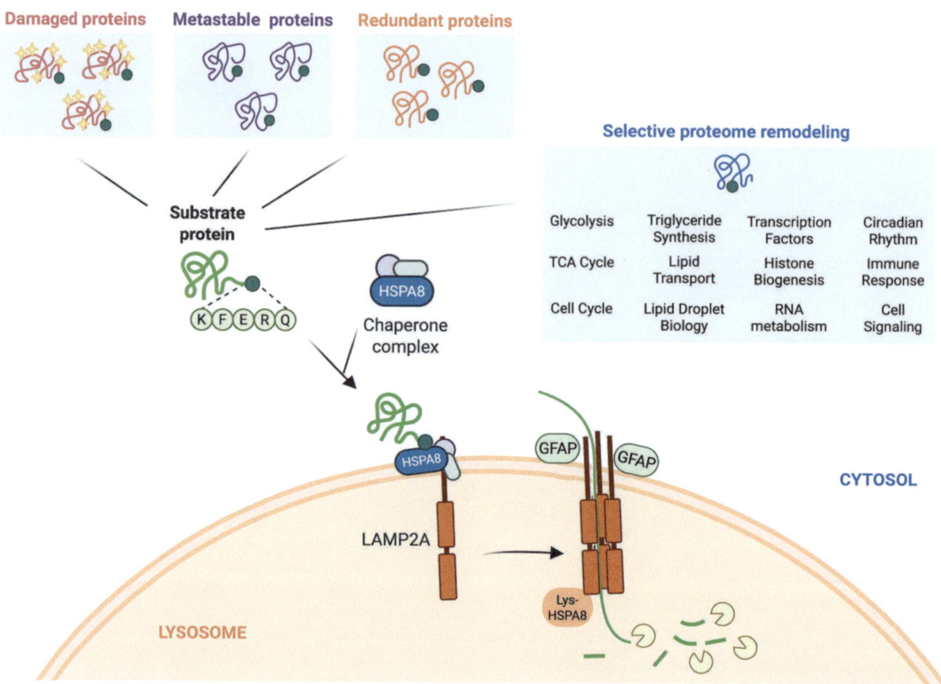



**Fig. 2** Functions of chaperone-mediated autophagy. CMA is an integral part of the protein quality control as it degrades damaged proteins and prone-to-aggregate metastable proteins. By degrading redundant proteins in times of extended starvation, CMA contributes amino acids. CMA plays a major role in selective proteome remodeling. Depending on the protein substrate degraded, CMA regulates several cellular processes such as the ones indicated here. Proteins involved in glycolysis, TCA cycle, triglyceride synthesis, lipid transport, lipid droplet biology, transcription, histone biogenesis, RNA biology, circadian rhythm, immune response, cell cycle and cell signaling are CMA substrates. (Created with BioRender.com)

the rate of the pathway that the substrate is involved in. For instance, by degrading glycolytic enzymes, many of which are CMA substrates in several different cell types, CMA can dampen the rate of glycolysis as per cellular need. By doing so, CMA can switch on/switch off fundamental cellular processes, as discussed in this chapter.

This chapter summarizes the molecular machinery that controls the CMA pathway, where this pathway is present, conditions that stimulate CMA and its degradative function, and the pathological consequences when CMA is dysfunctional.

## 2    CMA Substrates

Selectivity in the substrates that are degraded by CMA was known well before the discovery that the LAMP2A protein is the receptor for this pathway, and before the naming of the pathway itself [4, 5].

Selectivity of a given substrate for degradation by CMA is attained by the recognition of a consensus tag on the substrate protein by the cytosolic chaperone HSPA8. Only those proteins that have a specific KFERQ-like (Lys-Phe-Glu-Arg-Gln) targeting motif in their amino acid sequence are amenable to CMA degradation. The KFERQ-like motif is necessary and sufficient for targeting a protein for CMA degradation [5]. This pentapeptide motif is present in the primary sequence of the substrate and is compositional in nature, which means that the physicochemical properties of the motif sequence itself are more important than the exact sequence itself. The pentapeptide motif consists of glutamine flanked on either side by a tetrapeptide consisting of one or two positively charged residues (K and R), one or two hydrophobic residues (F, L, I or V) and a negatively charged residue (E or D) [4]. Canonical motifs are those that are present as the 'KFERQ' sequence in the substrate, while putative motifs are those that are generated by posttranslational modifications (such as phosphorylation or acetylation) to the protein (for example, KFSRQ upon phosphorylation of S), which then complete the motif [6].

Forty-six percent of the proteins in the human proteome bear at least one canonical KFERQ-like motif and 29% additional proteins harbor putative KFERQ-like motifs [6]. Furthermore, enrichment analysis revealed that canonical motifs are enriched in cytoskeleton-associated gene ontology terms (cytoskeleton organization, cell projection organization), proteins with phosphorylation-generated motifs frequently associate with transport-related terms, and proteins with acetylation-generated motifs associate with metabolism-related terms. It is, however, important to understand that the mere presence of a motif does not mean that each protein harboring the motif (75% of all proteins) is degraded by CMA at any given time. Whether or not a protein is a CMA substrate depends on the pathophysiological context of the cell and experimental validation is needed to determine if the protein is a substrate. As readers will note in the following sections, each CMA substrate protein undergoes degradation at a specific physiologically appropriate time to accommodate the needs of the cell.

Earlier studies identified CMA as a protein quality control pathway that degrades damaged or misfolded proteins (Fig. 2) [7]. The underlying idea being that partial or abnormal unfolding of such damaged proteins will likely expose the KFERQ-like motifs, rendering them visible to the CMA machinery and allowing their recognition by HSPA8, which precedes degradation of the damaged protein by CMA. Furthermore, CMA degrades a large fraction of prone-to-aggregate metastable proteins, thus preventing their intracellular accumulation as aggregates [8].

In addition to its role in maintaining protein quality control, it is being increasingly learnt that CMA also degrades functional proteins, and in this way, CMA remodels the cellular proteome according to the cellular requirements. CMA substrates identified till date are as diverse as the cellular processes they are involved in (Fig. 2). These include, among others, proteins involved in signaling pathways [9–11], transcriptional regulation [12–15], circadian rhythm [16], cell cycle [9, 17], glycolysis [9, 18, 19], lipolysis [20], triglyceride synthesis [19] and tricarboxylic acid cycle [19]. By degrading/eliminating proteins belonging to a certain pathway, CMA can effectively decrease the rate of that pathway in question. Because these functional proteins are properly folded, in often cases, the substrates are 'triggered' for degradation by posttranslational modifications in and around the KFERQ-like motifs or through exposure of motifs masked via protein/membrane binding or through translocation from a subcellular location to cytosol. For example, the KFERQ-like motif of membrane-associated SNCA/α-synuclein (protein linked to Parkinson disease; a CMA substrate) or SNCA fibers is masked [6], supporting the observations that only soluble forms of this protein are CMA substrates and once this pathogenic protein oligomerizes, it resists degradation via CMA [21].

> Question 1: For a protein to be a CMA substrate, it must have a specific motif present in the amino acid sequence. What is the motif?
>
> Question 2: Are damaged proteins the only substrates degraded by CMA?

# 3 The Molecular Components of CMA

## 3.1 Cytosolic Chaperones

The cytosolic chaperone HSPA8 binds to the KFERQ-like motif in a CMA substrate protein [22], while the other cochaperones, DNAJ/HSP40, STUB1/CHIP and DDIT3/CHOP, regulate the substrate-chaperone binding and delivery to lysosomes [23]. Contrary to macroautophagy, which relies on microtubules for autophagosome-lysosome fusion, there has till date been no documentation of the involvement of microtubules for the delivery of cargo proteins to lysosomes via CMA. It is interesting to note that cytosolic HSPA8 also targets KFERQ-like motif bearing proteins to late endosomes for a selective form of endosomal microautophagy; but in this case, HSPA8 binds phosphatidyl serine on the late endosomal membrane and substrate internalization occurs in vesicles forming from the surface of the endosome [24]. Endosomal microautophagy also has a non-selective arm that does not require HSPA8. Which leads to the question as to what determines if a substrate-HSPA8 complex is targeted to lysosomes for CMA or to late endosomes for selective endosomal microautophagy, and this remains an active area of research.

## 3.2 LAMP2A: The CMA Receptor

The key gate keeper for substrates to enter the CMA lysosomes is the LAMP2A receptor (Fig. 1). At the lysosomal membrane, substrate-HSPA8 complex interacts with the CMA receptor LAMP2A [25]. LAMP2A is essential for CMA and is the rate-limiting component for the pathway [25]. LAMP2A is an integral membrane protein with two lumenal domains (that make up 90% of the protein), a single transmembrane domain and a short (12 amino acid) C-terminal cytosolic tail. It is a heavily glycosylated protein with glycosylation exclusively found in its lumenal domains. LAMP2A is one of the three splice variants of *LAMP2* gene and is the only isoform involved in CMA. Substrate proteins bind the cytosolic tail of monomeric LAMP2A [26]. Four basic amino acid residues in the C terminus of LAMP2A are important for substrate binding [26]. Using NMR spectroscopy, it has been determined that the cytosolic tail of LAMP2A has a flexible conformation while the transmembrane domain has a coiled-coil helical conformation. Binding of the substrate-chaperone complex to the cytosolic tail triggers higher molecular weight multimer formation independent of the transmembrane and lumenal domains, which suggests that substrate recognition and targeting are likely coupled processes [27]. These LAMP2A multimers form part of the translocation complex via which the substrate proteins thread through to enter the lysosomal lumen [28]. LAMP2A protein, thus, shuttles between various states, monomer, trimer and multimer in the lysosome membrane, and undergoes cycles of assembly and disassembly into the translocation complex that are extensively controlled by several key CMA regulator proteins on the lysosomal surface as well as inside the lysosomes, as discussed below.

## 3.3 Lysosomal Chaperones

Lysosomal membrane-associated HSPA8 present on the cytosolic side of the lysosome, has a dual role in CMA (Fig. 1). HSPA8 in complex with other chaperones HSP90, DNAJ/HSP40, ST13/HIP and STIP1/HOP participates in the unfolding of the substrate protein, a prerequisite for its translocation into the lysosome [23]. Lysosomal membrane HSPA8 also participates in the disassembly of the CMA translocation complex once translocation is complete [28]. The two-domain architecture of the lumenal part of LAMP2A is important for the interaction with HSPA8 at the cytoplasmic surface of the lysosome [29]. Additionally, HSPA8 also exists as an acidic pH-stabilized protein in the lysosomal lumen [30, 31] and lysosomal HSPA8 is indispensable for the translocation of CMA substrates into the lysosomal lumen. In fact, the pool of lysosomes that do not have lumenal form of HSPA8, even though they contain LAMP2A, are CMA-incompetent (LAMP2A$^+$ HSPA8$^-$) and these lysosomes have greater affinity to fuse with autophagosomes [32]. The CMA-competent (LAMP2A$^+$ HSPA8$^+$) lysosomes are slightly more acidic than CMA-incompetent lysosomes [33]. A form of HSP90 associated with the lumenal side of the lysosomal membrane plays a key role in stabilizing LAMP2A as it transitions between its monomeric and multimeric forms [28].

## 4    Regulation of CMA

### 4.1    Lysosomal CMA Regulators

Many proteins associated to the lysosomal membrane act as CMA regulators, which either affect the assembly and disassembly of the LAMP2A translocation complex or regulate the levels of LAMP2A at the lysosomal membrane (Fig. 1). GFAP (glial fibrillary acidic protein) and its GTP-binding partner EEF1A/EF1α (eukaryotic translation elongation factor 1 alpha) regulate the LAMP2A translocation complex [34]. Non-phosphorylated GFAP stabilizes the multimeric complex, while GTP-driven release of EEF1A from the lysosomal membrane promotes self-dimerization of phosphorylated GFAP resulting in the disassembly of the multimeric translocation complex. A key nutrient and growth factor-sensitive kinase AKT1 plays a key role in inhibiting CMA activity by promoting multimer disassembly, in part by phosphorylating GFAP [35]. Consequently, the inhibition of AKT1 via class I PI3K inhibitors activates CMA by decreasing lysosomal levels of phospho-GFAP and increasing lysosomal HSPA8 levels [36]. AKT1 is, in turn, regulated by the kinase-phosphatase pair MTOR (mechanistic target of rapamycin kinase) complex 2 (MTORC2) and PHLPP1 (PH domain and leucine rich repeat protein phosphatase 1) with the former acting as an inhibitor for CMA [35]. In counterbalance, recruitment of PHLPP1 to the lysosomes (in a RAC1-dependent manner) leads to activation of CMA as it dephosphorylates AKT1 [35], as does PTEN, a lipid phosphatase, which also activates CMA by inactivating AKT1 [10]. Furthermore, levels of LAMP2A per se at the lysosomal membrane are also regulated by its degradation in lipid microdomains [37]. LAMP2A monomers when present in these cholesterol- and glycosphingolipid-rich lipid microdomains undergo cleavage by a metallopeptidase and cathepsin A [38] with their subsequent release and degradation in the lysosomal lumen. Hence, the degradation of LAMP2A in lysosomal membrane lipid microdomains plays an important part in maintaining its levels in the lysosome membrane.

### 4.2    Extra-Lysosomal CMA Regulators

Several studies have revealed extra-lysosomal mechanisms in the regulation of LAMP2A levels and CMA activity. Apart from proteins at the lysosomal surface, several proteins that are present elsewhere in the cell, can effectively regulate CMA activity. Many of these are transcription factors that either drive the expression of CMA effectors or decrease the expression of negative regulators of CMA. The first such regulator discovered was NFATC2/NFAT1 (nuclear factor of activated T cells 2), which was reported to transcriptionally upregulate *LAMP2A* expression during T cell activation [14]. NFE2L2/NRF2 (NFE2 like bZIP transcription factor 2), an important component of the oxidative stress response, also binds to antioxidant response elements in the *LAMP2* gene and activates *LAMP2A* expression in different cell types, in basal as well as oxidative stress conditions [39]. Yet another extra-lysosomal mechanism that regulates the lysosomal levels of LAMP2A is via targeting of LAMP2A to lysosomes. For instance, in cystinosis, a rare lysosomal storage disorder, LAMP2A is mistargeted resulting in impaired CMA, and this mistargeting of LAMP2A can be corrected by stimulating Golgi-to-lysosome transport of LAMP2A, which has been shown to depend on RAS-related protein RAB11 and the RAB7 effector RILP (Rab interacting lysosomal protein) [40]. In addition, we know that non-acetylated TPD52, known to confer stress resistance in prostate cancer, activates CMA by enhancing the substrate-HSPA8 binding while this absence of acetylation also prevents TPD52 from being delivered to lysosomes for degradation [41]. Conversely, RARA/RARα (retinoic acid receptor alpha) signaling is an extra-lysosomal mechanism that serves to inhibit CMA [42]. Indeed, chemical and genetic blockage of RARA revealed its robust inhibitory effect on *LAMP2A* expression and on the lysosomal trafficking mediated by RAB11 [40, 42]. In fact, currently available specific activators of CMA leverage their ability to block the negative impact of RARA signaling on CMA-related transcriptional program, without affecting other RARA-dependent genes [43] (see later sections). In addition, glucocorticoids [44] and growth hormone signaling [45] also play key roles in the inhibition of CMA, likely by impinging of AKT1 activity, as discussed earlier. Furthermore, as discussed below, BMAL1, a known driver of the circadian clock, was also recently shown to negatively regulate CMA by driving the expression of RARA, which then blocks *LAMP2A* expression [16].

> Question 3: Are all lysosomes able to perform CMA? Apart from the CMA receptor, LAMP2A, what is essential in a lysosome for CMA?

## 5  CMA Activation

All mammalian cells tested till date show CMA activity, albeit at different levels. Earlier studies relied on determining lysosomal levels of LAMP2A and on the functional uptake of CMA substrates in isolated lysosomes as a measure of CMA activity. The development of a CMA fluorescent reporter for use in cells [46] and mice [47] based on tagging the KFERQ-motif to a photoconvertible fluorescent protein has been a substantial forward-step for the study of CMA activity in intact cells and *in vivo*. There are currently several varieties of this reporter, which include KFERQ-photoswitchable-CFP2, KFERQ-photoactivable-mCherry1, KFERQ-photoswitchable-Dendra, with all reporters employing the same principle, i.e., visualizing the association of fluorescent reporter-protein construct with lysosomes as it is degraded by CMA. Using this CMA reporter, researchers are currently exploring new facets of tissue-specific differences in CMA activity, and examples of such spatiotemporal changes in CMA activity are highlighted in the following sections. In terms of physiological levels of CMA activity, it should also be noted that apart from basal levels of CMA, CMA is activated by various stressors. Mild oxidative or proteotoxic stress [7], lipotoxic stress [48], genotoxic stress [17], ER stress [49], hypoxia [15], and starvation [50] are each known to upregulate CMA. In addition to these stressors, a growing number of studies report the activation of CMA as an essential part of key physiological processes, for instance, T-cell activation [14], senescence [51], adipocyte differentiation [9], osteoblast differentiation [52], hematopoietic stem cell activation [53], and embryonic stem cell activation [54], as detailed in the following sections.

## 6    CMA in Circadian Rhythm

As with most physiological processes that are synchronous with the circadian clock, the basal activity of CMA is itself modulated across period as a function of time [16]. In fact, CMA was recently shown to be regulated by the circadian clock in a tissue-specific manner [16]. Intriguingly, in the liver (hepatocytes), maximal CMA activity is observed during the light phase in rodents. By contrast, mouse heart and kidney each display higher CMA activity during the dark phase. Additionally, these circadian oscillations of CMA activity are achieved by modulating the pool of CMA-competent lysosomes in liver and kidney, however, that is not the case in heart wherein the number of CMA-competent lysosomes remains unchanged with circadian rhythm [16]. Furthermore, these circadian changes in CMA activity reflect in the nature of the cargo that is degraded by CMA as a function of time; in other words, proteins degraded by CMA in the light phase are distinct from those degraded in the dark phase. For example, in the liver, nocturnal CMA substrates in mice are those enriched in pathways related to glucose and lipid metabolism, protein translation and processing, and endocytosis, while diurnal CMA substrates include proteins involved in quality control such as, ubiquitin-proteasome pathway, autophagy, cell death, drug metabolism and redox buffering [16]. Thus, by degrading distinct proteins at different circadian times, CMA remodels the subproteome according to the cellular requirements. In addition of inadequate proteome remodeling, CMA-deficient mice (whole body deletion of *lamp2a*) show disruption in both the central and peripheral clock and have aberrant circadian outputs because of failure to timely degrade clock proteins; in fact, young *lamp2a* null animals phenocopy the age-related dampening of the circadian clock [16].

## 7    CMA in Metabolism

From the initial years of its discovery, CMA has been well-characterized in liver. Liver is one of the metabolic tissues in which CMA displays higher basal activity, which is further induced in response to starvation [50]. With prolonged starvation, the pool of CMA-competent lysosomes (HSPA8$^+$) increase [33], and the ratio of assembly of the LAMP2A translocation complex [28] and reduced LAMP2A lysosomal degradation [37] tips towards greater assembly of the multimeric complex. Therefore, with fasting, one will observe greater uptake of CMA substrates and reduced degradation of the LAMP2A receptor in lysosome membrane lipid microdomains, as discussed previously. One important role that CMA plays in highly metabolic tissues such as liver, is in the regulation of cellular energetics. In fact, cellular and mouse models of CMA deficiency show substantially reduced ATP levels during starvation [19, 55]. In contrast, availability of nutrients leads to suppression of CMA, likely through activation of AKT1 signaling, as also discussed previously. This is best represented in livers of mice acutely fed a high cholesterol diet or chronically exposed to high fat diet, each of which blocks CMA [48]. In fact, CMA activity declines in these dietary models due to accelerated degradation of the LAMP2A receptor in lysosomes because of lipid composition changes induced by the diets [48].

It is also interesting to note that several proteins involved in glucose and lipid metabolic pathways, including those related to glycolysis, tricarboxylic acid cycle, lipid binding, lipid droplet coat proteins, triglyceride metabolism, steroid metabolism, fatty acid oxidation, and ketogenesis, are CMA substrates [19, 20]. Consequently, hepatocyte-specific *lamp2a* knockout (KO) mice show signs of impaired lipid utilization reflecting in steatosis, liver damage and reduced liver function, while relying heavily on glycogen reserves for energy [19]. In fact, *lamp2a* KO animals have reduced hepatic glycogen stores and massive accumulation of lipid droplets. Lipid droplet coat proteins, PLIN2 (perilipin 2) and PLIN3, assist in the storage of triglycerides. Because PLIN2 and PLIN3 are CMA substrates, in the absence of CMA (in *lamp2a* KO mice), there is an impairment in the removal of these coat proteins from the lipid droplets. Consequently, lipolysis via macrolipophagy machinery and cytoplasmic lipases fails to occur due to their inability to access the lipid droplet surface [20]. Consistent with what is observed in overnutrition models, *lamp2a* KO animals also phenocopy the age-related defects in metabolism, susceptibility of ROS and proteostasis failure [56]. A fundamental question in hepatology is why accumulation of lipids in liver sets the basis for development of metabolic dysfunction-associated steatohepatitis (MASH). From that angle, it is interesting to note that LAMP2A levels are substantially reduced in liver-resident macrophages from MASH mouse models [57]. While blocking CMA in myeloid cells (macrophages, monocytes, and granulocytes) in basal conditions results in hepatic macrophage infiltration, this *per se* does not lead to liver damage or lipid accumulation. By contrast, in the setting of MASH, dysfunctional CMA in macrophages, promotes rapid disease progression and fibrosis [57], pointing to diverse cell-specific hepatoprotective roles of CMA.

Insights into how different cell-types in the liver display changes in CMA during fasting and feeding were revealed using the CMA KFERQ-Dendra reporter mouse model [47]. Quite interestingly, under basal conditions, hepatocytes and Kupffer cells (resident macrophages) display higher CMA activity as compared to hepatic stellate cells [47]. With fasting, while hepatocytes and stellate cells activate CMA, Kupffer cells fail to do so. Furthermore, basal CMA activity appears to show zonal differences with periportal regions exhibiting lower CMA activity, while pericentral regions with lower nutrients and oxygen show higher CMA activity [47]. How these spatial differences in CMA affect liver function in health and disease remains to be seen.

In addition to the liver, the various fat depots play central roles in systemic metabolism and energy balance by serving to store neutral lipids in the fed state but releasing fatty acids during fasting. Both white and brown adipose tissue have been shown to display CMA activity, with the latter having ten-fold higher basal activity, while white fat shows robust induction of CMA with fasting [47]. Intriguingly, CMA is also upregulated with white adipocyte differentiation [9]. In fact, preadipocytes rely on CMA to degrade key metabolic enzymes (glycolytic enzymes), cell cycle factors (MYC) and signaling proteins (TGFB effector SMAD2-SMAD3) at distinct time-points in the multi-step commitment-to-differentiation program [9]. By degrading different proteins at various times in the differentiation process, CMA modulates the functional levels of each of the pathways in which these substrate proteins are involved. For example, at early steps, CMA degrades SMAD2-SMAD3, an adipogenesis inhibitor, which leads to the transcriptional upregulation of adipocyte gene expression in preadipocytes. In contrast, at later steps, by degrading glycolytic enzymes after the energy burst that precedes differentiation, CMA ensures the shutting down of glycolysis so that the cells can become quiescent and commence the differentiation program [9]. Blocking CMA in preadipocytes (by deleting *lamp2a*) led to defects in adipocyte differentiation resulting in aberrant visceral white adipose tissue with accumulation of preadipocytes. Additionally, the white adipose tissue showed smaller depot size, macrophage accumulation and fibrosis [9].

## 8    CMA in Cardiovascular Function

Recent studies have identified a role of CMA in maintaining cardiovascular function. A systematic analysis of the aorta revealed high activity in the middle media layer (vascular smooth muscle cells) and lower activity in the inner intima layer (endothelial cells) when examined under basal conditions [58, 59]. It has also been observed that CMA activity declines with atherosclerosis [58, 59]. In fact, in an atherosclerosis mouse model, as the disease progresses, neither vascular smooth muscle cells nor macrophages in the aorta show any detectable levels of CMA activity [58] and LAMP2A levels are decreased in the plaque. In support that reduced CMA contributes to vascular disease, mice with systemic CMA blockage (total-body *lamp2a* KO) when subjected to pro-atherogenic diets display aggravation of the disease phenotype, in terms of plaque size, larger necrotic plaque cores and elevated loss of macrophages and smooth muscle cells [58]. Although part of the higher severity of the atherosclerosis disease in these mice can be a consequence of the systemic changes in lipid metabolism in this model, studies with selective blockage in different cellular vascular components related with atherosclerosis support also a protective cell-autonomous effect of CMA. Indeed, we now know that loss of *lamp2a* in vascular smooth muscle cells result in their dedifferentiation, inflammation, and death [58]. Vascular disease in this model is also due to impaired local immune function considering the key role macrophage foam cells play in the fate of atherosclerotic plaques. Loss of *lamp2a* in macrophages renders a proinflammatory phenotype [58] with macrophage-specific loss of *lamp2a* resulting in more atherosclerotic plaque formation and increased vascular inflammation, associated with NLRP3 inflammasome activation and secretion of proinflammatory factors [59]. Comparative proteomics have recently demonstrated that CMA contributes to macrophage function by turning over key protein substrates, and that accumulation of these substrates in CMA-null macrophages impairs its roles in immune response and cell adhesion leading to inflammation and leukocyte activation [58]. Interestingly, under conditions of lipopolysaccharide-induced macrophage activation, CMA degrades proteins that are broadly involved in nitric oxide synthesis (NOS and five other stimulators of nitric oxide synthesis), immune response, neutrophil degradation, inflammation (NLRP3), and transendothelial migration (for cell adhesion, cell localization) [58, 59]. Although direct measurement of CMA activity in humans is still not possible, analysis of changes in the expression of all components that participate in CMA (CMA network) allows inferring differences in CMA activity outcome [8]. Using this CMA predictor in human cohorts, it has been confirmed that CMA activity in plaque macrophages and smooth muscle cells decreases with disease stage and that unfavorable plaque characteristics (inflammation and hemorrhage) correlate with reduced CMA activity. In fact, low plaque LAMP2A protein levels at the time of first cardiovascular event correlates with higher risk of a future secondary cardiovascular event [58].

In addition to its aforementioned role in vascular health, LAMP2A protein levels were also shown to be elevated with heart failure [60]. Compared to control subjects, the hearts of patients with ischemic heart failure and other causes of heart failure, including nonischemic cardiomyopathy, dilated cardiomyopathy, and chronic systolic heart failure, displayed elevated levels of LAMP2A. This increase in CMA is likely a part of the compensatory stress response to hypoxia, which is a known activator for CMA, as hypoxically stressed primary cardiomyocytes also show increased CMA activity [60].

## 9    CMA in Neurological Function

Proteostasis plays a critical role in central nervous system. Given the idea that CMA could be neuroprotective, great effort is going into understanding the pathophysiology of CMA in relation to diseases of the nervous system. Initial studies showed that CMA is active basally in all brain cell types, neurons, and glial cells, and as with liver, the level of its activity varies between regions, and between the various cell types. Using the CMA reporter mouse model (KFERQ-Dendra), it was discovered that neuronal CMA activity differs across various brain regions [47]. For instance, basal CMA activity was found to be higher in the CA1 and CA2 regions of the hippocampus, whereas dentate gyrus displayed lower activity. The cortex also displays high CMA activity with visible KFERQ-Dendra puncta observed in neuronal cell bodies and in their projections [47].

Understandably, the protein quality control function of CMA, which deters the prone-to-aggregate proteins from forming toxic aggregates [8], is immensely vital in post-mitotic tissues such as the brain. In fact, CMA failure has been linked with many neurodegenerative disorders, including Parkinson disease (PD) and in tauopathies such as frontotemporal dementia (FTD), and Alzheimer disease (AD) (reviewed in [3, 61]). Indeed, several unmodified neurodegeneration-related proteins have been validated as CMA substrates, including SNCA, GBA, LRRK2, PARK7, UCHL1 and VPS35 (PD), MAPT/tau and TARDBP/TDP-43 (related to tauopathies such as AD and FTD), and HTT (huntingtin; Huntington disease) [3]. However, as commonly observed, the pathogenic variants of these proteins tend to inhibit CMA and thus, these pathogenic variants accumulate in the cells resulting in neurotoxicity. For example, even though mutant SNCA (A53T or A30P) [21] or posttranslationally-modified SNCA (dopamine-modified) [62] or acetylated MAPT/tau [63] are recognized by cytosolic HSPA8 and targeted to LAMP2A on lysosomes, these toxic proteins fail to get internalized and form oligomers at the lysosomal membrane, which disrupt the dynamics of assembly/disassembly of LAMP2A translocation complex, in turn, inhibiting CMA. Although most studies have been performed in experimental mouse models, in case of PD, isolation of lysosomes from patients' brains confirmed the CMA blockage and disruption of LAMP2A dynamics [64]. Similarly, in case of Alzheimer disease (AD), brains of AD patients show neuron-specific CMA inhibition, which was reported to be due to transcriptional downregulation of CMA components and its regulators, showing strong correlation with the stages of the pathology. Consistently, a mouse model of tauopathy that expresses human MAPT/tau displays marked reduction in neuronal CMA activity in hippocampus [8]. In addition, loss of CMA in neurons also accelerates the pathology and proteotoxicity in AD mouse model. CMA blockage resulted in increased release of extracellular acetylated MAPT/tau and in fact, MAPT/tau spreading was enhanced in CMA-deficient mice, promoting thus, MAPT/tau pathology propagation [63]. Due to reduced CMA activity in AD and PD and the aggravating effect of CMA blockage on the pathology of several neurodegenerative diseases, approaches used to activate CMA as a preventative measure are an active area of research (see Sect. 13).

Genetic loss of CMA in neurons, using CaMKII neuron-specific deletion of *lamp2a*, results in disruption of neuronal proteostasis and function with accumulation of diverse cellular protein debris such as protein inclusions, lipofuscin, ubiquitinated and oxidized proteins [8]. In this case, because CMA substrates cannot be degraded, they tend to entrap within protein aggregates leading to loss of their physiological functions. Many of these protein inclusions include CMA substrates involved in metabolism, cation homeostasis, protein targeting, endocytosis, and cytoskeleton organization [8]. Similarly, an acute loss of *lamp2a* in dopaminergic neurons results in accumulation of SNCA, proteostasis failure, and neurodegeneration [65]. Given these developments, needless to say, a better understanding of how we can restore CMA activity will help develop novel approaches to prevent or delay age-related neurodegeneration.

## 10 CMA in Immune Cells

Besides the previously discussed role of CMA in macrophage polarization and function, as briefly alluded to earlier, CMA is activated in response to T-cell receptor (TCR) engagement [14]. Upon T-cell activation, expression of *LAMP2A* is increased owing to abundant TCR-induced ROS production that triggers NFAT-induced expression of the CMA receptor. CMA maintains T-cells in activated stage by degrading two inhibitors of TCR signaling, ubiquitin ligase ITCH and PPP3/calcineurin inhibitor RCAN1. Consequently, deletion of *lamp2a* in T-cells causes reduced spleen and thymus size, decreased numbers of total thymocytes, reduced percentage of peripheral CD4+ and CD8+ T-cells, impaired T-cell response, and diminished response to immunization or Listeria infection [14].

CMA also plays important roles in diverse cells that directly or indirectly affect the immune system. For example, mesenchymal stromal cells are multifaceted immunomodulatory cells that regulate immune cells to alleviate inflammation. It has been shown that CMA negatively regulates inflammation-induced immunosuppression by mesenchymal stem cells [66]. Proinflammatory factors (IFNG/IFNγ and TNF/TNFα) decrease LAMP2A levels in mesenchymal stromal cells and blockage of CMA increases the immunosuppressive capacity of stromal cells on T-cell proliferation. In fact, CMA inhibition improved the therapeutic effect of mesenchymal stromal cells in concanavalin A-induced liver injury model of T-cell mediated hepatitis [66]. Similarly, pericytes (perivascular stromal cells) are part of the immunological defense response; however, these cells get hijacked by glioblastoma to favor tumor growth. Glioma tumor cells tend to induce abnormal CMA upregulation in these surrounding pericytes, which causes an anti-inflammatory phenotype that prevents T-cell activation, thereby inhibiting their antitumor function and inducing immunotolerance to cancer [67, 68]. In fact, blockage of CMA in pericytes maintains their proinflammatory function and supports activation of antitumor T-cell response and promotes tumor cell death. Indeed, glioblastoma mouse models grafted with CMA-deficient pericytes exhibit reduced tumor proliferation and improved immune response [67].

Yet another example of how CMA activity is important for an overall maintenance of immune function is through its regulation of B cell function. Interestingly, CMA activity is increased in splenic B cells in mouse models of lupus [69]. In fact, under these conditions, there is marked expansion of the lysosomal compartment in terms of its size and acidity with elevated levels of CMA effector proteins, including LAMP2A and HSPA8. Overexpression of *lamp2a* in antigen presenting cells augmented cytosolic autoantigen presentation possibly priming autoreactive T-cells [70]. Finally, CMA appears to regulate many additional immune response related proteins. For example, STING1, an important component of the innate immune response, is degraded by CMA in its unsumoylated form [71]. Similarly, levels of NFKB/NF-κB, a proinflammatory transcription factor, appear to be regulated by CMA since CMA was shown to degrade the NFKB inhibitor [13].

## 11 CMA in Stem Cells

Recent studies have revealed new roles of CMA in stem cell function including embryonic stem cells (ESCs) [54], hematopoietic stem cells (HSCs) [53], and mesenchymal stem cells (MSCs) [52]. It has been shown that CMA controls the balance between self-renewal and differentiation in ESCs [54]. While low CMA activity promotes quiescent ESC self-renewal, upregulation of CMA occurs during ESC differentiation. Quite intriguingly, the expression of *LAMP2A per se* is suppressed by POU5F1/OCT4 and SOX2, which are critical factors for pluripotency, until ESC differentiation is initiated. Conversely, CMA degrades key enzymes of tricarboxylic acid cycle (isocitrate dehydrogenases IDH1 and IDH2) in these cells that convert isocitrate to α-ketoglutarate and hence, decreases the levels of intracellular α-ketoglutarate, which promotes cell differentiation [54].

CMA also plays a role in sustaining HSCs function [53]. In the quiescent stage, HSCs have high CMA activity compared to myeloid progenitors, and CMA was shown to be further activated upon HSCs activation. Using *lamp2a* KO mice, it was found that although CMA is dispensable for multilineage blood cell production; CMA is necessary for maintaining HSC protein quality control in quiescent stage and facilitate part of the proteome remodeling required during its activation [53]. Moreover, CMA-deficient HSCs had elevated ROS levels, higher levels of oxidized proteins, and protein inclusions, all of which indicate loss of proteostasis. CMA loss in HSCs also resulted in reduced ATP production and reduced glycolysis. Upon HSC activation, CMA is required for the metabolic switch toward lipid-generated energy as it increases linoleic acid metabolism by modulating the ratio of active/inactive FADS2, limiting enzyme in this metabolic pathway [53].

CMA is also an important regulator for MSC differentiation into osteoblasts by its ability to degrade SOX9 and MYD88, which are known inhibitors of osteogenesis [52]. CMA activity increases with osteoblast differentiation and this increase is controlled by VANGL2, a regulator of planar cell polarity and organ development [52].

## 12 CMA in Aging

Loss of proteostasis is one of the hallmarks of aging [72]. Aging and many age-related morbidities are associated with defective proteostasis, resulting in accumulation of damaged, oxidized, glycated, ubiquitinated, and misfolded proteins that are more prone to form intracellular aggregates and inclusions or extracellular plaques. Links of CMA with aging have been established from the early years of discovery of the pathway. In fact, a decline in LAMP2A levels has been observed in most cell types and tissues, including, liver [73], skeletal muscle [74], fibroblasts [73], hematopoietic stem cells [53], mesenchymal stem cells [52], and T-cells [14]. However, this decrease in LAMP2A with age is not always the case, since tissues such retina [75] and cardiac muscle [74] show preserved total levels of LAMP2A until late in life. It is noteworthy that measuring the total cellular

**lamp2a KO phenotype**

| | | |
|---|---|---|
| Neurons | Proteotoxicity, neurodegeneration, pathology aggravation | [8, 63] |
| Hepatocytes | Metabolic alterations, steatosis, liver damage, metabolic syndrome. With aging – susceptibility to oxidative stress, proteostasis failure | [19, 56] |
| Preadipocytes | Smaller depot, reduces adipogenesis, adipose dysfunction, inflammation, fibrosis | [9] |
| Smooth muscle cells | Atherosclerotic lesions, inflammation and dedifferentiated vascular smooth muscle cells | [58, 59] |
| Macrophages | Aggravated recruitment of liver targeted monocyte, promoted steatosis, fibrosis; promoted inflammation, increased atherosclerotic plaque formation | [57-59] |
| T-cells | T-cell inactivation, decreased adaptive immune response, decreased T-cell response | [14] |
| Hematopoietic stem cells | Loss of stem cells, impaired HSC activation | [53] |
| Systemic | Neurodegeneration, disruption of circadian rhythm, metabolic syndrome, susceptibility to atherosclerosis challenges | [8, 16, 19, 58] |

**Fig. 3** Diverse physiological outcomes of loss of CMA. The phenotypes of the *lamp2a* KO mice in different tissues. (Created with BioRender.com)

levels of LAMP2A does not necessarily reflect net CMA activity and in most aging tissues, the status of CMA activity remains undetermined. The best characterized organ in relation to aging of CMA is liver. With aging, in liver, LAMP2A levels decrease at the lysosomal membrane with the decline starting at mid-life [73]. This decrease is not transcriptional but attributed to reduced stability of this membrane receptor with age, resulting in diminished uptake/degradation of CMA substrates [76].

Extension of healthspan exhibits a positive correlation with elevated LAMP2A/CMA activity. For example, long-lived *pou1f1/Snell dwarf* and *ghr* (growth hormone receptor) knockout mice display increased hepatic CMA activity [45]. In addition, transgenic mice expressing an inducible extra copy of LAMP2A in the liver display improved liver function and proteostasis when compared to normally aged controls [77] (Sect. 13). Expectedly, young LAMP2A-deficient mice phenocopy many of the age-related traits (Fig. 3). Indeed, as discussed in preceding sections, CMA loss causes neurodegeneration and proteotoxicity in brain [8, 63], metabolic alterations and metabolic syndrome in liver [19, 56, 57], inflammation and impaired adipocyte differentiation [9], atherosclerotic lesions [58, 59], T-cell inactivation [14], loss of hematopoietic stem cells [53], and disruption in circadian rhythm [16] and low CMA activity associates with photoreceptor degeneration in retina [43], Consequently, restoring levels of LAMP2A to keep CMA activated with aging holds great promise in improving healthspan and decreasing the onset of age-related diseases.

## 13    Restoration of CMA

Genetic proof-of-concept approaches to restore CMA activity have been utilized to reverse liver aging and disease. By using a transgenic inducible mouse model in which *lamp2a* is specifically expressed in hepatocytes, it was found that maintaining constant LAMP2A levels after mid-age, when lysosomal levels of LAMP2A begin to decline, resulted in restoration of CMA activity in old transgenic mice livers when compared to levels in young mice [77]. Consequently, preserving CMA activity until late in life associated with decreased oxidized and damaged proteins, reduced liver damage, decreased cell death, and improved proteostasis and overall liver function [77].

Similarly, restoring CMA activity in old HSCs preserved their function [53]. As with other cell types and tissues (and as discussed in Sect. 11), genetic CMA blockage in young mice HSCs phenocopied the changes observed in old HSCs. Using mice overexpressing an inducible copy of *lamp2a*, to prevent the decline in LAMP2A levels with age, reversed these age-related phenotypes seen in old HSCs, by reducing the expansion of non-functional stem cells, restoring stem cell function, lowering ROS levels, maintaining proteostasis, and preserving glucose and fatty acid metabolism [53]. As also mentioned in Sects. 8 and 9, CMA activity is reduced with atherosclerosis progression and with neurodegeneration. Consequently, normalizing LAMP2A levels, using the whole-body inducible *lamp2a*-overexpressing mice, after plaque development has occurred, halted plaque size and improved metabolic profile. This supports the therapeutic potential of reactivating CMA in slowing atherosclerosis disease progression [58]. Consistent with this idea, adenovirus-based *lamp2a* overexpression in substantia nigra in rats was shown to reduce human SNCA aggregation, increased neuronal survival, and protect dopaminergic neurons from degeneration [78].

Finally, approaches to pharmacologically activate CMA have been recently described. Indeed, CMA activators (CAs) have been developed that serve to stabilize the interaction between RARA (a known CMA inhibitor) and its corepressor, NCOR1 (nuclear receptor corepressor 1), resulting in selectively activating CMA [43]. These CA molecules have favorable oral bioavailability and show brain penetrance. Owing to the efficacy of these small molecules, they have been successfully utilized *in vivo* to reduce stem cell aging [53] and to ameliorate phenotypes of neurodegeneration [8] and retinal degeneration [43]. Indeed, pharmacologically activating CMA via oral route rejuvenated aged HSCs from old mice and even from old people and improved HSC function and proteostasis [53]. Similarly, CA administration in two different mouse models of MAPT/tau-induced neurodegeneration (mutant MAPT/tau expressing mice and AD-model mice) led to robust beneficial effects in terms of alleviating multiple signature traits of neurodegeneration, including behavioral deficiencies, and histopathology and biochemistry evidence of proteotoxicity [8]. By the same token, CA delivered systemically or intravitreally in a mouse model of retinitis pigmentosa led to preservation of retinal histology and visual function and reduction of photoreceptor cell death [43]. These pieces of evidence provide support for the idea that restoration of CMA holds great potential in preventing or delaying diverse age-related diseases.

> Question 4: Which two strategies have been used experimentally to activate CMA in aging, and age-related diseases such as neurodegeneration?

## 14    Introduction to Cardinal Articles

## 14.1    Cardinal Articles

**Cardinal article 1** [21]: Cuervo AM, Stefanis L, Fredenburg R, Lansbury PT, Sulzer D. Impaired degradation of mutant alpha-synuclein by chaperone-mediated autophagy. *Science*. 2004;305(5688):1292–5.

Before this publication, very little was known about the contribution of CMA to degradation of any neuronal protein or possible alteration of this pathway in disease. This landmark study [21] was the first to demonstrate that: (a) wild-type SNCA is a CMA substrate in normal conditions, and that (b) pathogenic mutant SNCA (A53T and A30P that cause familial form of Parkinson disease) are inefficiently degraded by CMA and that these mutant forms of SNCA block CMA activity. This work [21] demonstrated that although mutant forms of SNCA have high affinity for the CMA receptor LAMP2A, these mutant forms are poorly degraded by CMA because they fail to cross the lysosomal membrane. Indeed, high affinity binding of the mutant SNCA to LAMP2A blocks the uptake and degradation of other CMA substrates by disrupting the dynamics of the CMA receptor at the lysosomal membrane. Thus, the study concluded that impaired turnover of pathogenic forms of SNCA by CMA resulted in accumulation of these toxic proteins, which favor aggregation of proteins or promote additional modifications that may contribute to Parkinson disease progression.

This cardinal paper [21] helped move the field forward because it directly connected CMA malfunction to a human disease i.e., neurodegeneration, and laid the foundation for subsequent studies investigating the roles of CMA in Parkinson disease. For instance, subsequent research based on this initial study showed that malfunction of CMA was not limited to familial forms of PD (those caused by mutations in SNCA) but that also extended to idiopathic PD, since posttranslational modification of SNCA by dopamine prevents its efficient degradation by CMA and inhibits CMA of other proteins through similar mechanisms to the one described for the mutant forms [62]. The predicted negative consequences of this SNCA-induced CMA blockage have been confirmed experimentally by showing that acute loss of *LAMP2A* in dopaminergic neurons results in accumulation of SNCA causing neurodegeneration [65] and that genetic loss of neuronal CMA also disrupts the metastable prone-to-aggregate proteome, promoting a shift toward loss of solubility and contributing to neurodegeneration [8]. Accordingly, *lamp2a* overexpression via adenoviral vectors in substantia nigra [78] or pharmacological upregulation of CMA [8] each decreases human SNCA aggregation and protects against neurodegeneration. Since this cardinal study [21] was the first to link impaired CMA with human disease, it raised many questions at that time. For instance, are other prone-to-aggregate proteins also degraded by CMA, and whether CMA dysfunction is causally linked to additional human diseases, including other neurodegenerative diseases. Strikingly, two decades after this initial work, many studies have discovered that indeed many neurodegeneration-related proteins are degraded by CMA, and that CMA dysfunction is broadly associated with diverse human diseases, as detailed in this chapter.

## 14.2    Take Note Of

It is noteworthy to mention that substrate proteins must be in soluble form to be amenable to degradation by CMA, and that aggregated proteins cannot undergo degradation through CMA. Take note of the fact that, unlike macroautophagy or microautophagy, CMA does not degrade protein aggregates and organelles. In fact, CMA relies on soluble substrate proteins being unfolded and translocated across the multimerized LAMP2A receptor complex into the lysosome. However, even if CMA cannot degrade full organelles, it can contribute to degrade organelle proteins when in transit through the cytoplasm toward their final destination. Readers must specially understand that increased binding of substrate protein to LAMP2A is not by itself synonymous with increased substrate uptake and degradation. In many pathological conditions, such as neurodegenerative disorders, mutant forms of pathogenic proteins will bind LAMP2A, but these are not translocated across the lysosome for degradation. These pathogenic proteins form oligomers at the lysosomal membrane that disrupt the dynamics of assembly/disassembly of LAMP2A translocation complex, thereby inhibiting CMA. Similarly, higher LAMP2A levels are not always equivalent to more CMA, understating the importance of LAMP2A lysosomal membrane dynamics.

**Cardinal article 2** [77]: Zhang C, Cuervo AM. Restoration of chaperone-mediated autophagy in aging liver improves cellular maintenance and hepatic function. *Nat Med.* 2008;14(9):959–65.

A link between impaired CMA and aging had been established from the initial years of CMA studies. Early work demonstrated that CMA decreases with age due to a decline in levels of the LAMP2A receptor at the lysosomal membrane [73]. It was therefore proposed that loss of proteostasis that occurs with aging, is partly due to defective CMA activity and that upregulation of CMA may potentially serve as an anti-aging strategy. This landmark article [77] demonstrated that preventing the decline in LAMP2A with age is sufficient to preserve liver function until late in life, and that CMA activation plays a key role in eliminating oxidized proteins in liver. The mouse model utilized in this study was an inducible transgenic model that expressed an extra copy of *LAMP2A* in liver, and using this approach, the authors prevented the decline in LAMP2A levels, and of CMA activity, starting at mid-age. Strikingly, maintaining CMA activity until later in life led to considerably less damaged proteins, reduced cell death, improved cellular and tissue homeostasis, and improved organ function. In fact, aged transgenic *lamp2a* overexpressing mice exhibited the levels of youthfulness noted in young wild-type mice, at least in each of the parameters studied. This proof-of-concept study [77] showed that indeed, maintaining optimal CMA activity throughout life leads to improved proteostasis, improved cellular and organ function, and delayed aging. This study provides strong support to the idea that preventing the decline of CMA is a promising strategy to extend healthspan and lifespan and could be included among the sought after gerotherapies. Additionally, this study paved the way for future studies testing the possibility that increasing LAMP2A levels serves as a therapeutic approach and is *per se* sufficient to upregulate CMA in diseases in which this pathway is impaired.

## 14.3 Take Note Of

CMA restoration via genetic approaches is achieved by increasing LAMP2A levels at the lysosome membrane. Please note that because the enhancement is at the receptor level (rate-limiting step of the pathway; steps 2 and 3 in Fig. 1), it does not result in aberrantly increased substrate recognition (step 1 in Fig. 1). In comparison, increasing LAMP2A levels is synonymous with opening up of additional customs counter (LAMP2A) at a busy airport to clear up the backlog of travelers (substrates).

**Take-Home Message**
- CMA is a highly selective mechanism for lysosomal degradation of soluble proteins containing a specific KFERQ tag by mediating their internalization into lysosomes through a membrane receptor, LAMP2A.
- CMA is a central player in the proteostasis network. It clears damaged, unfolded, and metastable proteins that are prone to form toxic aggregates.
- By context- and time-dependent degradation of fully functional proteins, CMA controls many fundamental and physiological processes in the cell.
- CMA-defective mice show accelerated diverse age-related phenotypes, for instance, neurodegeneration, fatty liver, adipose fibrosis and inflammation, lower resistance to atherosclerosis, defective T-cell activation, loss of stem cells and dysregulated circadian rhythm.
- Pharmacological activation CMA has been used successfully to ameliorate neurological and retinal degeneration and rejuvenate cellular aging, conditions in which CMA activity is reduced.

**Answer Guide to Questions**

Answer 1: For a protein to be a CMA substrate, it must contain KFERQ-like motif in its amino acid sequence. The physicochemical properties of the amino acid residues of the motif are more important than the exact sequence itself. The motif can be (a) canonical—established in the unmodified protein (KFERQ) and (b) putative—those that get generated by posttranslational modifications that impart the missing charge (i.e., KFSRQ—when phosphorylated at serine residue (S) will impart the missing negative charge and make the motif complete).

Answer 2: No, in addition to damaged proteins, undamaged functional proteins are also degraded by CMA. In this way, CMA selectively remodels the cellular proteome to change cell functions according to the cellular conditions.

Answer 3: No, only those lysosomes that contain LAMP2A and HSPA8 are able to perform CMA. HSPA8 inside the lysosomes is necessary for substrate translocation. Lysosomes that are LAMP2A$^+$ HSPA8$^+$ are called CMA-competent or CMA-active lysosomes. The two populations of lysosomes (CMA-competent and CMA-incompetent) are interchangeable, and the cells increase the pool of lysosomes competent for CMA depending on the cellular conditions.

Answer 4: Two strategies that have been used to activate CMA:

(a) Genetic upregulation of CMA: proof-of-principle strategy of activating CMA by expressing an extra copy of *lamp2a* in mice. The old *lamp2a* overexpressing mice had improved proteostasis, and better cellular and tissue health that was comparable to young mice [77].

(b) Chemical upregulation of CMA: small molecule CMA activator (CA) stimulates CMA *in vivo*. The pharmacological activator is based on inhibiting one of the inhibitors of CMA (RARA). The CA molecules have been used successfully to amend phenotypes of neurodegeneration [8].

## Definitions

**Autophagy**  'Auto' means self and 'phagy' means eat. Autophagy is the degradation of intracellular components in lysosomes. Autophagy is the intracellular recycling program.

**Chaperones**  Chaperones are proteins that assist other proteins to fold properly during or after synthesis, to refold after partial denaturation, to stabilize partially folded proteins and prevent aggregation and that can target proteins to specific compartments or to the degradation systems.

**LAMP2A**  Lysosomal associated membrane protein 2A is one of the three isoforms of the *LAMP2* gene. It is single-span membrane glycoprotein localized at the lysosome membrane with C terminus towards the cytosol and most of the protein facing the lysosomal lumen. The three isoforms have identical lumenal domains but distinct transmembrane and cytoplasmic domains.

**Lysosome**       Lysosomes are membrane-enclosed acidic organelles that contain various hydrolytic enzymes capable of breaking down all types of biomolecules—proteins, carbohydrates, lipids and nucleic acids.

**Proteostasis**   Protein homeostasis or proteostasis is the maintenance of the health and functionality of the cellular proteins. Proteostasis network includes integrated pathways that ensure optimal protein folding upon synthesis, trafficking and degradation.

# References


1. Levine B, Klionsky DJ. Autophagy wins the 2016 Nobel Prize in Physiology or Medicine: breakthroughs in baker's yeast fuel advances in biomedical research. Proc Natl Acad Sci USA. 2017;114(2):201–5.
2. Mizushima N. The ATG conjugation systems in autophagy. Curr Opin Cell Biol. 2020;63:1–10.
3. Kaushik S, Cuervo AM. The coming of age of chaperone-mediated autophagy. Nat Rev Mol Cell Biol. 2018;19(6):365–81.
4. Dice JF. Peptide sequences that target cytosolic proteins for lysosomal proteolysis. Trends Biochem Sci. 1990;15(8):305–9.
5. Dice JF. Altered degradation of proteins microinjected into senescent human fibroblasts. J Biol Chem. 1982;257(24):14624–7.
6. Kirchner P, Bourdenx M, Madrigal-Matute J, Tiano S, Diaz A, Bartholdy BA, et al. Proteome-wide analysis of chaperone-mediated autophagy targeting motifs. PLoS Biol. 2019;17(5):e3000301.
7. Kiffin R, Christian C, Knecht E, Cuervo AM. Activation of chaperone-mediated autophagy during oxidative stress. Mol Biol Cell. 2004;15(11):4829–40.
8. Bourdenx M, Martin-Segura A, Scrivo A, Rodriguez-Navarro JA, Kaushik S, Tasset I, et al. Chaperone-mediated autophagy prevents collapse of the neuronal metastable proteome. Cell. 2021;184(10):2696–714 e25.
9. Kaushik S, Juste YR, Lindenau K, Dong S, Macho-Gonzalez A, Santiago-Fernandez O, et al. Chaperone-mediated autophagy regulates adipocyte differentiation. Sci Adv. 2022;8(46):eabq2733.
10. Zhang KK, Burns CM, Skinner ME, Lombard DB, Miller RA, Endicott SJ. PTEN is both an activator and a substrate of chaperone-mediated autophagy. J Cell Biol. 2023;222(9):e202208150.
11. Liu H, Wang P, Song W, Sun X. Degradation of regulator of calcineurin 1 (RCAN1) is mediated by both chaperone-mediated autophagy and ubiquitin proteasome pathways. FASEB J. 2009;23(10):3383–92.
12. Yang Q, She H, Gearing M, Colla E, Lee M, Shacka JJ, et al. Regulation of neuronal survival factor MEF2D by chaperone-mediated autophagy. Science. 2009;323(5910):124–7.
13. Cuervo AM, Hu W, Lim B, Dice JF. IkappaB is a substrate for a selective pathway of lysosomal proteolysis. Mol Biol Cell. 1998;9(8):1995–2010.
14. Valdor R, Mocholi E, Botbol Y, Guerrero-Ros I, Chandra D, Koga H, et al. Chaperone-mediated autophagy regulates T cell responses through targeted degradation of negative regulators of T cell activation. Nat Immunol. 2014;15(11):1046–54.
15. Hubbi ME, Hu H, Kshitiz, Ahmed I, Levchenko A, Semenza GL. Chaperone-mediated autophagy targets hypoxia-inducible factor-1alpha (HIF-1alpha) for lysosomal degradation. J Biol Chem. 2013;288(15):10703–14.

16. Juste YR, Kaushik S, Bourdenx M, Aflakpui R, Bandyopadhyay S, Garcia F, et al. Reciprocal regulation of chaperone-mediated autophagy and the circadian clock. Nat Cell Biol. 2021;23(12):1255–70.

17. Park C, Suh Y, Cuervo AM. Regulated degradation of Chk1 by chaperone-mediated autophagy in response to DNA damage. Nat Commun. 2015;6:6823.

18. Aniento F, Roche E, Cuervo AM, Knecht E. Uptake and degradation of glyceraldehyde-3-phosphate dehydrogenase by rat liver lysosomes. J Biol Chem. 1993;268(14):10463–70.

19. Schneider JL, Suh Y, Cuervo AM. Deficient chaperone-mediated autophagy in liver leads to metabolic dysregulation. Cell Metab. 2014;20(3):417–32.

20. Kaushik S, Cuervo AM. Degradation of lipid droplet-associated proteins by chaperone-mediated autophagy facilitates lipolysis. Nat Cell Biol. 2015;17(6):759–70.

21. Cuervo AM, Stefanis L, Fredenburg R, Lansbury PT, Sulzer D. Impaired degradation of mutant alpha-synuclein by chaperone-mediated autophagy. Science. 2004;305(5688):1292–5.

22. Chiang HL, Terlecky SR, Plant CP, Dice JF. A role for a 70-kilodalton heat shock protein in lysosomal degradation of intracellular proteins. Science. 1989;246(4928):382–5.

23. Agarraberes FA, Dice JF. A molecular chaperone complex at the lysosomal membrane is required for protein translocation. J Cell Sci. 2001;114(Pt 13):2491–9.

24. Sahu R, Kaushik S, Clement CC, Cannizzo ES, Scharf B, Follenzi A, et al. Microautophagy of cytosolic proteins by late endosomes. Dev Cell. 2011;20(1):131–9.

25. Cuervo AM, Dice JF. A receptor for the selective uptake and degradation of proteins by lysosomes. Science. 1996;273(5274):501–3.

26. Cuervo AM, Dice JF. Unique properties of lamp2a compared to other lamp2 isoforms. J Cell Sci. 2000;113(Pt 24):4441–50.

27. Rout AK, Strub MP, Piszczek G, Tjandra N. Structure of transmembrane domain of lysosome-associated membrane protein type 2a (LAMP-2A) reveals key features for substrate specificity in chaperone-mediated autophagy. J Biol Chem. 2014;289(51):35111–23.

28. Bandyopadhyay U, Kaushik S, Varticovski L, Cuervo AM. The chaperone-mediated autophagy receptor organizes in dynamic protein complexes at the lysosomal membrane. Mol Cell Biol. 2008;28(18):5747–63.

29. Ikami Y, Terasawa K, Sakamoto K, Ohtake K, Harada H, Watabe T, et al. The two-domain architecture of LAMP2A regulates its interaction with Hsc70. Exp Cell Res. 2022;411(1):112986.

30. Cuervo AM, Gomes AV, Barnes JA, Dice JF. Selective degradation of annexins by chaperone-mediated autophagy. J Biol Chem. 2000;275(43):33329–35.

31. Agarraberes FA, Terlecky SR, Dice JF. An intralysosomal hsp70 is required for a selective pathway of lysosomal protein degradation. J Cell Biol. 1997;137(4):825–34.

32. Koga H, Kaushik S, Cuervo AM. Altered lipid content inhibits autophagic vesicular fusion. FASEB J. 2010;24(8):3052–65.

33. Cuervo AM, Dice JF, Knecht E. A population of rat liver lysosomes responsible for the selective uptake and degradation of cytosolic proteins. J Biol Chem. 1997;272(9):5606–15.

34. Bandyopadhyay U, Sridhar S, Kaushik S, Kiffin R, Cuervo AM. Identification of regulators of chaperone-mediated autophagy. Mol Cell. 2010;39(4):535–47.

35. Arias E, Koga H, Diaz A, Mocholi E, Patel B, Cuervo AM. Lysosomal mTORC2/PHLPP1/Akt regulate chaperone-mediated autophagy. Mol Cell. 2015;59(2):270–84.

36. Endicott SJ, Ziemba ZJ, Beckmann LJ, Boynton DN, Miller RA. Inhibition of class I PI3K enhances chaperone-mediated autophagy. J Cell Biol. 2020;219(12):e202001031.

37. Kaushik S, Massey AC, Cuervo AM. Lysosome membrane lipid microdomains: novel regulators of chaperone-mediated autophagy. EMBO J. 2006;25(17):3921–33.

38. Cuervo AM, Dice JF. Regulation of lamp2a levels in the lysosomal membrane. Traffic. 2000;1(7):570–83.

39. Pajares M, Rojo AI, Arias E, Diaz-Carretero A, Cuervo AM, Cuadrado A. Transcription factor NFE2L2/NRF2 modulates chaperone-mediated autophagy through the regulation of LAMP2A. Autophagy. 2018;14(8):1310–22.

40. Zhang J, Johnson JL, He J, Napolitano G, Ramadass M, Rocca C, et al. Cystinosin, the small GTPase Rab11, and the Rab7 effector RILP regulate intracellular trafficking of the chaperone-mediated autophagy receptor LAMP2A. J Biol Chem. 2017;292(25):10328–46.

41. Fan Y, Hou T, Gao Y, Dan W, Liu T, Liu B, et al. Acetylation-dependent regulation of TPD52 isoform 1 modulates chaperone-mediated autophagy in prostate cancer. Autophagy. 2021;17(12):4386–400.

42. Anguiano J, Garner TP, Mahalingam M, Das BC, Gavathiotis E, Cuervo AM. Chemical modulation of chaperone-mediated autophagy by retinoic acid derivatives. Nat Chem Biol. 2013;9(6):374–82.

43. Gomez-Sintes R, Xin Q, Jimenez-Loygorri JI, McCabe M, Diaz A, Garner TP, et al. Targeting retinoic acid receptor alpha-corepressor interaction activates chaperone-mediated autophagy and protects against retinal degeneration. Nat Commun. 2022;13(1):4220.

44. Sato M, Ueda E, Konno A, Hirai H, Kurauchi Y, Hisatsune A, et al. Glucocorticoids negatively regulates chaperone mediated autophagy and microautophagy. Biochem Biophys Res Commun. 2020;528(1):199–205.

45. Endicott SJ, Boynton DN Jr, Beckmann LJ, Miller RA. Long-lived mice with reduced growth hormone signaling have a constitutive upregulation of hepatic chaperone-mediated autophagy. Autophagy. 2021;17(3):612–25.

46. Koga H, Martinez-Vicente M, Macian F, Verkhusha VV, Cuervo AM. A photoconvertible fluorescent reporter to track chaperone-mediated autophagy. Nat Commun. 2011;2:386.

47. Dong S, Aguirre-Hernandez C, Scrivo A, Eliscovich C, Arias E, Bravo-Cordero JJ, et al. Monitoring spatiotemporal changes in chaperone-mediated autophagy in vivo. Nat Commun. 2020;11(1):645.

48. Rodriguez-Navarro JA, Kaushik S, Koga H, Dall'Armi C, Shui G, Wenk MR, et al. Inhibitory effect of dietary lipids on chaperone-mediated autophagy. Proc Natl Acad Sci USA. 2012;109(12):E705–14.

49. Li W, Zhu J, Dou J, She H, Tao K, Xu H, et al. Phosphorylation of LAMP2A by p38 MAPK couples ER stress to chaperone-mediated autophagy. Nat Commun. 2017;8(1):1763.

50. Cuervo AM, Knecht E, Terlecky SR, Dice JF. Activation of a selective pathway of lysosomal proteolysis in rat liver by prolonged starvation. Am J Phys. 1995;269(5 Pt 1):C1200–8.

51. Rovira M, Sereda R, Pladevall-Morera D, Ramponi V, Marin I, Maus M, et al. The lysosomal proteome of senescent cells contributes to the senescence secretome. Aging Cell. 2022;21(10):e13707.

52. Gong Y, Li Z, Zou S, Deng D, Lai P, Hu H, et al. Vangl2 limits chaperone-mediated autophagy to balance osteogenic differentiation in mesenchymal stem cells. Dev Cell. 2021;56(14):2103–20 e9.

53. Dong S, Wang Q, Kao YR, Diaz A, Tasset I, Kaushik S, et al. Chaperone-mediated autophagy sustains haematopoietic stem-cell function. Nature. 2021;591(7848):117–23.

54. Xu Y, Zhang Y, Garcia-Canaveras JC, Guo L, Kan M, Yu S, et al. Chaperone-mediated autophagy regulates the pluripotency of embryonic stem cells. Science. 2020;369(6502):397–403.

55. Massey AC, Kaushik S, Sovak G, Kiffin R, Cuervo AM. Consequences of the selective blockage of chaperone-mediated autophagy. Proc Natl Acad Sci USA. 2006;103(15):5805–10.

56. Schneider JL, Villarroya J, Diaz-Carretero A, Patel B, Urbanska AM, Thi MM, et al. Loss of hepatic chaperone-mediated autophagy accelerates proteostasis failure in aging. Aging Cell. 2015;14(2):249–64.

57. Zhang M, Tian SY, Ma SY, Zhou X, Zheng XH, Li B, et al. Deficient chaperone-mediated autophagy in macrophage aggravates inflammation of nonalcoholic steatohepatitis by targeting Nup85. Liver Int. 2023;43(5):1021–34.

58. Madrigal-Matute J, de Bruijn J, van Kuijk K, Riascos-Bernal DF, Diaz A, Tasset I, et al. Protective role of chaperone-mediated autophagy against atherosclerosis. Proc Natl Acad Sci USA. 2022;119(14):e2121133119.

59. Qiao L, Ma J, Zhang Z, Sui W, Zhai C, Xu D, et al. Deficient chaperone-mediated autophagy promotes inflammation and atherosclerosis. Circ Res. 2021;129(12):1141–57.

60. Ghosh R, Gillaspie JJ, Campbell KS, Symons JD, Boudina S, Pattison JS. Chaperone-mediated autophagy protects cardiomyocytes against hypoxic-cell death. Am J Physiol Cell Physiol. 2022;323(5):C1555–C75.

61. Fleming A, Bourdenx M, Fujimaki M, Karabiyik C, Krause GJ, Lopez A, et al. The different autophagy degradation pathways and neurodegeneration. Neuron. 2022;110(6):935–66.

62. Martinez-Vicente M, Talloczy Z, Kaushik S, Massey AC, Mazzulli J, Mosharov EV, et al. Dopamine-modified alpha-synuclein blocks chaperone-mediated autophagy. J Clin Invest. 2008;118(2):777–88.

63. Caballero B, Bourdenx M, Luengo E, Diaz A, Sohn PD, Chen X, et al. Acetylated tau inhibits chaperone-mediated autophagy and promotes tau pathology propagation in mice. Nat Commun. 2021;12(1):2238.

64. Kuo SH, Tasset I, Cheng MM, Diaz A, Pan MK, Lieberman OJ, et al. Mutant glucocerebrosidase impairs alpha-synuclein degradation by blockade of chaperone-mediated autophagy. Sci Adv. 2022;8(6):eabm6393.

65. Xilouri M, Brekk OR, Polissidis A, Chrysanthou-Piterou M, Kloukina I, Stefanis L. Impairment of chaperone-mediated autophagy induces dopaminergic neurodegeneration in rats. Autophagy. 2016;12(11):2230–47.

66. Zhang J, Huang J, Gu Y, Xue M, Qian F, Wang B, et al. Inflammation-induced inhibition of chaperone-mediated autophagy maintains the immunosuppressive function of murine mesenchymal stromal cells. Cell Mol Immunol. 2021;18(6):1476–88.

67. Valdor R, Garcia-Bernal D, Riquelme D, Martinez CM, Moraleda JM, Cuervo AM, et al. Glioblastoma ablates pericytes antitumor immune function through aberrant up-regulation of chaperone-mediated autophagy. Proc Natl Acad Sci USA. 2019;116(41):20655–65.

68. Molina ML, Garcia-Bernal D, Salinas MD, Rubio G, Aparicio P, Moraleda JM, et al. Chaperone-mediated autophagy ablation in pericytes reveals new glioblastoma prognostic markers and efficient treatment against tumor progression. Front Cell Dev Biol. 2022;10:797945.

69. Macri C, Wang F, Tasset I, Schall N, Page N, Briand JP, et al. Modulation of deregulated chaperone-mediated autophagy by a phosphopeptide. Autophagy. 2015;11(3):472–86.

70. Zhou D, Li P, Lin Y, Lott JM, Hislop AD, Canaday DH, et al. Lamp-2a facilitates MHC class II presentation of cytoplasmic antigens. Immunity. 2005;22(5):571–81.

71. Hu MM, Yang Q, Xie XQ, Liao CY, Lin H, Liu TT, et al. Sumoylation promotes the stability of the DNA sensor cGAS and the adaptor STING to regulate the kinetics of response to DNA virus. Immunity. 2016;45(3):555–69.

72. Lopez-Otin C, Blasco MA, Partridge L, Serrano M, Kroemer G. Hallmarks of aging: an expanding universe. Cell. 2023;186(2):243–78.

73. Cuervo AM, Dice JF. Age-related decline in chaperone-mediated autophagy. J Biol Chem. 2000;275(40):31505–13.

74. Zhou J, Chong SY, Lim A, Singh BK, Sinha RA, Salmon AB, et al. Changes in macroautophagy, chaperone-mediated autophagy, and mitochondrial metabolism in murine skeletal and cardiac muscle during aging. Aging (Albany NY). 2017;9(2):583–99.

75. Rodriguez-Muela N, Koga H, Garcia-Ledo L, de la Villa P, de la Rosa EJ, Cuervo AM, et al. Balance between autophagic pathways preserves retinal homeostasis. Aging Cell. 2013;12(3):478–88.
76. Kiffin R, Kaushik S, Zeng M, Bandyopadhyay U, Zhang C, Massey AC, et al. Altered dynamics of the lysosomal receptor for chaperone-mediated autophagy with age. J Cell Sci. 2007;120(Pt 5):782–91.
77. Zhang C, Cuervo AM. Restoration of chaperone-mediated autophagy in aging liver improves cellular maintenance and hepatic function. Nat Med. 2008;14(9):959–65.
78. Xilouri M, Brekk OR, Landeck N, Pitychoutis PM, Papasilekas T, Papadopoulou-Daifoti Z, et al. Boosting chaperone-mediated autophagy in vivo mitigates alpha-synuclein-induced neurodegeneration. Brain. 2013;136(Pt 7):2130–46.

## Further Reading

Web-based resource to check if your favorite protein has a canonical or putative KFERQ-like motif: https://rshine.einsteinmed.edu.

CMA is increased in most cancer cell lines and CMA receptor LAMP2A is elevated in a wide variety of human tumors. Role of CMA in cancer biology and progression has been extensively reviewed in:

Kaushik S, Cuervo AM. The coming of age of chaperone-mediated autophagy. Nat Rev. Mol Cell Biol. 2018;19(6):365–81.

Arias E, Cuervo AM. Pros and cons of chaperone-mediated autophagy in cancer biology. Trends Endocrinol Metab. 2020;31(1):53–66.

# A Phenotypic Screening Routine for the Identification of Autophagic Flux Inducers

Marion Leduc, Sabrina Forveille, Allan Sauvat, Giulia Cerrato, Guido Kroemer, and Oliver Kepp

**Part 1: What Will You Learn in This Chapter**

The phenotypic assessment of autophagy facilitates the establishment of workflows applicable to high throughput chemical drug screening campaigns by means of robotized automation platforms. Here, we describe a phenotypic screening routine for the identification of novel autophagic flux inducers from chemical libraries of small compounds. This screening routine encompasses fluorescent biosensor cells, automated liquid handling, robotized bioimaging as well as a multistep analysis pipeline. In this setting fluorescent biosensor cells expressing green fluorescent protein (GFP) conjugated to MAP1LC3/LC3 (microtubule associated protein 1 light chain 3), as well as mCherry-GFP-LC3 tandem reporter cells are cocultured for the phenotypic assessment of LC3 puncta formation and lysosomal turnover. Image analysis and data processing facilitates the phenotypic separation of autophagic flux inducers from drugs that affect lysosomal acidification or fusion of lysosomes with the autophagosome.

M. Leduc · S. Forveille · A. Sauvat · G. Cerrato · O. Kepp (✉)
Metabolomics and Cell Biology Platforms, Gustave Roussy Cancer Center, Université Paris Saclay, Villejuif, France

Centre de Recherche des Cordeliers, Equipe labellisée par la Ligue contre le cancer, Université Paris Cité, Sorbonne Université, Inserm U1138, Institut Universitaire de France, Paris, France

G. Kroemer (✉)
Metabolomics and Cell Biology Platforms, Gustave Roussy Cancer Center, Université Paris Saclay, Villejuif, France

Centre de Recherche des Cordeliers, Equipe labellisée par la Ligue contre le cancer, Université Paris Cité, Sorbonne Université, Inserm U1138, Institut Universitaire de France, Paris, France

Institut du Cancer Paris CARPEM, Department of Biology, APHP, Hôpital Européen Georges Pompidou, Paris, France

                    311
B. Loos, D. J. Klionsky (eds.), *Autophagy - From Molecular Mechanisms to Flux Control in Health and Disease*, Learning Materials in Biosciences,
https://doi.org/10.1007/978-3-031-88121-3_12

**Part 2: Introduction to Cardinal Articles**

Macroautophagy (from here on referred to as autophagy) is a cellular response to energetic crisis, translational dysfunction and pathogenic infection that targets cellular content for degradation within double membraned cytoplasmic vesicles to facilitate the clearance of potentially harmful cellular matter such as invading microorganisms and pathogenic protein aggregates. Moreover, autophagy is implicated in the turnover of aged and dysfunctional organelles thus controlling cellular rejuvenation and organismal equilibria altogether limiting aging-related diseases [1–3].

The induction of autophagic flux is a tightly regulated process that relies on several multiprotein complexes including the ULK1/ULK2, PtdIns3K and ATG9 complexes as well as the ATG12 and LC3 conjugation systems that serve initiation, vesicle nucleation and phagophore expansion, respectively [4]. The lipidation of LC3 into the membranes of such nascent phagophores serves the recruitment of ubiquitinated SQSTM1 (sequestosome 1) attached to autophagic substrates. Subsequent autophagosome maturation and lysosomal fusion leads to the generation of autolysosomes and the proteolytic degradation of autophagic cargo molecules by lysosomal hydrolyses thus allowing for the recycling of metabolic building blocks and the re-establishment of energetic homeostasis [5–8].

Harnessing the beneficial effect of autophagy on organismal health by pharmaceutical inducers offers a tempting approach to address the need for extending health- and potentially life span emerging from ageing populations of modern societies. Thus, autophagy-mediated bacterial clearance, limitation of inflammatory processes and (re)-establishment of cellular redox homeostasis facilitates the prevention of cancer, has the capacity to limit a decline of immune functions and halt tissue degeneration [9–12]. Pharmaceutical autophagy inducers can be employed to complement the positive health effects of caloric restriction, to overcome fasting-related impacts on productivity that are incompatible with socioeconomic standards [13, 14]. In addition to prophylactic effects, autophagy induction also has therapeutic potential. Particularly in neurodegenerative diseases autophagy was shown to be able to halt and partially revert disease symptoms, which unveils an unmet medical need that awaits urgent exploration.

**Questions**

1. How can small molecules with autophagy-inducing capacity be identified by automated drug discovery?
2. How can autophagic flux inducers be differentiated from agents that interfere with lysosomal functions?

# 1    Drug Discovery for Pharmaceutical Autophagy Induction

Here we detail a screening workflow employing an LC3-biosensor co-culture assay that can be employed on robotized high throughout (HT) drug discovery platforms which encompass automated sample preparation and high content bioimaging. Subsequent image processing and phenotypic clustering facilitates the identification of novel, non-toxic autophagic flux inducers with potentially diverse mode of actions.

# 2    Step-by-Step Protocol

## 2.1    Consumables and Equipment

### 2.1.1    Standard Plastic Ware
- Standard cell culture vessels (in our case T175 cm² flasks with filter cap, Corning; 431080)
- Sterile plastic pipettes (5 mL, 10 mL, 25 mL)
- Disposable cell counting chamber slides for automated cell counting and viability assessment (in our case by a Countess FL2 from ThermoFisher Scientific, AMQAF2001) (*see* **Answer 1**)

### 2.1.2    Lab Automation Compatible Plastic Ware
- AP384 P30XL pre-sterilized polypropylene automation-grade tips (Beckman Coulter, A22288)
- Barcoded µClear® 384-well polystyrene cell culture microplates with black walls and thin foil bottom (Greiner Bio-one, 781091)
- V-bottom, clear polystyrene 96- and 384-well microplates with conical bottom (Greiner Bio-one, 781281)
- Disposable spinner flasks (500 mL; 1 L) (Corning; 3153) (*see* **Answer 2**)
- Nalgene™ disposable polypropylene robotic reservoir (ThermoFisher Scientific, 1200-1301)
- Disposable polypropylene automation reservoir with pyramid bottom (Porvair Sciences, 390001)
- Adhesive aluminum seal for storage of microtiter plates before image acquisition.

### 2.1.3    Equipment
- Standard humidified cell culture incubator at 37 °C with a 5% $CO_2$ atmosphere; in our case Heracell™ 150i (ThermoFisher Scientific, 51032719)
- Sterile class II biological safety cabinets such as the MSC-Advantage™ (ThermoFisher Scientific)
- Temperature-controlled magnetic stirring plate
- Automated liquid handling platform in our case equipped with a FX$^P$ laboratory automation workstation (Beckman Coulter) equipped with a dual arm system encompass-

ing a multichannel (compatible with 200-µL 96-channel and 30-µL 384-channel heads) and a span-8 pipettor (A31844) as well as a gripper for on deck labware transport; in our case integrated with a 24-inch servo shuttle (Beckman Coulter) for robotic access and located in a sterile class II biological safety cabinet (Noroit, Bouaye, France); an industrial grade 6 axis Motoman HP3JC robot (Yaskawa, Kitakyushu, Japan) equipped with a servo-electric gripper (PTM mechatronics), a Cytomat™ 24 automated plates hotel, a Cytomat™ 6000 automated cell culture incubator (37 °C with 5% $CO_2$) and a Cytomat™ 2 automated 4 °C plate storage (all from ThermoFisher Scientific), vacuum operated delidder (Beckman Coulter), multiple barcode readers for sample tracking, a standard automated reagent dispenser (MultiDrop™ combi; ThermoFisher Scientific, 5,840,300) all operated with Sami EX scheduling software (Beckman Coulter)

- Automated fluorescence bioimager such as in our case two sister-built IXM-XL automated widefield microscopes (Molecular Devices) (*see* **Answer 3**) equipped with Spectra X light sources (Lumencor), and Zyla sCMOS cameras (Andor), 20 X PlanAPO objectives (Nikon), and multiband filters for the excitation and emission of mCherry, GFP and Hoechst 33342

- Freely available R software (https://www.r-project.org) employing R-packages as detailed below for image analysis, data curation, quality control and statistical evaluation (*see* **Answer 4**)

### 2.1.4 Cell Lines and Reagents

- Human osteosarcoma U2OS biosensor cells stably GFP-LC3 as produced by transduction with Lentibrite (Merck-Millipore, 17-10193) lentiviral particles or stably transfected with a GFP-LC3-coding plasmid [15, 16] or the mCherry-GFP-LC3 tandem reporter [17]

- Cell culture media and consumables such as Dulbecco's Modified Eagle's Medium (DMEM) supplemented with 1% non-essential amino (ThermoFisher Scientific, 11140-035) and 10% heat inactivated fetal calf serum (Merck-Millipore, F7524-500ML), phosphate-buffered saline (PBS) pH 7.4 (Gibco, ThermoFisher Scientific, 10010023) and TrypLE™ enzyme (ThermoFisher Scientific, 12604039)

- 36.5–38% formaldehyde (Merck-Millipore, F8775) diluted with PBS to a final concentration of 4%

- 10 mg/mL Hoechst 33342 (ThermoFisher Scientific, H3570) diluted to a final concentration of 2 µg/mL

- Dimethylsulfoxide (DMSO; Merck-Millipore, D8418)

- Chemical inducers and inhibitors of autophagic flux such as torin1 (R&D Systems, 4247/10) and bafilomycin $A_1$ (R&D Systems, 1334/10 U)

- Suitable compound libraries arrayed in either 96-well or in 384-well format. In our case chemicals came in 384-well microplates at 10 mM (with border columns suitable for controls) and were stored at −80 °C until use

## 3          Methods

### 3.1          Cell Culture

- Cells are kept in a humidified, temperature-controlled cell culture incubator at 37 °C in 5% $CO_2$ atmosphere and are routinely passaged using supplemented cell culture media (*see* **Answer 5**).
- Cells are adapted to culture for a minimum of 1 week in complete medium before being used for experimentation.
- Cells are passaged 3 times per week and confluency is kept between 40% and 80% (*see* **Answers 6** and **7**).
- For experimental purposes GFP-LC3 and tandem reporter cells are seeded in a 1:1 ratio at a total density of $1.5\times10^3$ cells per well of a 384-well plate.

### 3.2          Scheduled Automation Workflow (for a Detailed Description Please Refer to Reference [15])

- In short, the programmed schedule looks as follows: plates are transported form the plate hotel to the automated dispenser for cell seeding. Then cells are transferred to the incubator and are let adapt to culture conditions for 24 h (*see* **Answer 8**). Next cells are placed on the deck of the liquid handler together with pre-diluted compounds arrayed in a 384-well plate at 4 °C. Cells are treated with pre-diluted compounds by means of disposable tips mounted to a 384-channel head. Then the cells are incubated for 24 h at 37 °C and 5% $CO_2$, before media removal and addition of 4% formaldehyde containing Hoechst 33342. After 30 min the fixative is removed, PBS is being added and the plates are sealed and stored at 4 °C until image acquisition (*see* **Answer 9**).

### 3.3          Preparation of Drugs and Solutions

- The MTORC1 and MTORC2 inhibitor torin1 was used as positive control for autophagy induction on each plate for normalization. Torin1 stocks are diluted at a concentration of 300 μM in DMSO and are kept at −20 °C protected from light exposure.
- The lysosomal acidification inhibitor bafilomycin $A_1$ (BafA1) is employed as a positive control for autophagic flux inhibition. BafA1 is diluted in DMSO at a concentration of 100 μM and are kept at −20 °C protected from light exposure (*see* **Answer 10**).
- One day before treatment "working plates" of all compounds from the chemical libraries are prepared at 10× of the final concentration in medium and are stored at 4 °C overnight. (*see* **Answers 11–14**).
- Controls such as PBS, DMSO, torin1 and BafA1 (as untreated-, vector-, autophagic flux inducer- and autophagic flux inhibitor control, respectively) are prepared in the

border columns of a 384 V-shape plate and are then distributed at 10x of the final concentration (torin1 3 µM; BafA1 1 µM) to all "working plates" containing chemical drugs.

## 3.4 Cell Seeding and Treatment

- Cells ($1.5 \times 10^3$ cells; tandem reporter and GFP-LC3 cells in a 1:1 ratio) are seeded in 45 µl DMEM per well into imaging compatible µClear® 384-well cell culture microplates ("cell plates") and allowed to adapt for 24 h at 37 °C and 5% $CO_2$ in the Cytomat™ 6000 automated cell culture incubator (*see* **Answers 15** and **16**).
- "Working plates" containing compounds and controls at 10× the final concentration are placed in the Cytomat™ 2 at 4 °C.
- "Cell plates" and "working plates" are transferred to the liquid handler and cells are treated at final concentration by addition of 5 µL (10× concentrated chemicals and controls) from the "working plates" to the cell plates. The final volume then is 50 µL (*see* **Answer 17**).
- Treated "cell plates" are incubated at 37 °C and 5% $CO_2$ in the Cytomat™ 6000 automated cell culture incubator.
- Following incubation cells are fixed with 4% formaldehyde containing 2 mg/L Hoechst 33342 for 30 min at room temperature (*see* **Answer 18**).
- Then formaldehyde is removed and 50 µL of PBS are added to each well before plates are sealed with adhesive aluminum foil.

## 3.5 Image Acquisition and Analysis

- Images are acquired using two sister-built automated fluorescence microscopes ImageXpress IXM XL microscopes or equivalent bioimagers, equipped with 20X PlanApo objectives and DAPI, FITC and TRITC triple bandpass filters. MetaXpress (Molecular Devices) or equivalent software is used for image acquisition. A minimum of 4 view fields are captured to ensure an adequate number of cells for subsequent statistical evaluation. The DAPI channel is used for visualizing Hoechst-stained nuclei, the FITC channel for detecting GFP-LC3 dots, and the TRITC channel for assessing autophagic flux in complement to the FITC filter (*see* **Answers 19** and **20**).
- A robotic arm is employed to expedite the acquisition process and handle large batches of plates.
- Laser autofocus settings are adjusted manually and are consistently used for the same batch of plates in one experiment (*see* **Answer 21**).

## 3.6    Image Processing and Segmentation

- Images are processed and segmented using the open-access R software with several packages, including EBImage (available on the Bioconductor repository), RBioFormats (https://github.com/aoles/RBioFormats), MetaxpR (https://github.com/asauvat/MetaxpR), and MorphR (https://github.com/kroemerlab/MorphR) (Fig. 1).

- Images are normalized plate-wise based on the pixel intensity distribution of controls. The image histograms are scaled between the 0.0005% and 99.9995% intensity quantiles to remove extreme intensities and enhance contrast.

- For nuclei, background subtraction is applied using a Gaussian convolution kernel to denoise the images, which are then rescaled using a sigmoidal transformation. A threshold computed with Otsu's method [18] is used for binarization, followed by three morphological operations to clean the obtained mask.

- For images acquired with the FITC and TRITC filter sets, normalization within an automatically calculated range of intensities enhances the cytoplasmic signal. After scaling with a sigmoid transformation, a median filter is applied, and a binarized mask is obtained using Otsu's method.

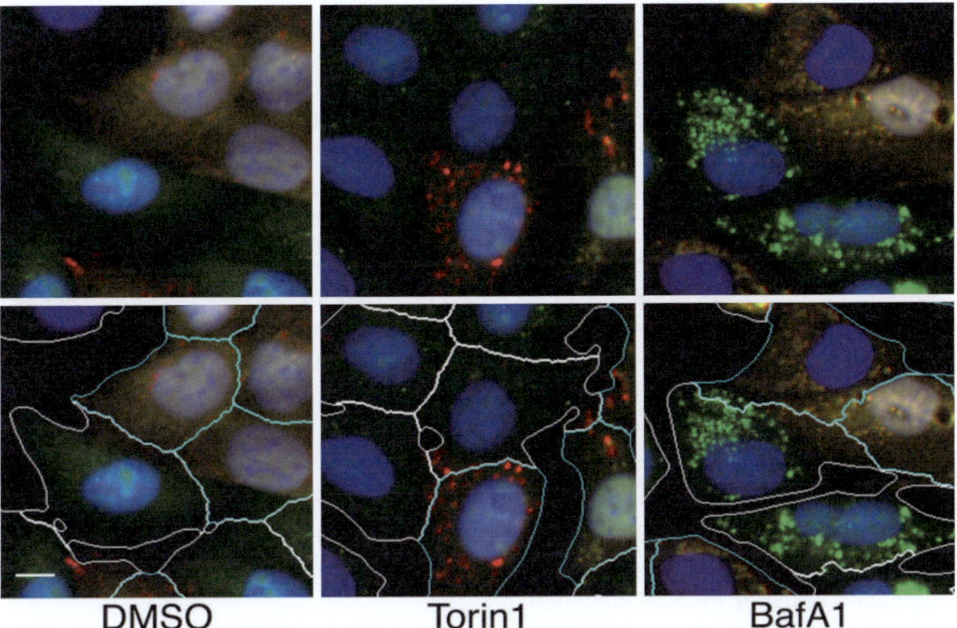


**Fig. 1** Automated image segmentation. A co-culture of human osteosarcoma U2OS cell stably expressing GFP-LC3 and mCherry-GFP-LC3 were treated with vector control (dimethylsulfoxide, DMSO), the MTORC inhibitor torin1 (300 nM) as prototype autophagic flux inducer or the lysosomal acidification inhibitor bafilomycin A$_1$ (BafA1; 100 nM) for 24 h. Representative images of cocultures (upper row) alongside their segmentation masks (lower row) are depicted. Nuclear and cytoplasmic regions were segmented. White segmentation masks define GFP-LC3-expressing cells and blue masks mark mCherry-GFP-LC3 biosensors. Size bar: 10 μm

- To quantify LC3 aggregation and fluxes, cytoplasmic regions with higher GFP and mCherry fluorescence intensities are identified. Two sets of operations are applied to optimize the detection of aggregates of different sizes: a maximum projection of top hat filters of various sizes combined with a sigmoidal transformation and Otsu thresholding detects small dots, whereas larger areas of GFP aggregates are identified using a combination of top hat filters of larger sizes and adaptive thresholding.
- Cellular parameters are computed and extracted using distinct masks. The nuclear mask enables the computation of the nuclear area in square-µm and the intensity of Hoechst signal by measuring the pixel intensity within the labeled regions. The cytoplasmic area, GFP and mCherry intensities are calculated based on the cytoplasmic mask. Additionally, the total surface area of dots per cell and the number of dots per cell are computed to assess autophagy induction. For a more comprehensive characterization of different autophagy phenotypes, the average distance from dots to the centroid of dots and nucleus is determined to assess the dot distribution within the cell. All parameters are then compiled into a data frame with unique identifiers.

## 3.7    Data Analysis

- Data extracted from image segmentation are analyzed with analytic R packages freely available at https://github.com/kroemerlab.
- Fluorescence intensities from Hoechst and GFP signals are normalized by plate using the maximum of distribution (MOD) in controls.
- Gating steps based on intensity and size parameters are applied to detect and exclude dead cells and debris from the data set.
- A random forest model trained on a dataset of 7680 manually labeled images is used to control data quality, excluding potentially problematic data (arising from technical limitations, physical interferences or biological complications).
- Data are aggregated by well and normalized for viability and autophagy data using appropriate methods. Quality control is conducted by plotting boxplots of positive and negative controls and heatmaps of all plates to verify dataset consistency.
- A flux inhibition score is then determined using a linear regression of the surface of autophagosomes compared to the surface of autophagosomes and autolysosomes in control data (for a detailed description of the method please refer to *reference* [15]).
- Statistical indicators are used for data reduction such as computing the average and standard deviation for each condition.
- LC3 aggregation and autophagic flux are evaluated using GFP dot number and surface measurements.
- Based on their effects on viability, flux inhibition scores, as well as the number of GFP-LC3 dots and the GFP-LC3 surface, screened compounds are classified as toxic, autophagic flux inhibitors, autophagy inducers, or inactive, facilitating hit selection (Fig. 2).

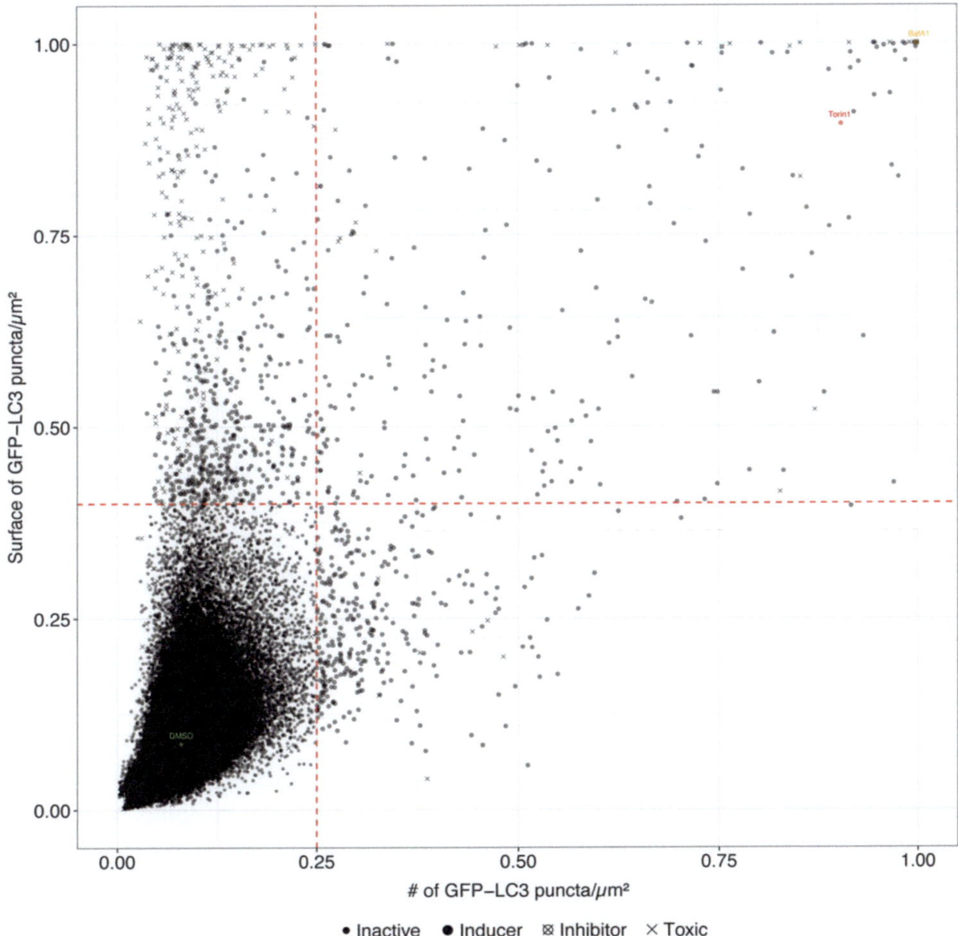


**Fig. 2** Phenotypic clustering and data visualization for hit identification. A representative chemical library (of about $2{\times}10^5$ agents) was screened using a co-culture assay of human osteosarcoma U2OS cell stably expressing GFP-LC3 and mCherry-GFP-LC3. Automated fluorescence microscopy coupled with image segmentation algorithms were employed for the assessment of viability, autophagic flux inhibition and autophagy induction. The surface of GFP-LC3$^+$ puncta was plotted against the number GFP-LC3$^+$ puncta. Dimethylsulfoxide (DMSO) was used as vector control torin1 (300 nM) served as autophagic flux inducer and bafilomycin A$_1$ (100 nM; BafA1) was employed as flux inhibitor. Small dots mark inactive compounds, large dots represent autophagy inducing agents, crosses mark toxic conditions and crossed circles indicate autophagic flux inhibitors. Arbitrary red lines were set to indicate thresholds for hit selection, aiding the identification of potentially active compounds

## 4    Advantages and Disadvantages

- Coculture assays combining complementary biosensors expressing a distinct set of fluorescent proteins that allow for retrospective signal segregation allow for a cost-effective multiplex analysis that yield rich data sets for quality control, multifactorial data mining and hit identification. In combination with lab automation and robotized screening workflows this assay offers a robust way for screening large chemical libraries for autophagic flux inducers. Moreover, the present workflow can be flexibly adapted for validation and mode of action studies that mandatorily follow the hit identification from the primary screen. Furthermore, as compared to plate reader-based methods or flow cytometry, image based phenotypic drug screening offers the possibility to extract additional features such as morphological changes or subcellular signal distribution, altogether offering additional means of result classification.
- The outlined approach is limited by the initial expenses for lab automation and fluorescent confocal bioimagers, which can be challenging for academia. Moreover, the sheer volume of data necessitates high volume (raid-configured) data storage capacity in the petabyte range and high-performance computing for secure data storage and timely data analysis.

## 5    Concluding Remarks

Drug discovery is often hampered by the lack of appropriate assays compatible with large-scale screening campaigns either due to issues with sensitivity, reproducibility or throughput. Moreover, most drug screening approaches are limited to only one assessed parameter. Here we developed an automation friendly workflow that is based on a multicolor biosensors coculture assay facilitating multiplex image-based analysis. Image analysis routines coupled with a sophisticated data processing pipeline allow for phenotypic clustering and for the identification of autophagic flux within large chemical compound collections. Compared to the manual assessment or other screening protocols, the outlined method offers advantages in throughput, reproducibility, and the possibility to extract multiple phenotypic features, while limiting batch difference due to standardized and robotized screening routines, rigorous quality control algorithms and sophisticated normalization routines.

**Take Home Message**
All together, the present protocol describes the essential components of a drug discovery pipeline that can be used for both academic purposes as well as for the pharmaceutical industry.

**Alternative Reagents, Approaches and Considerations**

1. Alternatively, environmentally friendly, economical Countess reusable slide can be employed.
2. Alternatively, environmentally friendly, reusable, autoclavable glass Spinner flasks can be employed.
3. Alternative automated microscopes such as an IXM-C automated spinning disc confocal microscope (Molecular Devices, San Jose, CA, USA) equipped with an Auro II light source (Lumencor, Beaverton, OR, USA), and a Zyla sCMOS cameras (Andor, Belfast, Northern Ireland) can be employed for image acquisition.
4. Alternative image analysis software can be used for image segmentation and feature extraction.
5. A large number of stocks from an early passage number are stored in liquid nitrogen and those are used to frequently restart cultures in order to keep the total passage number low.
6. Low- or high-density cell culture can affect cellular metabolism and might interfere increase the baseline of autophagy thus decreasing the signal to noise ratio and negatively impacting the sensitivity of the assay
7. Optimal cell culture conditions might differ for other cell lines. The guidelines available from the furnisher or repository should be strictly regarded.
8. It is important to let cells sediment by gravity and adhere to the plasticware before transfer to the cell culture incubator.
9. Autophagy can be monitored at different time points to visualize autophagsome formation (typically analyzed upon 6 h of incubation) and autophagic flux (prominently visible at later time points).
10. Stocks of positive controls should yield for the total screen to avoid batch dependent variations.
11. If drugs are prepared long ahead of screening diluted agents in "working plates" can be stored at −20 °C for several weeks.
12. V-bottom plates are used to decrease dead volume.
13. Plates should be sealed to avoid evaporation and contamination.
14. Alternative to conventional liquid transfer contactless dispensers such the Echo™ acoustic dispenser can be used.
15. Cells are seeded in media complemented with FBS and non-essential amino acids. Antibiotics should be limited but can be added to avoid bacterial contamination in non-sterile conditions.
16. An excessive number of cells that would lead to more than 70–80% confluence in untreated controls at the end of the experiment needs to be avoided as this would impinge on baseline autophagy.
17. Please *see*.

18. Formaldehyde is a carcinogen that needs to be handled according to safety regulations.
19. Z-stacks were acquired and collapsed into a composite image employing a proprietary algorithm within the MetaXpress acquisition software to reduce data volume.
20. The exposure time needs to be set according to signal intensity for each filter set.
21. HTS screening campaigns are sensitive to batch variations. It is recommended to use plasticware as well as media and supplements from one batch.

**Acknowledgments** O.K. receives funding by Association pour la recherche sur le cancer (ARC) and Institut National du Cancer (INCa). G.K. is supported by the Ligue contre le Cancer (équipe labellisée); Agence National de la Recherche (ANR)—Projets blancs; AMMICa US23/CNRS UMS3655; ARC; Cancéropôle Ile-de-France; Fondation pour la Recherche Médicale (FRM); a donation by Elior; Equipex Onco-Pheno-Screen (ANR 21-ESRE-0028); European Joint Programme on Rare Diseases (EJPRD); European Research Council Advanced Investigator Award (ERC-2021-ADG, ICD-Cancer, Grant No. 101052444), European Union Horizon 2020 Projects Oncobiome, Prevalung (grant No. 101095604) and Crimson (No. 101016923); Fondation Carrefour; INCa; Institut Universitaire de France; LabEx Immuno-Oncology (ANR-18-IDEX-0001); a Cancer Research ASPIRE Award from the Mark Foundation; the RHU Immunolife; Seerave Foundation; SIRIC Stratified Oncology Cell DNA Repair and Tumor Immune Elimination (SOCRATE); and SIRIC Cancer Research and Personalized Medicine (CARPEM). M.M. and Y.P. were supported by the China Scholarship Council (CSC, file n°202106320287, n°2021069100).

**Conflicts of Interest** O.K. is a scientific co-founder of Samsara Therapeutics. GK has been holding research contracts with Daiichi Sankyo, Eleor, Kaleido, Lytix Pharma, PharmaMar, Osasuna Therapeutics, Samsara Therapeutics, Sanofi, Sutro, Tollys, and Vascage. GK is on the Board of Directors of the Bristol Myers Squibb Foundation France. GK is a scientific co-founder of everImmune, Osasuna Therapeutics, Samsara Therapeutics and Therafast Bio. GK is in the scientific advisory boards of Hevolution, Institut Servier, Longevity Vision Funds and Rejuveron Life Sciences. GK is the inventor of patents covering therapeutic targeting of aging, cancer, cystic fibrosis and metabolic disorders. GK's brother, Romano Kroemer, was an employee of Sanofi and now consults for Boehringer-Ingelheim. GK's wife, Laurence Zitvogel, has held research contracts with Glaxo Smyth Kline, Incyte, Lytix, Kaleido, Innovate Pharma, Daiichi Sankyo, Pilege, Merus, Transgene, 9 m, Tusk and Roche, was on the on the Board of Directors of Transgene, is a cofounder of everImmune, and holds patents covering the treatment of cancer and the therapeutic manipulation of the microbiota. The funders had no role in the design of the study, in the writing of the manuscript, or in the decision to publish the results.

# References

1. Rubio-Tomas T, Sotiriou A, Tavernarakis N. The interplay between selective types of (macro) autophagy: mitophagy and xenophagy. Int Rev Cell Mol Biol. 2023;374:129–57. https://doi.org/10.1016/bs.ircmb.2022.10.003.

2. Lopez-Otin C, Blasco MA, Partridge L, Serrano M, Kroemer G. Hallmarks of aging: an expanding universe. Cell. 2023;186:243–78. https://doi.org/10.1016/j.cell.2022.11.001.

3. Hofer SJ, Kroemer G, Kepp O. Autophagy-inducing nutritional interventions in experimental and clinical oncology. Int Rev Cell Mol Biol. 2022;373:125–58. https://doi.org/10.1016/bs.ircmb.2022.08.003.

4. Annis MG, et al. Endoplasmic reticulum localized Bcl-2 prevents apoptosis when redistribution of cytochrome c is a late event. Oncogene. 2001;20:1939–52. https://doi.org/10.1038/sj.onc.1204288.

5. Gatica D, Lahiri V, Klionsky DJ. Cargo recognition and degradation by selective autophagy. Nat Cell Biol. 2018;20:233–42. https://doi.org/10.1038/s41556-018-0037-z.

6. Galluzzi L, Bravo-San Pedro JM, Levine B, Green DR, Kroemer G. Pharmacological modulation of autophagy: therapeutic potential and persisting obstacles. Nat Rev Drug Discov. 2017;16:487–511. https://doi.org/10.1038/nrd.2017.22.

7. Chen G, et al. 3,4-Dimethoxychalcone induces autophagy through activation of the transcription factors TFE3 and TFEB. EMBO Mol Med. 2019;11:e10469. https://doi.org/10.15252/emmm.201910469.

8. Klionsky DJ, et al. Guidelines for the use and interpretation of assays for monitoring autophagy (4th edition)(1). Autophagy. 2021;17:1–382. https://doi.org/10.1080/15548627.2020.1797280.

9. Menzies FM, Fleming A, Rubinsztein DC. Compromised autophagy and neurodegenerative diseases. Nat Rev Neurosci. 2015;16:345–57. https://doi.org/10.1038/nrn3961.

10. Liu GY, Sabatini DM. mTOR at the nexus of nutrition, growth, ageing and disease. Nat Rev Mol Cell Biol. 2020;21:183–203. https://doi.org/10.1038/s41580-019-0199-y.

11. Bravo-San Pedro JM, Pietrocola F. Fasting and cancer responses to therapy. Int Rev Cell Mol Biol. 2022;373:107–23. https://doi.org/10.1016/bs.ircmb.2022.08.002.

12. Castoldi F, Kroemer G, Pietrocola F. Spermidine rejuvenates T lymphocytes and restores anti-cancer immunosurveillance in aged mice. Onco Targets Ther. 2022;11:2146855. https://doi.org/10.1080/2162402X.2022.2146855.

13. Jia K, Levine B. Autophagy is required for dietary restriction-mediated life span extension in C. elegans. Autophagy. 2007;3:597–9. https://doi.org/10.4161/auto.4989.

14. Madeo F, Zimmermann A, Maiuri MC, Kroemer G. Essential role for autophagy in life span extension. J Clin Invest. 2015;125:85–93. https://doi.org/10.1172/JCI73946.

15. Kabeya Y, et al. LC3, a mammalian homologue of yeast Apg8p, is localized in autophagosome membranes after processing. EMBO J. 2000;19:5720–8. https://doi.org/10.1093/emboj/19.21.5720.

16. Forveille S, et al. High throughput screening for autophagy. Methods Cell Biol. 2021;165:89–101. https://doi.org/10.1016/bs.mcb.2020.12.011.

17. Agrotis A, Pengo N, Burden JJ, Ketteler R. Redundancy of human ATG4 protease isoforms in autophagy and LC3/GABARAP processing revealed in cells. Autophagy. 2019;15:976–97. https://doi.org/10.1080/15548627.2019.1569925.

18. Andrade AR, et al. Recent computational methods for white blood cell nuclei segmentation: a comparative study. Comput Methods Prog Biomed. 2019;173:1–14. https://doi.org/10.1016/j.cmpb.2019.03.001.

## Further Reading

Forveille S, Leduc M, Sauvat A, Cerrato G, Kroemer G, Kepp O. High throughput screening for autophagy. Methods Cell Biol 2021;165:89–101, Elsevier.

https://github.com/kroemerlab

https://www.r-project.org

https://www.bioconductor.org

# Index


**A**

Acetylation-generated motifs, 287
Acidification, 258
Acquisition process, 54
Active reading, 5, 6
Acute kidney injury (AKI), 213
Aggregation-prone proteins, 128, 134, 143, 149, 152
Aggrephagy, 35
Aging, 171, 172
  autophagy declines with age, 172, 173
  autophagy modulators
    MTOR inhibitor, 189
    NAD⁺, 190
    spermidine, 191
    TFEB modulator, 189, 190
    Urolithin A, 191
  CMA in, 298, 299
  impaired autophagy, 174
  lifespan and healthy life expectancy, 177
    genetic activation, 175
    inflammaging, 180–183
    lipophagy, 179
    lysophagy, 180
    mitophagy, 177, 179
    nucleophagy, 180
    selective autophagy, 177
  lifestyle interventions, 187
    dietary interventions, 187, 188
    exercise interventions, 188, 189
  longevity paradigm and autophagy, 174
  non-canonical autophagy, 185–187
  organ- specific autophagy, 183–185
  overactivation of autophagy, 175, 176
  suppression of autophagic activity, 194
  upregulation of RUBCN promotes autosis, 195, 196
Aging of immune system, 184
AKT1, 290
Alzheimer disease (AD), 177, 296
Amino-acid starvation, 31, 35
AMP-activated protein kinase (AMPK), 217, 257–258, 268
Amphisomes, 30, 36, 38, 39, 141, 142
  autophagosomes fuse with multivesicular endosomes, 35
  fuse with lysosomes, 37
Amyloid beta precursor protein (APP), 60
Apermidine, 60
Apoptosis, 265
Assessment, 24–26
Astrocytes, 132, 151
Atg11, 87
ATG12–ATG5-ATG16L1 complex, 217
ATG2, 109
ATG2A/ATG2B, 74
ATG5, 106, 268
ATG7, 249
ATG7 mutations, 136
ATG7-deficient patients, 249
Atg8 family, 47, 109
ATG8 turnover assay, 262, 263
Atg8-family interacting motif (AIM), 74, 75, 77
Atg8-family protein, 74, 75, 77, 118, 119, 182, 185, 252



© The Editor(s) (if applicable) and The Author(s), under exclusive license to
Springer Nature Switzerland AG 2025
B. Loos, D. J. Klionsky (eds.), *Autophagy - From Molecular Mechanisms to Flux
Control in Health and Disease*, Learning Materials in Biosciences,
https://doi.org/10.1007/978-3-031-88121-3

Atg8-family protein conjugation reactions,
        116, 117
Atg8ylated phagosomal and endosomal
        membranes, 186
Atg8ylation, 185
Atg32, 87
Atherosclerosis disease, 294
ATL (atlastin) proteins, 140
ATP, 214
Autolysosome, 9, 15, 26, 30, 38, 39, 46, 59, 63,
        197, 263
Autolysosome size, analysis of, 263
Automated fluorescence bioimager, 314
Autophagic activation, 183
Autophagic compartment, 37
Autophagic flux inducers, 173, 258, 312
    advantages and disadvantages, 320
    harnessing beneficial effect of, 312
    methods
        cell culture, 315
        cell seeding and treatment, 316
        data analysis, 318
        drugs and solutions, preparation of,
            315, 316
        image acquisition and analysis, 316
        image processing and segmentation,
            317, 318
        scheduled automation workflow, 315
    identifying defects in, 246
    pharmaceutical autophagy induction, drug
        discovery for, 313
    by protein aggregates, impairment of,
        142, 143
    step-by-step protocol
        cell lines and reagents, 314
        equipment, 313, 314
        lab automation compatible plastic
            ware, 313
        standard plastic ware, 313
Autophagic induction, 247, 249, 250
Autophagic loss, 128
Autophagic lysosome reformation, defects in,
        141, 142
Autophagic vacuolar myopathies (AVMs), 235
Autophagic vacuole, 37, 104
Autophagic-lysosomal proteolysis, 147, 152
Autophagolysosome, 26
Autophagosomal membrane, 51
    orgin of, 102

early biochemical studies failed to
        resolve provenance of, 104
    early morphological studies, 102
    phagophore formation, 104
Autophagosomal structures, 112–113
Autophagosome biogenesis, 129, 137,
        141–143, 147, 151, 152, 176
    defects in, 134–137
Autophagosome biosynthesis, 197, 198
Autophagosome flux
    definition of, 58
    in pathology and autophagy dysfunction, 60
Autophagosome formation, 173
    biochemical reactions and structural
        arrangements, 111
        Atg8-family protein conjugation
            reactions, 116, 117
        closing of phagophore membrane,
            117, 118
        feedback regulation of ULK and
            PIK3C3/VPS34 complexes, 116
        lipid transfer and scrambling feed
            forming autophagosome, 117
        PIK3C3/VPS34, activation of, 114, 116
        PtdIns3P formation, 114, 116
        signalling via MTOR and ULK, 113
        ULK complex, scaffolding func-
            tion of, 113
    early biochemical studies, 105
    in mammalian cells, 103
    morphological studies, 105, 106
        ER and other membranes implication,
            110, 111
        initiation, 106, 107
        nucleation, 107–109
        phagophores and sealing off of double
            membranes, 109, 110
    non-selective and selective autophagy at the
        initiation stage, 120
    pathways, 118–120
Autophagosome maturation, 140, 141, 197
Autophagosome tethering compounds
        (ATTEC), 146
Autophagosomes, 9, 26, 27, 30, 34–35, 45, 46,
        55, 63, 72, 75, 101, 102, 104, 106,
        109, 111, 117, 118, 121, 151,
        222, 232
    autophagosomes/amphisomes fuse with
        lysosomes, 37

and cargo, 46
cytoplasmic cargo, 33, 35
EM morphology of, 102
degradation of, 51
fuse with multivesicular endosomes, 35
fuse with lysosomes, 37
fusion inhibitor, 48
LC3-II, 47
lipid transfer and scrambling, 117
pool sizes, 59
simultaneous detection of, 263
spherical, limited by two lipid bilayers, 33, 35
and super-resolution techniques, 53, 54
Autophagy, 5, 6, 26, 30, 31, 39, 43, 85, 276,
        284, 304, 312
    activity, 45, 47, 58, 62
    addiction, 60
    assessment
        CLEM, 54–56
        electron microscopy, 46
        fluorescence microscopy, 49, 51, 52
        super-resolution techniques, 53, 54
        Western blotting, 47, 48
    Atg8-family proteins, 74, 75, 77
    cardiac ischemia models and reperfu-
            sion, 264
        cell culture, 264, 265
        ex vivo organ perfusion, 265, 266
        *in vivo* models, 266
        slice culture, 265
        working heart model, 266
    cessation, 258, 259
    components and processes of, 257
    concepts and methodology, 261
        flux assessment, 262, 263
        immunoblotting, 261, 262
        immunohistology/immunofluores-
                cence, 262
    core machinery, 72–74
    degradation, 258
    dysfunction, 45
    dysregulation, 217
    engulfment, 258
    implication of, 267
        cardiac hypertrophy, 268, 269
        diabetic cardiomyopathy, 267
        heart failure, 269
        myocardial ischemia/reperfusion
                injury, 268

    initiation of, 257
    mitochondrial mass, assessment of,
            263, 264
    MitoTimer, 264
    modulation of, 270
    monitoring, 44, 45
    selective, 71, 72
    steps of, 273
Autophagy flux, 49, 62, 275
    assessment, 58
    skeletal muscle diseases, by decreased, 235
        colchicine and vincristine toxicity, 239
        CQ and HCQ, 239
        Danon disease, 236, 237
        Pompe disease, 238, 239
        XMEA, 235
Autophagy flux measurement, 58, 60, 62
Autophagy induction, skeletal muscle dis-
            eases, 239
    congenital myopathies, 241, 242
    muscular dystrophies, 241
    mutations in machinery, 240, 241
Autophagy machinery with proteinaceous
            cargo, 60
Autophagy modulators, 193
    MTOR inhibitor, 189
    NAD$^+$, 190
    spermidine, 191
    TFEB modulator, 189, 190
    Urolithin A, 191
Autophagy pathway intermediates, 46, 51, 65
Autophagy related (ATG) proteins, 217
Autophagy targeting chimeras (AUTAC), 147
Autophagy-mediated bacterial clearance, 312
AUTOphagy-TArgeting Chimera
            (AUTOTAC), 147
Autosis, 176


**B**
Bafilomycin A$_1$ (BafA1), 51, 258, 265,
            266, 315
*Becn1*, 48, 268
    degradation, 135
    overexpression, 269
BECN1-BCL2 complex formation, 175
Beta-propeller-protein-associated neurodegen-
            eration (BPAN), 136
Bethlem myopathy, 241

Blood vessels, 259
Blood-derived autophagy assessment, 62
Bloom's taxonomy, 18, 19
BNIP3L, 87

**C**
C9orf72, 134
Calorie restriction, 187
Calpastatin, 146
Calpeptin, 146
CaMKII, 296
Canonical motifs, 287
Carbamazepine (CBZ), 146
Cardiac hypertrophy, 276
Cardiac ischemia models, 264
Cardiac-specific Drp1 gene, 269
Cardinal articles, 301–303
Cardiomyocytes, 237, 259, 274
    function, 277
Cardioprotection, 270
Cardiovascular disorders, autophagy in, 267
    cardiac hypertrophy, 268, 269
    diabetic cardiomyopathy, 267
    heart failure, 269
    modulation, 270
    myocardial ischemia/reperfusion injury, 268
Cardiovascular function, CMA in, 294, 295
Cargo degradation, 232
Cargo recognition and sequestration, 232
Cargo recruitment, defects in, 137–140
CASP1 (caspase 1), 181
Cells, 264
Cell culture
    autophagic flux inducers, 315
    cardiac ischemia models and reperfusion,
        264, 265
Cell plates, 316
Cell seeding, 316
Cell-based response, 102
Cellular parameters, 318
Centronuclear myopathy, 253
Cessation, 258, 259
CGAS-STING1 pathway, 181, 182
Chaperone lysosomal chaperones, 290
Chaperone-assisted selective autophagy
        (CASA), 244, 253
Chaperone-mediated autophagy
        (CMA), 284–286

activation, 291, 292
    in aging, 298, 299
    cardinal articles, 301–303
    in cardiovascular function, 294, 295
    in circadian rhythm, 292
    functions of, 286
    in immune cells, 297
    in metabolism, 292–294
    in neurological function, 295, 296
    in stem cells, 298
    molecular components
        cytosolic chaperones, 288
        LAMP2A, 288, 289
        lysosomal chaperones, 289
    physiological outcomes of loss, 299
    regulation
        extra-lysosomal CMA regulators,
            290, 291
        lysosomal chaperones, 290
    restoration of, 300, 301
    substrates, 286–288
Chaperones, 304
Chloroquine (CQ), 239, 265, 266
Chronic kidney disease (CKD), 213
Ciliogenesis, 221
Circadian clock, 292
Circadian rhythm, CMA in, 292
Claw domain, 80
CMA-competent lysosomes, 292
CMA-deficient HSCs, 298
Colchicine, 239, 246
Collagen type VI (COL6), 241
Colocalization, 54
Comparative proteomics, 295
Compensatory hypertrophy, 268
Confocal microscopes, 56
Congenital myopathies, 241, 242
Conjugation of LC3, 258
Correlative light and electron microscopy
        (CLEM), 54, 55, 60, 62, 64
    autophagy with multidimensional detail, 55
    cargo clearance study, 65
    vs. fluorescence microscopy, 65
    vs. SR-SIM, 65
    workflow, 56
Correlative microscopy, 60–66
COX8 (cytochrome c oxidase subunit 8), 263
Cristae, 39
Cutting-edge microscopy techniques, 62

Cvt pathway, 4, 86, 87
Cytoplasmic cargo, 33, 35, 40
Cytoplasmic chromatin fragments (CCF), 182
Cytoplasm-to-vacuole targeting (Cvt)
       pathway, 77
Cytosolic chaperones, 288
Cytosolic fatty acids, 214
Cytosolic HSPA8, 296

**D**
Danon disease, 236, 237
Data analysis, 318
*De novo* formation, 104
Degradation, 258
Diabetic cardiomyopathy (DCM), 267, 276
Diabetic syndrome models, 267
Diffuse distribution, 262
3D approach, 65
Direct stochastic optical reconstruction
       microscopy (dSTORM), 53
Disabled autophagy, 171, 172, 175
Double membrane structure, 46
Double-membrane phagophore, 232
Double-membraned structures, 63
Drift correction, 54
Drug discovery, for pharmaceutical autophagy
       induction, 313
DsRed-E5, 264
dSTORM, 62
Dynamics of early autophagy
       responses, 108
Dynamics of LC3 during autophagy, 110
Dynein-driven transport of neuronal autophago-
       somes, 140

**E**
Eat me signal, 120
Electron microscopy (EM), 31, 46, 63, 65
    acquisition, 56
    applications, 46
    artifacts, 46
    of membranous vacuoles, 102
    morphology of autophagosomes, 102
Electron tomography, 111
Endolysosomal vesicles, 104
Endoplasmic reticulum (ER), 101, 110–113
Endosomal microautophagy, 288

Endosomal sorting complexes required for
      transport (ESCRT) machinery, 141, 219
Energy balance, 294
Engulfment, 258
Everolimus, 143
Ex vivo organ perfusion, 265, 266
Exercise-induced autophagy, 231
Extra-lysosomal CMA regulators, 290, 291

**F**
Feedback loop, 143
Ferree 2013, 272
Ferritinophagy, 136
Fibroblasts, 259
FLAG, 222
Fluid flow mechanosensing, 216
Fluorescence microscopy, 49, 56, 60
    advantages, 52
    vs. CLEM, 65
    disadvantages, 52
    fluorescent probes, 49
    GFP, 49
    GFP-LC3, 49, 51
    GFP-LC3-RFP-LC3ΔG, 51
    mRFP-GFP-LC3 fluorescent construct, 51
    pool size analysis and pathway intermediate
      counts, 51
    RFP-LC3ΔG chimera, 51
Fluorescence-based microscopy approaches, 44
Fluorescent confocal bioimagers, 320
Fluorescent microscopes, 316
Fluorescent probes, 49, 53
Fluorescent timer protein, 264
Fluorochromes, 49
Flux assessment, autophagy and
      mitophagy, 262
    ATG8 turnover assay, 262, 263
    autolysosome size, analysis of, 263
    autophagosomes and autolysosomes,
      simultaneous detection of, 263
Flux measurement techniques, 275
Formaldehyde, 316
Formative assessments, 23
Fourier transform, 53
FOXA (forkhead box A), 187
Frontotemporal dementia (FTD), 134
Functional autophagy, 62
Fusion, 258

**G**

Gap junction/connexin proteins (GJ), 137
Genetic loss of CMA, 296
Genetic proof-of-concept approaches, 300
Glia, functions of autophagy in, 129–132
Glial autophagic impairment, 131
Glial autophagy, 131
Glomerular filtration rate (GFR), 216
Glomerulotubular balance, 216
Glucocerebrosidase, 141
Glucocorticoids, 291
Gluconeogenesis, 214
Golgi complex, 104
Golgi-to-lysosome transport, 291
G-protein-coupled receptors (GPCR), 217
Graphical organizer, 9
Green fluorescent protein (GFP), 31, 49
Growth hormone signaling, 291
GTP-bound RAB7, 140

**H**

Hamacher-Brady 2006, 271
Haploinsufficiency, 141, 269
HDAC inhibitors, 270
Healthy longevity, 172
Heart failure (HF), 269, 277
Heavy membrane compartment, 266
Hematoxylin and eosin (H&E), 235
Heterophagy, 102
Hit identification, 320
Homeostasis, 45
Homotypic vacuole fusion and protein sorting
      (HOPS) complex, 219
HOPS complex, 140
Hoshino 2013, 272
HSPA8, 289, 304
HTT, 143, 148
Huang 2011, 272
Human proteome bear, 287
Huntington disease (HD), 128
Hydrophobic pockets, 80
Hydroxychloroquine (HCQ), 239, 246

**I**

Ibidi® Pump system, 221
Image acquisition, 316
Image processing and segmentation, 317, 318
Immune cells, CMA in, 297
Immunoblotting, 261–263, 265

Immuno-electron microscopy, 37
Immunofuorescence microscopy, 112–113
Immunohistology/immunofluorescence,
      autophagy and mitophagy, 262
Impaired autophagy, 130, 174, 276
   diseases of human skeletal muscle caused
      by, 231, 232, 235
*In vitro* models, 116
*In vivo* models, 266
In-chamber ultramicrotome, 57
In-class learning activities, 23
Induction, 232
Ineffective strategies, 2
Inflammaging, 177, 180, 181, 185
   CGAS-STING1 pathway, 181, 182
   NLRP3 inflammasome, 181
   T cell-mediated inflammaging, 182, 183
Inflammasome, 260
Inflammation, 260
Inflammation-induced immunosuppression, 297
Interactive teaching, 21, 22
   assessment, 24–26
   learning activities, 23, 24
   learning goals, 22
Ischemia, 268, 276
Ischemia-reperfusion (I/R) injury, 260
Ischemic preconditioning, 260
Isolation membranes, 31

**K**

KFERQ-like motif, 287, 288, 291, 304
Kidney proximal cells, 214
Klionsky's Guidelines, 261

**L**

Lab automation compatible plastic ware, 313
*Lamp2*, 236, 268
LAMP2A, 284–286, 288, 289, 291, 292, 295,
      297, 304
*lamp2a* knockout, 293
*lamp2a* KO animals, 293, 298
LAMP2A monomers, 290
Langendorff model, 266
LC3 aggregation, 251, 274, 318
LC3 lipidation, 135
LC3-II, 222, 225
LC3-interacting region (LIR) motif, 74, 75, 77
Learning strategies, 3, 23, 24
   active reading, 5, 6

Bloom's taxonomy, 18, 19
flashcards and free recall, 14
graphs, figures and diagrams, 7, 8
homework, 10
    answer key, 11, 12
    make time, space and motivation, 12, 13
    quiz, 10, 11
needs, 2, 3
practice examination, 16, 17
previewing, 3, 4
question everything, 15, 16
study anytime, anywhere, without missing
    anything, 15
study strategies, 13
summarizing, 8, 9
teaching, 19, 20
test yourself, 14
using time wisely, 20, 21
Lentiviral transduction, 222
Lifespan extension mechanism, 171
Lipid phosphatidylinositol 3-phosphate
    (PtdIns3P), 105
Lipid stains, 222
Lipid synthesis, 122
Lipidation of LC3, 312
Lipidation reaction, 117
Lipophagy, 35, 178, 179, 217, 224
Lipopolysaccharide-induced macrophage
    activation, 295
Liquid-liquid phase separation (LLPS), 72
LIR docking site (LDS), 77
Loss-of-function mutations, 137
Lymphocytes, 198
Lysophagy, 178, 180
Lysosomal acidification inhibitor, 315
Lysosomal blockade, 275
Lysosomal chaperones, 289
Lysosomal dysfunction, 45, 189
Lysosomal enzyme activity, 198, 199
Lysosomal fusion, 312
Lysosomal GBA, 141
Lysosomal homeostasis, 132
Lysosomal HSPA8, 289
Lysosomal inhibitors, 51
Lysosomal lipolysis, 179
Lysosomal membrane, 288
Lysosomal protease inhibitors, 105
Lysosome, 46, 59, 104, 117, 119, 122, 284, 304, 305
Lysosomotropic amines, 105
LysoTracker, 50, 51

**M**
Macroautophagy, 5, 15, 16, 30, 43, 71, 172,
    217, 231, 232, 257, 312
Mammalian cell core autophagy machinery, 72
Mammalian macroautophagy, 25, 233
MAP1LC3/LC3, 47
MAPT/tau pathology propagation, 296
MAPT/tau spreading, 296
MAPT/tau-induced neurodegeneration, 300
McCrudden, 4
McGuire, 3–5
mCherry, 31
mCherry-EGFP tandem tag, 86
Mechanistic target of rapamycin (MTOR), 105,
    113, 118, 122, 149, 174, 189, 242
Membrane permeabilization (MOMP), 259
Membrane-associated SNCA/α-synuclein, 288
Membrane-bound SARs, 81, 83, 84
Metabolic control analysis, 58
Metabolic feedback, 45
Metabolic poisons, 265
Metabolic syndrome models, 267
Metabolism, CMA in, 292–294
Metacognition, 18
Metformin, 192, 270, 277
Microglia, 131, 132, 151
Microlysophagy, 186
Microvilli, 216
Mihaylova 2011, 272
Mitochondria, 259, 261
Mitochondrial DNA (mtDNA), 260
Mitochondrial mass analysis, 263, 264
Mito-Keima, 263, 275
Mitophagy, 4, 35, 72, 83, 178, 217, 259, 260,
    269, 274
and aging, 177, 179
cardiac ischemia models and
    reperfusion, 264
    cell culture, 264, 265
    ex vivo organ perfusion, 265, 266
    *in vivo* models, 266
    slice culture, 265
    working heart model, 266
concepts and methodology, 261
    flux assessment, 262, 263
    immunoblotting, 261, 262
    immunohistology/immunofluorescence, 262
inflammation, 260
mitochondrial mass, assessment of, 263, 264
MitoTimer, 264

MitoTimer, 264, 275
Mizushima 2004, 271
Modulating autophagy, 276
Moiré fringes, 53
Monomeric red fluorescent protein (mRFP), 31
MSP1, 243
MTOR complex 1 (MTORC1), 79, 106
MTOR complex 2 (MTORC2), 290
MTORC1, 268
Multi-dimensional approach, 60
Multisystem proteinopathies (MSPs), 243, 244
Muscle-specific ATG7 deficiency, 184
Muscular dystrophies, 241
Myocardial ischemia, 268
Myofibrillar myopathies, 244, 245

**N**
Narendra 2010, 271–272
NBR1, 82
NCOR1, 300
Neoglucogenesis, 224
Nephron, 214–215
Neurodegeneration, 128, 129, 149, 299, 301
Neurodegenerative diseases, 129, 144–145,
        149, 172
    autophagic flux by protein aggregates,
        impairment of, 142, 143
    autophagic lysosome reformation, defects
        in, 141, 142
    autophagosome biogenesis, 134–137
    autophagosome maturation, defects in, 140, 141
    autophagy as therapeutic target for,
        143, 145–148
    cargo recruitment, defects in, 137–140
    functions of autophagy in neurons and
        glia, 129
        microglial autophagy modulates neuronal
            health and survival, 131, 132
        neurodevelopmental pathways require
            autophagy, 130, 131
        oligodendrocyte and Schwann cell
            autophagy, 132
        synaptic function, learning and memory
            require autophagy, 130
    neurons vulnerable to, impaired autophagic
        function, 128
        efficient mitophagy to enable their high
            metabolic activity, 129
        morphology of neurons exacerbates
            cellular trafficking defects, 129
        post-mitotic and cannot dilute autopha-
            gic substrates through cell divi-
            sion, 128
Neurodevelopmental disease, 127, 134
Neurodevelopmental pathways, 130, 131
Neurological function, CMA in, 295, 296
Neuronal autophagy, 128
NFATC2/NFAT1, 290
Nicotinamide adenine dinucleotide
        (NAD+), 190
NLRP3 (NLR family pyrin domain contain-
        ing 3), 180
NLRP3 inflammasome, 181
Non-alcoholic steatohepatitis (NASH), 293
Non-canonical autophagy, 171, 172, 185–187
Non-selective autophagy, 120
Nucleation, 232
    of autophagosome formation, 107–109
Nucleophagy and aging, 180
Numerical aperture (NA), 53

**O**
Oka 2012, 272
Oligodendrocyte, `, 132, 151
OM phosphoproteins, 260
Omegasome, 107, 108, 112–113
OPCA motif, 82
OPTN, 83, 139
Organ-specific autophagy, 171, 172, 183–185
Osmiophilic pre-autophagosomal struc-
        tures, 102
Osmium tetroxide, 33
Osteoblasts, 298
Outer membrane (OM), 260
Overactivation of autophagy, 175, 176
Oxidative phosphorylation (OXPHOS), 214
Oxidative stress, 290
Oxygenated buffer, 265

**P**
Parkinson disease, 134, 177
PARL (presenilin associated rhomboid
        like), 260
Passive mode of reading, 5
Pathological hypertrophy, 276

Pentapeptide motif, 287
Pexophagy, 4, 35, 87
Phagophore, 9, 26, 27, 219
  closure, 232
  formation, 104
  membrane, 51, 63, 136
    closing of, 117, 118
Phagophores, 31–33, 102, 122, 217, 222
  and sealing off of double membranes,
    109, 110
Pharmaceutical autophagy induction, drug
    discovery for, 312, 313
Phosphatidylethanolamine (PE), 111, 185, 258
Phospholipid phosphatidylethanolamine, 74
Phosphorylations, 118
Photobleaching, 52
Physiological hypertrophy, 276
Physiological shear stress, 225
PIK3C3/VPS34, 111, 114, 116
  feedback regulation of, 116
  and inhibitors, 114
  PIK3C3/VPS34 complex I-specific
    activator, 122
  PIK3C3/VPS34 lipid kinase, 105
PINK1, 259
PINK1-PRKN mitophagy, 138
Polyubiquitinated proteins, 149
Pomodoro technique, 12
Pompe disease, 238, 239
Posttranslationally-modified SNCA, 296
Primary cell culture, 264
Primary cilium, 216
Primary cilium-dependent regulation of
    autophagy, 219, 220
  metabolism in kidney epithelial cells,
    220, 221
  and metabolism in tubular cells, 218–219
  MTORC1, 219
  YAP1, 220
Primary cilium-dependent regulation of
    intracellular pathways, 216
PRKN, 137, 260, 266
*PRKN* deletion, 260
Proapoptotic factors, 137
Problem-solving, 11
Proinflammatory factors, 297
Proteasome-targeting chimeras
    (PROTACs), 146
Protein aggregates, 217

Proteostasis, 295, 298, 305
Proteotoxicity, 299
Proximal tubule cells, metabolism of, 213–216
PSEN1 (presenilin 1), 142
PtdIns3K complex, 134
PtdIns3K-C1 activity, 135
PtdIns3P formation, 114, 116, 117


**Q**
Quantitative data, 7


**R**
R software, 314
Radaforolimus, 143
Rapalogs, 143, 145
Rapamycin, 105, 143, 149, 152, 189, 269,
    270, 277
RARA, 300
RARA/RARα, 291
Ratio-metric imaging, 51
RB1CC1, 80
RB1CC1/FIP200-interacting region (FIR), 80
Reactive oxygen species (ROS), 258
Reading, 3
Rejuvenate, 198
Reperfusion injury, 264–268, 276
Resin embedding, 57
Restoration, CMA, 300, 301
Resveratrol, 270, 277
Reticulophagy, 4, 35, 79
Retinitis pigmentosa, 300
RETREG1/FAM134B, mutation in, 139
Retrograde perfusion, 265
Retrotransposons, 182
RNAseq analysis, 174
ROS-damaged mitochondria, 260
Rubicon autophagy regulator (RUBCN), 173,
    183, 194–196, 219, 225


**S**
SAHA, 273
*Salmonella*-containing vacuoles (SCV), 83
Scheduled automation workflow, 315
Schwann cells, 132, 151
Second brain, 184
Selective autophagy, 71, 72, 78, 79, 83,
    120, 177

Selective autophagy receptors (SARs), 77–79
 Cvt pathway, 77
 definition, 77
 essential function of, 79
 membrane-bound, 83, 84
 oligomerization, 79
 soluble, 81–83
sIBM, 245
siRNA, 222, 225
SIRT1 (sirtuin1), 176
Skeletal muscle diseases, 234
 assessment of autophagy defects
  in autophagic flux, 246
  in autophagic induction, 247, 249, 250
 clearance of specific autophagy cargo,
   defects in, 242
  MSPs, 243, 244
  myofibrillar myopathies, 244, 245
  secondarily impaired, 245
 by decreased autophagy flux, 235
  colchicine and vincristine toxicity, 239
  CQ and HCQ, 239
  Danon disease, 236, 237
  Pompe disease, 238, 239
  XMEA, 235
 by decreased autophagy induction, 239
  congenital myopathies, 241, 242
  muscular dystrophies, 241
  mutations in machinery, 240, 241
 impaired autophagy, caused by, 231,
   232, 235
Skeletal muscle homeostasis, 232
Skeletal myopathies, 234
Slice culture, cardiac ischemia models and
  reperfusion, 265
Smith-Magenis syndrome, 134
SNCA, 142, 301
Soluble N-ethylmaleimide-sensitive-factor
  attachment protein receptor
  (SNARE)-dependent process, 219
Soluble SARs, 81–83
Spastic paraplegia, 140
Spermidine, 62, 191
SQSTM1, 48, 76, 82, 85, 86, 138, 234,
  246–248, 250, 251
SQSTM1/p62, 47, 86, 197, 198, 262, 266
Standard plastic ware, 313
Starvation-induced macroautophagy, 79
Steady state, 58

Stem cells, CMA in, 298
STING1, 186, 297
STORM, 109
Stress granule clearance in skeletal muscle, 253
Structured illumination microscopy (SIM), 53
Study strategies, 13
Subsequent autophagosome maturation, 312
Substrate-HSPA8 complex, 288, 291
Summative assessments, 23
Superoxide dismutase, 37
Super-resolution microscopy, 106, 107
Super-Resolution Structured Illumination
   Microscopy (SR-SIM), 54, 55,
   60, 62, 65
Super-resolution techniques, 52, 53, 60–66
 acquisition process, 54
 disadvantages, 54
 dSTORM, 53
 SIM, 53
Synaptic function, 130
Systemic metabolism, 294


**T**
T cell-mediated inflammaging, 182, 183
Tandem fluorescent, 30
Tandem fluorescent LC3 constructs, 275
Tandem mCherry-GFP constructs, 275
Tauopathy, 296
TBK1 (TANK binding kinase 1), 139
T-cell-specific TFAM deficiency
   model, 184–185
Teaching, 19, 20
Teaching strategies, 21
 interactive teaching, 21, 22
  assessment, 24–26
   learning activities, 23, 24
   learning goals, 22
Temsirolimus, 143
TFEB, 147, 189, 190
Time-restricted feeding (TRF), 188
Tissue-resident macrophages, 259
Tissue-specific autophagy-deficient mouse
   models, 175
Transmembrane protein, 106
Transmission electron microscopy
 amphisomes, 36
 autolysosomes, 38
 autophagosomes, 34–35

issues, 37, 39
   phagophores, 32
Transverse aortic constriction (TAC) surgery, 269
Trehalose, 192
Tricarboxylic acid cycle, 298
TUNEL staining, 265
Type 2 diabetes, 270

**U**
Ubiquitin-dependent degradation of depolarized mitochondria, 83
Ubiquitin-dependent selective autophagy, 120
ubiquitin-independent selective autophagy, 120
Ubiquitin-like conjugation, 111
Ubiquitous loss, 128
ULK complex, 106, 108, 120
   feedback regulation of, 116
   proteins, 106
   scaffolding function of, 113
ULK1, 258
ULK1 (unc-51 like autophagy activating kinase 1)-BECN1-ATG16L1, 181
ULK1/ULK2 complex, 72
Ullrich congenital muscular dystrophy, 241
Ultramicrotome sectioning, 56
Urolithin A, 191

**V**
V249M, 135
Vacuolar apparatus (VA), 102
Vacuolar myopathy, 252, 253
Vascular disease, 294
Vascular health, 295

V-ATPase, 258
Vinblastine treatment, 104
Vincristine toxicity, 239
Volume electron microscopy, 56
Volume EM, 62
VPS13D, 138

**W**
Wang 2014, 273
Weak bases, 258
Western blotting (WB), 222, 262
   advantages, 48
   limitations, 48
   molecular footprint, 47, 48
   protein abundance, 47, 48
   proteins of interest, 64
WIPI/WDR family, 135
WNT signalling, 138
WNT-CTNNB1/β-catenin signaling, 217

**X**
Xenophagy, 35
Xie 2014, 273–278
X-linked myopathy with excessive autophagy (XMEA), 235, 236

**Y**
YAP1 (Yes1 associated transcriptional regulator), 220, 222, 224

**Z**
z-stacking, 56